Communications
in Computer and Information Science   61

Dominik Ślęzak   Sankar K. Pal
Byeong-Ho Kang   Junzhong Gu
Hideo Kuroda   Tai-hoon Kim (Eds.)

Signal Processing, Image Processing, and Pattern Recognition

International Conference, SIP 2009
Held as Part of the Future Generation
Information Technology Conference, FGIT 2009
Jeju Island, Korea, December 10-12, 2009
Proceedings

 Springer

Volume Editors

Dominik Ślęzak
University of Warsaw and Infobright Inc., Poland
E-mail: slezak@infobright.com

Sankar K. Pal
Indian Statistical Institute, India
E-mail: sankar@isical.ac.in

Byeong-Ho Kang
University of Tasmania, Australia
E-mail: bhkang@utas.edu.au

Junzhong Gu
East China Normal University (ECNU), China
E-mail: jzgu@ica.stc.sh.cn

Hideo Kuroda
Nagasaki University, Japan
E-mail: kuroda@cis.nagasaki-u.ac.jp

Tai-hoon Kim
Hannam University, South Korea
E-mail: taihoonn@hnu.kr

Library of Congress Control Number: 2009939694

CR Subject Classification (1998): C.3, H.5.5, I.5.4, G.1.2, F.2.1, J.1, J.6, J.7

ISSN   1865-0929
ISBN-10   3-642-10545-9 Springer Berlin Heidelberg New York
ISBN-13   978-3-642-10545-6 Springer Berlin Heidelberg New York

springer.com

© Springer-Verlag Berlin Heidelberg 2009
Printed in Germany

Typesetting: Camera-ready by author, data conversion by Scientific Publishing Services, Chennai, India
Printed on acid-free paper   SPIN: 12805372   06/3180   5 4 3 2 1 0

Foreword

As future generation information technology (FGIT) becomes specialized and fragmented, it is easy to lose sight that many topics in FGIT have common threads and, because of this, advances in one discipline may be transmitted to others. Presentation of recent results obtained in different disciplines encourages this interchange for the advancement of FGIT as a whole. Of particular interest are hybrid solutions that combine ideas taken from multiple disciplines in order to achieve something more significant than the sum of the individual parts. Through such hybrid philosophy, a new principle can be discovered, which has the propensity to propagate throughout multifaceted disciplines.

FGIT 2009 was the first mega-conference that attempted to follow the above idea of hybridization in FGIT in a form of multiple events related to particular disciplines of IT, conducted by separate scientific committees, but coordinated in order to expose the most important contributions. It included the following international conferences: Advanced Software Engineering and Its Applications (ASEA), Bio-Science and Bio-Technology (BSBT), Control and Automation (CA), Database Theory and Application (DTA), Disaster Recovery and Business Continuity (DRBC; published independently), Future Generation Communication and Networking (FGCN) that was combined with Advanced Communication and Networking (ACN), Grid and Distributed Computing (GDC), Multimedia, Computer Graphics and Broadcasting (MulGraB), Security Technology (SecTech), Signal Processing, Image Processing and Pattern Recognition (SIP), and u- and e-Service, Science and Technology (UNESST).

We acknowledge the great effort of all the Chairs and the members of advisory boards and Program Committees of the above-listed events, who selected 28% of over 1,050 submissions, following a rigorous peer-review process. Special thanks go to the following organizations supporting FGIT 2009: ECSIS, Korean Institute of Information Technology, Australian Computer Society, SERSC, Springer LNCS/CCIS, COEIA, ICC Jeju, ISEP/IPP, GECAD, PoDIT, Business Community Partnership, Brno University of Technology, KISA, K-NBTC and National Taipei University of Education.

We are very grateful to the following speakers who accepted our invitation and helped to meet the objectives of FGIT 2009: Ruay-Shiung Chang (National Dong Hwa University, Taiwan), Jack Dongarra (University of Tennessee, USA), Xiaohua (Tony) Hu (Drexel University, USA), Irwin King (Chinese University of Hong Kong, Hong Kong), Carlos Ramos (Polytechnic of Porto, Portugal), Timothy K. Shih (Asia University, Taiwan), Peter M.A. Sloot (University of Amsterdam, The Netherlands), Kyu-Young Whang (KAIST, South Korea), and Stephen S. Yau (Arizona State University, USA).

We would also like to thank Rosslin John Robles, Maricel O. Balitanas, Farkhod Alisherov Alisherovish, and Feruza Sattarova Yusfovna – graduate students of Hannam University who helped in editing the FGIT 2009 material with a great passion.

October 2009

Young-hoon Lee
Tai-hoon Kim
Wai-chi Fang
Dominik Ślęzak

Preface

We would like to welcome you to the proceedings of the 2009 International Conference on Signal Processing, Image Processing and Pattern Recognition (SIP 2009), which was organized as part of the 2009 International Mega-Conference on Future Generation Information Technology (FGIT 2009), held during December 10–12, 2009, at the International Convention Center Jeju, Jeju Island, South Korea.

SIP 2009 focused on various aspects of advances in signal processing, image processing and pattern recognition with computational sciences, mathematics and information technology. It provided a chance for academic and industry professionals to discuss recent progress in the related areas. We expect that the conference and its publications will be a trigger for further related research and technology improvements in this important subject.

We would like to acknowledge the great effort of all the Chairs and members of the Program Committee. Out of 140 submissions to SIP 2009, we accepted 42 papers to be included in the proceedings and presented during the conference. This gives roughly a 30% acceptance ratio. Four of the papers accepted for SIP 2009 were published in the special FGIT 2009 volume, LNCS 5899, by Springer. The remaining 38 accepted papers can be found in this CCIS volume.

We would like to express our gratitude to all of the authors of submitted papers and to all of the attendees, for their contributions and participation. We believe in the need for continuing this undertaking in the future.

Once more, we would like to thank all the organizations and individuals who supported FGIT 2009 as a whole and, in particular, helped in the success of SIP 2009.

October 2009

Dominik Ślęzak
Sankar K. Pal
Byeong-Ho Kang
Junzhong Gu
Hideo Kuroda
Tai-hoon Kim

Organization

Organizing Committee

General Chair   Sankar K. Pal (Indian Statistical Institute, India)

Program Chair   Byeong-Ho Kang (University of Tasmania, Australia)

Publicity Chairs   Junzhong Gu (East China Normal University, China)
                         Hideo Kuroda (Nagasaki University, Japan)

Program Committee

Andrzej Dzielinski
Andrzej Kasinski
Antonio Dourado
Chng Eng Siong
Dimitris Iakovidis
Debnath Bhattacharyya
Ernesto Exposito
Francesco Masulli
Gérard Medioni
Hideo Kuroda
Hong Kook Kim
Janusz Kacprzyk

Jocelyn Chanussot
Joonki Paik
Joseph Ronsin
Junzhong Gu
Kenneth Barner
Mei-Ling Shyu
Miroslaw Swiercz
Mototaka Suzuki
N. Jaisankar
N. Magnenat-Thalmann
Nikos Komodakis
Makoto Fujimura

Paolo Remagnino
Roman Neruda
Rudolf Albrecht
Ryszard Tadeusiewicz
Salah Bourennane
Selim Balcisoy
Serhan Dagtas
Shu-Ching Chen
Tae-Sun Choi
William I. Grosky
Xavier Maldague
Yue Lu

Table of Contents

A Blind Image Wavelet-Based Watermarking Using Interval Arithmetic

Teruya Minamoto[1], Kentaro Aoki[1], and Mitsuaki Yoshihara[2]

[1] Saga University, Saga, Japan
{minamoto,aoki}@ma.is.saga-u.ac.jp
[2] Kobe Steel, Ltd., Japan

Abstract. In this paper, we present a new blind digital image watermarking method. For the intended purpose, we introduce the interval wavelet decomposition which is a combination between discrete wavelet transform and interval arithmetic, and examine its properties. According to our experimental results, this combination is a good way to produce a kind of redundancy from the original image and to develop new watermarking methods. Thanks to this property, we can obtain specific frequency components where the watermark is embedded. We describe the procedures of our method in detail and the relations with the human visual system (HVS), and give some experimental results. Experimental results demonstrate that our method give the watermarked images of better quality and is robust against attacks such as clipping, marking, JPEG and JPEG2000 compressions.

1 Introduction

Many digital image watermarking methods have been proposed over the last decade [2,3]. According to whether the original signal is needed or not during the watermark detection process, digital watermarking methods can be roughly categorized in the following two types : non-blind and blind. Non-blind method requires the original image in the detection end, whereas blind one does not. The blind methods are more useful than non-blind, because the original image may not be available in actual scenarios.

If we understand correctly, almost all existing watermarking methods utilize the redundancies of the original image to embed the digital watermark. Therefore, it is important to find or produce the redundant parts of the original image based on the characteristics of the human visual system (HVS).

While interval arithmetic which are mainly used in the field where rigorous mathematics is associated with scientific computing including computer graphics and computer-aided design [6,8]. Interval arithmetic has the property that the width of the interval which includes the original data expands in proportional to the number of arithmetic computations. This property is sometimes called "Interval expansion". Interval expansion is, so to speak, the redundancy produced by the original data and interval arithmetic. This implies that there is a possibility to apply interval arithmetic to digital watermarking.

D. Ślęzak et al. (Eds.): SIP 2009, CCIS 61, pp. 1–8, 2009.

Based on this idea, we proposed a new non-blind digital image watermarking method in Ref. [7] where we used discrete wavelet transforms and interval arithmetic. In this paper we call such a combination between discrete wavelet transform and interval arithmetic "Interval wavelet decomposition".

Up to now, we neither investigate the properties of the interval wavelet decomposition, nor consider the blind type in Ref. [7]. Therefore, our goal in this paper is to examine the properties of the interval wavelet decomposition and the relations with the HVS, and to propose a new blind digital watermarking method using the interval wavelet decomposition.

This paper is organized as follows: in Section 2, we briefly describe the basics of interval arithmetic. In Section 3, we introduce the interval wavelet decomposition, and propose a blind digital watermarking method in Section 4. Then, we describe how to choose the parameter appearing in our experiments, and the relationship between our method and the HVS in Section 5. We demonstrate experimental results in Section 6, and conclude this paper in Section 7.

2   Interval Arithmetic

An interval is a set of the form $A = [a_1, a_2] = \{t | a_1 \leq t \leq a_2, a_1, a_2 \in \mathbb{R}\}$, where $\mathbb{R}$ denotes the set of real numbers. We denote the lower and upper bounds of an interval A by $\inf(A) = a_1$ and $\sup(A) = a_2$, respectively, and the width of any non-empty interval A is defined by $w(A) = a_2 - a_1$. The four basic operations, namely, addition$(+)$, subtraction$(-)$, multiplication$(*)$ and division $(/)$ on two intervals $A = [a_1, a_2]$ and $B = [b_1, b_2]$ are defined as follows:

$$A + B = [a_1 + b_1, a_2 + b_2], \quad A - B = [a_1 - b_2, a_2 - b_1],$$
$$A * B = [\min\{a_1 b_1, a_1 b_2, a_2 b_1, a_2 b_2\}, \max\{a_1 b_1, a_1 b_2, a_2 b_1, a_2 b_2\}] \qquad (1)$$
$$A / B = [a_1, a_2] * [1/b_2, 1/b_1], \quad 0 \notin B.$$

For interval vectors and matrices whose elements consist of intervals, these operations are executed at each element.

From the basic operations (1), the width of interval expands in proportional to the number of computations in general. For examples, let $A = [-1, 2]$, $B = [1, 4]$, and $C = [-2, 1]$, then $A + B = [0, 6]$, $A + B - C = [-1, 8]$. The widths of A, B and C are 3, while the widths of $A + B$ and $A + B - C$ are 6 and 9, respectively. This phenomena is sometimes called "Interval expansion". This is the typical demerit of interval arithmetic, but we regard this interval expansion as a useful tool to produce the redundant part from the original image in this paper.

3   Interval Wavelet Decomposition

Let us denote the original image by S. It is well known from Refs. [4,5] that the usual Daubechies wavelet decomposition formulae for images with support width $2N - 1$ are given as follows:

$$C(i,j) = \sum_{m,n=0}^{2N-1} p_m p_n S(m+2i, n+2j), \quad D(i,j) = \sum_{m,n=0}^{2N-1} p_m q_n S(m+2i, n+2j),$$

$$E(i,j) = \sum_{m,n=0}^{2N-1} q_m p_n S(m+2i, n+2j), \quad F(i,j) = \sum_{m,n=0}^{2N-1} q_m q_n S(m+2i, n+2j). \tag{2}$$

Here p_n and q_n are real parameters which has the relation $q_n = (-1)^n p_{2N-1-n}$, and the indices i and j are the locations in horizontal and vertical directions, respectively. The components C, D, E and F indicate low frequency components, high frequency components in vertical, in horizontal, and in diagonal directions, respectively.

To accelerate the interval expansion, we define the interval wavelet decomposition based on (2) by

$$IC(i,j) = \sum_{m,n=0}^{2N-1} (1 + I(\Delta_{m,n})) \widehat{S}_{p_m,p_n}^{i,j}, \quad ID(i,j) = \sum_{m,n=0}^{2N-1} (1 + I(\Delta_{m,n})) \widehat{S}_{p_m,q_n}^{i,j},$$

$$IE(i,j) = \sum_{m,n=0}^{2N-1} (1 + I(\Delta_{m,n})) \widehat{S}_{q_m,p_n}^{i,j}, \quad IF(i,j) = \sum_{m,n=0}^{2N-1} (1 + I(\Delta_{m,n})) \widehat{S}_{q_m,q_n}^{i,j}, \tag{3}$$

where $\widehat{S}_{p_m,q_n}^{i,j} = p_m q_n S(m+2i, n+2j)$, $I(\Delta_{m,n}) = [-\Delta_{m,n}, \Delta_{m,n}]$, and $\Delta_{m,n}$ are positive real numbers.

4 Watermarking Algorithm

We assume that the binary-valued watermark W consists of -1 and 1. To develop a new blind method, we modify one of the best known block-based blind watermark scheme described in Chapter 3 in Ref.[2] and apply it to the non-blind method in Ref.[7]. In our method, we use a slide window instead of a block, because the slide window based scheme is superior to the block-based one in many cases according to our experience. Then, the embedding procedures are as follows:

1. Choose one among the four interval components IC, ID, IE and IF and decide the position where the watermark is embedded. For simplicity, we assume that we choose IF and decide $F' = \sup(IF)$ in this section.
2. Replace other components as floating point ones. In this case, replace IC, ID, and IE as C, D, E, respectively.
3. Compute the following sequence by a slide window of $2k+1$:

$$\overline{F}(i,j) = \text{sgn}(F'(i,j)) \cdot \frac{1}{(2k+1)} \sum_{l=-k}^{k} |F'(i+l,j)|, \tag{4}$$

where k is a fixed natural number and $\text{sgn}(x)$ is the usual signum function of a real number x.

4. Embed the watermark W by computing

$$F_w(i,j) = \overline{F}(i,j)\{1 + \alpha W(i,j)\}. \qquad (5)$$

Here $0 < \alpha < 1$ is a given hiding factor which adjusts the robustness.
5. Reconstruct the image using $C \sim E$, F_w and the inverse wavelet transform, then the watermarked image S_w is obtained.

The following steps describe the detection procedures:

1. Decompose S_w into four components C_{S_w}, D_{S_w}, E_{S_w}, and F_{S_w} by the wavelet transform.
2. Compute

$$W_e(i,j) = \text{sgn}\big(|F_{S_w}(i,j)| - |\overline{F_{S_w}}(i,j)|\big) \qquad (6)$$

to extract the watermark. Here we note that we must consider the absolute values of F_{S_w} and $\overline{F_{S_w}}$ to avoid changing the sign of $\alpha W \approx (F_{S_w} - \overline{F_{S_w}})/\overline{F_{S_w}}$ corresponding to $(F_w - \overline{F})/\overline{F}$ in (5). For example, assuming that $W = 1$, $(F_w - \overline{F})/\overline{F} < 0$ when $\overline{F} < 0$, whereas $(F_w - \overline{F})/\overline{F} > 0$ when $\overline{F} > 0$.

5   Considerations on the Proposed Algorithms

In this section, we consider how to choose the parameters described in the previous section and the relationship between the proposed redundancy and the HVS.

5.1   The Choice of Parameters

Digital watermarking methods involve the trade-off between the robustness and the quality of the watermarked image. The values $\Delta_{m,n}$ in (3) represents this relation. In fact, the robustness is proportional to the values $\Delta_{m,n}$, but the quality of the watermarked image is inversely proportional to the values $\Delta_{m,n}$, and vice versa. Since the hiding parameter α in (5) also decides the robustness of our watermarking method, it is important to find the good values of $\Delta_{m,n}$ and α. Furthermore, a natural number k in (4) will affect the robustness.

We decide the parameters so as not to decrease considerably the value of the peak signal to noise ratio (PSNR) in decibels computed by

$$PSNR = 20 \log_{10} \left(\frac{255}{\sqrt{\frac{1}{N_x N_y} \sum_{i=1}^{N_x} \sum_{j=1}^{N_y} (S_w(i,j) - S(i,j))^2}} \right),$$

where N_x and N_y are the size of image in horizontal and in vertical directions, respectively.

In our experiment, we require that the PSNR value exceeds 30. According to our preliminary experiments, we must set $\Delta_{m,n} \leq 0.008$ to fill this requirement, if we use the same constant $\Delta_{m,n}$ at all pixels. While k in (4) does not affect the PSNR too much, but we confirmed that it affects the quality of extracted watermark in our simulations. Moreover, α should be chosen as large as possible in order to maintain the robustness.

As a result, we choose $\Delta_{m,n} = 0.008$ in (3), $k = 10$ in (4), and $\alpha = 0.9$ in (5) as initial values.

5.2 Relationship between Our Method and the HVS

In this subsection, we investigate the relationship between the produced redundancy by the interval wavelet decomposition and the HVS. For the intended purpose, we prepare the following theorem.

Theorem 1. *The interval high frequency components ID, IE and IF contain the low frequency component C. More precisely, the following relations hold.*

$$ID(i,j) = D(i,j) + \theta_D \left(IC(i,j) - C(i,j) \right) + R_D(i,j),$$
$$IE(i,j) = E(i,j) + \theta_E \left(IC(i,j) - C(i,j) \right) + R_E(i,j), \qquad (7)$$
$$IF(i,j) = F(i,j) + \theta_F \left(IC(i,j) - C(i,j) \right) + R_F(i,j).$$

Here θ_D, θ_E and θ_F are certain constants depending on p_n in (2), and $R_D(i,j)$, $R_E(i,j)$ and $R_F(i,j)$ are certain components which values are expectatively small.

Proof. We only prove the first equality in (7), because the other equalities can be shown in the same manner, and we abbreviate $S(i,j)$ to $S_{i,j}$ in this proof to save space.

From (3), we first note that $ID(i,j) = D(i,j) + \sum_{m,n=0}^{2N-1} I(\Delta_{m,n}) p_m q_n S_{m+2i,n+2j}$.

Using the relation $q_n = (-1)^n p_{2N-1-n}$ appearing in Ref. [4], the second term of the right-hand side is denoted by

$$\sum_{m,n=0}^{2N-1} I(\Delta_{m,n}) p_m q_n S_{m+2i,n+2j} = \sum_{m,n=0}^{2N-1} I(\Delta_{m,n}) \frac{p_{2N-1-n}}{p_n} p_m p_n S_{m+2i,n+2j}.$$

Next, putting $c_n = p_{2N-1-n}/p_n$ and $\theta_D = \min_n |c_n|$, the right-hand side term above is represented by

$$\sum_{m,n=0}^{2N-1} \left(I(\Delta_{m,n}) \theta_D p_m p_n S_{m+2i,n+2j} + I(\Delta_{m,n})(c_n - \theta_D) p_m p_n S_{m+2i,n+2j} \right)$$

$$= \theta_D \sum_{m,n=0}^{2N-1} I(\Delta_{m,n}) p_m p_n S_{m+2i,n+2j} + R_D(i,j),$$

where $R_D(i,j) = \sum_{m,n=0}^{2N-1} I(\Delta_{m,n})(c_n - \theta_D) p_m p_n S_{m+2i,n+2j}$.

Then, we obtain

$$\sum_{m,n=0}^{2N-1} I(\Delta_{m,n}) p_m p_n S_{m+2i,n+2j} = \sum_{m,n=0}^{2N-1} (1 + I(\Delta_{m,n}) - 1) p_m p_n S_{m+2i,n+2j}$$

$$= \sum_{m,n=0}^{2N-1} (1 + I(\Delta_{m,n})) p_m p_n S_{m+2i,n+2j} - \sum_{m,n=0}^{2N-1} p_m p_n S_{m+2i,n+2j}.$$

Therefore, $ID(i,j) = D(i,j) + \theta_D(IC(i,j) - C(i,j)) + R_D(i,j)$ holds. $\square$

Barni et al. proposed the weighing function using the masking characteristics of the HVS [1]. To compare their method with our method, we set $Y_l^0 = D_l$, $Y_l^1 = F_l$, $Y_l^2 = E_l$ and $Y_l^3 = C_l$, where the numerical subscript $l(l = 0, 1, 2, 3)$ stands for the resolution level. The components C_0, D_0, E_0 and F_0 are identical with C, D, E and F, respectively in Section 3.

They embed the watermark $w^\theta(i, j)$ according to the rule $\tilde{Y}_0^\theta(i, j) = Y_0^\theta(i, j) + \beta q^\theta(i, j) w^\theta(i, j)$ using the weighing function $q^\theta(i, j) = \Theta(\theta)\Lambda(i, j)\Xi(i, j)^{0.2}$.

According to Ref. [1], the functions $\Theta(\theta)$, $\Lambda(i, j)$, and $\Xi(i, j)^{0.2}$ correspond to the following considerations, respectively:

1. The eye is less sensitive to noise in high resolution bands.
2. The eye is less sensitive to noise in those areas of the image where brightness is high or low.
3. The eye is less sensitive to noise in highly textured areas, but, among these, more sensitive near the edge.

Since F is the high resolution bands in general, the first item corresponds with what we embed the watermark into F in our method. To consider the second item, Barni et al. used the low component C_3 of the image. This choice corresponds to $\bar{F}$ in our method, because $\bar{F}$ depends on IF, and IF contains the low frequency component by Theorem 1. Moreover, they used the product of the local mean square values of Y_l^θ in all components and the local variance of the low frequency component Y_3^3 to take into account of the third item. These things also corresponds to $\bar{F}$ in our method, because $\bar{F}$ is produced by the slide window which takes into account of the neighborhood of the pixel, and $\bar{F}$ contains both high and low frequency factors.

From these discussions, we may conclude that our method takes the HVS into consideration. In particular, the high frequency component $\bar{F}$ has the information of the low frequency component thanks to the interval wavelet decomposition, and the mathematical expressions are simpler and more sophisticated than the rule proposed by Barni et al. This is the new merit of interval arithmetic.

6   Experimental Results

To evaluate the performance of the proposed method, we adopt the 256-grayscale Lenna image of size 256×256 and a binary watermark of size 128×128 as shown in Fig.1 (a) and (b). We implemented our method using INTLAB [9] which is the MATLAB toolbox and supports interval arithmetic.

At first, we set $\Delta_{m,n} = 0.008$ and $N = 3$ in (3), $k = 10$ in (4), and $\alpha = 0.9$ in (5). In this experiment, we compute the $mean(S)$ which is the arithmetic mean of S, and varied $\Delta_{m,n}$ from 0.0012 to 0.0163 depending on the ratio $S(i, j)/mean(S)$ at each pixel.

Fig.1 (c) shows the watermarked image with PSNR=33.5774 obtained by the proposed blind method and the watermark extracted from the watermarked

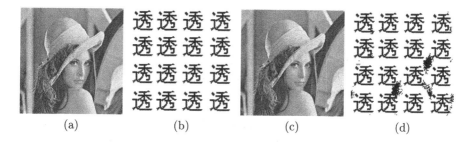

Fig. 1. (a) Original image, (b) watermark, (c) watermarked image with PSNR=33.5774 and (d) extracted watermark without any attack

Fig. 2. (a) Watermarked image with marked areas, (b) extracted watermark, (c) 188 × 210 fragment of watermarked image and (d) extracted watermark

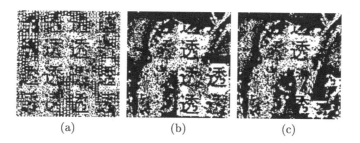

Fig. 3. Extracted watermarks from (a) the watermarked JPEG image with the file size ratio 34%, the watermarked JPEG2000 images with the file size ratio (b) 10.6% and (c) 9.7%

image without any attack. Figs.2~3 illustrate the watermarks extracted from the attacked watermarked images under attacks such as marking, clipping, JPEG and JPEG2000 compressions. The extracted images are damaged, but we are able to identify the existence of watermark at a single glance in Figs.2 (b), (d) and 3 (b), and do barely in Fig.3 (a) and (c).

7   Conclusion

We proposed a blind digital image watermarking method. It seems that our method are relatively easier to implement than other methods based on the frequency domain [1,2]. To develop our method, we introduced the interval wavelet decomposition formulae and investigate its properties. Using the interval wavelet decomposition formulae, we realized that every high frequency component contains the low frequency component. Since the low component is the main part of the original image, these high frequency components also contain it. This means that every high frequency component has a certain redundancy. We also mentioned the relationship between this redundancy and the HVS, and realized that our method takes a kind of the HVS into consideration automatically. Of course, this conclusion is only an interpretation of the relation between the HVS and our method, because the HVS is certainly one of the most complex biological devices. We would like to remark this point.

Experimental results demonstrate that our method gives the better-quality watermarked images and have the robustness against some attacks such as clipping, marking, JPEG and JPEG2000 compressions.

At present, our digital watermarking method may not have enough robustness in practice, and we do not compare our method with other up-to-date methods [3]. However, this work gives a novel technique of digital watermarking methods, besides it may open up new possibilities for interval arithmetic. In this sense, we believe that our approaches are very important in the field of both digital watermarking and interval arithmetic.

References

1. Barni, M., Bartolini, F., Piva, A.: Improved wavelet-based watermarking through pixel-wise masking. IEEE Trans. Image Processing 10(5), 783–791 (2001)
2. Cox, I.J., Miller, M.L., Bloom, J.A.: Digital Watermarking. Morgan Kaufmann Publishers, San Francisco (2002)
3. Cox, I.J., Miller, M.L., Bloom, J.A., Fridrich, J., Kalker, T.: Digital Watermarking and Steganography. Morgan Kaufmann Publishers, San Francisco (2008)
4. Daubechies, I.: Orthonormal bases of compactly supported wavelets. Comm. Pure Appl. Math. 41, 909–996 (1988)
5. Daubechies, I.: Ten Lectures on Wavelets. SIAM, Philadelphia (1992)
6. Minamoto, T.: Numerical method with guaranteed accuracy of a double turning point for a radially symmetric solution of the perturbed Gelfand equation. Journal of Computational and Applied Mathematics 169/1, 151–160 (2004)
7. Minamoto, T., Yoshihara, M., Fujii, S.: A Digital Image Watermarking Method Using Interval Arithmetic. IEICE Trans. Fundamentals E90-A(12), 2949–2951 (2007)
8. Moore, R.E., Kearfott, R.B., Cloud, M.J.: Introduction to Interval Analysis. SIAM, Philadelphia (2009)
9. S.M. Rump: INTLAB - Interval Laboratory,
   http://www.ti3.tu-harburg.de/rump/intlab/

Hand Gesture Spotting Based on 3D Dynamic Features Using Hidden Markov Models

Mahmoud Elmezain, Ayoub Al-Hamadi, and Bernd Michaelis

Institute for Electronics, Signal Processing and Communications
Otto-von-Guericke-University Magdeburg, Germany
{Mahmoud.Elmezain,Ayoub.Al-Hamadi}@ovgu.de

Abstract. In this paper, we propose an automatic system that handles hand gesture spotting and recognition simultaneously in stereo color image sequences without any time delay based on Hidden Markov Models (HMMs). Color and 3D depth map are used to segment hand regions. The hand trajectory will determine in further step using Mean-shift algorithm and Kalman filter to generate 3D dynamic features. Furthermore, k-means clustering algorithm is employed for the HMMs codewords. To spot meaningful gestures accurately, a non-gesture model is proposed, which provides confidence limit for the calculated likelihood by other gesture models. The confidence measures are used as an adaptive threshold for spotting meaningful gestures. Experimental results show that the proposed system can successfully recognize isolated gestures with 98.33% and meaningful gestures with 94.35% reliability for numbers (0-9).

Keywords: Gesture spotting, Gesture recognition, Pattern recognition, Computer vision.

1 Introduction

Although automatic hand gesture recognition technologies have been successfully applied to real-world applications, there are still several problems that need to be solved for wider applications of Human-Computer Interaction (HCI). One of such problems, which arise in real-time hand gesture recognition, is to extract meaningful gestures from the continuous sequence of hand motions. Another problem is caused by the fact that the same gesture varies in shape, trajectory and duration, even for the same person. In the last decade, several methods of potential applications [1], [2], [3], [4], [5] in the advanced gesture interfaces for HCI have been suggested but these differ from one another in their models. Some of these models are Neural Network (NN) [3], HMMs [5] and Dynamic Time Warping (DTW) [6]. Lee *et al.* [7] proposed an Ergodic model using adaptive threshold to spot the start and the end points of input patterns, and also classify the meaningful gestures by combining all states from all trained gesture models using HMMs. Yang *et al.* [2] presented a method to recognize the whole-body key gestures in Human-Robot Interaction (HRI) by designing the garbage model for non-gesture patterns. Yang *et al.* [8] introduced a method for designing threshold model using the weight of state and transition of the original

D. Ślęzak et al. (Eds.): SIP 2009, CCIS 61, pp. 9–16, 2009.
© Springer-Verlag Berlin Heidelberg 2009

Conditional Random Fields (CRF). This model performs an adaptive threshold to distinguish between signs and non-sign patterns.

Previous approaches mostly used the backward spotting technique that first detects the end point of gesture. Moreover, they track back to discover the start point of the gesture and then the segmented gesture is sent to the recognizer for recognition. So, there is an inevitable time delay between the meaningful gesture spotting and recognition. This time delay is not suitable for on-line applications.

The main contribution of this paper is to propose a gesture spotting system that executes hand gesture segmentation and recognition simultaneously using HMMs. A Markov Model is is capable of modeling spatio-temporal time series of gestures effectively and can handle non-gesture patterns rather than NN and DTW. To spot meaningful gesture accurately, a sophisticated method of designing a non-gesture model is proposed. The non-gesture model is a weak model for all trained gesture models where its likelihood is smaller than that the dedicated model for a given gesture. The start and end points of gestures are based on the competitive differential observation probability value, which is determined by the difference of observation probability value of maximal gesture models and non-gesture model. Each isolated gesture number is based on 60 video sequences for training and testing, while the continuous gestures are based on 280 video sequences for spotting meaningful gestures. The achievement recognition rates on isolated and meaningful gestures are 98.33% and 94.35% respectively.

2   Hidden Markov Models

Hidden Markov Model is a mathematical model of stochastic process, which generates random sequences of outcomes according to certain probabilities [5], [9], [10], [11], [12], [13]. A stochastic process is a sequence of feature extraction codewords, the outcomes being the classification of hand gesture path.

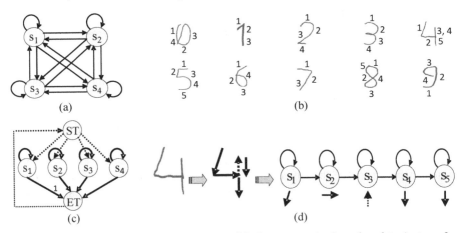

Fig. 1. (a) Ergodic topology with 4 states. (b) Gesture paths from hand trajectory for numbers (0-9) with its segmented parts. (c) Simplified Ergodic with fewer transitions. (d) LRB topology with its straight-line segmentation for gesture path 4.

Evaluation, Decoding and Training are the main problems of HMMs and they can be solved by using Forward-Backward, Viterbi and Baum-Welch (BW) algorithms respectively. Also, HMMs has three topologies; Fully Connected (i.e. Ergodic model) where any state in it can be reached from other states, Left-Right (LR) model such that each state can go back to itself or to the following states and Left-Right Bannded (LRB) model in which each state can go back to itself or the next state only (Fig. 1).

3   3D Dynamic Feature Extraction

There is no doubt that selecting good features to recognize the hand gesture path play significant role in system performance. A gesture path is spatio-temporal pattern that consists of centroid points (x_{hand}, y_{hand}). There are three basic features; location, orientation and velocity. We consider the center of gravity to compute the location features because different location features are generated for the same gesture according to the different starting points (Eq. 2).

$$L_t = \sqrt{(x_{t+1} - C_x)^2 + (y_{t+1} - C_y)^2} \quad ; t = 1, 2, ..., T - 1 \tag{1}$$

$$(C_x, C_y) = \frac{1}{n}\left(\sum_{t=1}^{n} x_t, \sum_{t=1}^{n} y_t\right) \tag{2}$$

where T represents the length of gesture path and (C_x, C_y) refer to the center of gravity at the point n. To verify the real-time implementation in our system, the center of gravity is computed after each image frame. Moreover, the orientation θ_t is determined between the current point and the center of gravity by Eq. 3.

$$\theta_t = \arctan\left(\frac{y_{t+1} - C_y}{x_{t+1} - C_x}\right) \quad ; t = 1, 2, ..., T - 1 \tag{3}$$

The velocity is based on the fact that each gesture is made at different speeds where the velocity of the hand decreases at the corner point of a gesture path. The velocity is calculated as the Euclidean distance between the two successive points divided by the time in terms of the number of video frames as follows;

$$V_t = \sqrt{(x_{t+1} - x_t)^2 + (y_{t+1} - y_t)^2} \quad ; t = 1, 2, ..., T - 1 \tag{4}$$

Furthermore, the feature vector F is obtained by the union of location, orientation and velocity features (Eq. 5).

$$F = \{(L_1, \theta_1, V_1), (L_2, \theta_2, V_2), ..., (L_{T-1}, \theta_{T-1}, V_{T-1})\} \tag{5}$$

The extracted features are quantized so as to obtain discrete symbols to apply to the HMMs. This done using k-means clustering algorithm [11], which classify the gesture pattern into K clusters in the feature space. We evaluate the gesture recognition according to different clusters number from 28 to 37. Therefore, Our experiments showed that the optimal number of clusters K is equal to 33.

4   Gesture Spotting and Recognition System

To spot meaningful gestures, we mention how to model gesture patterns discriminately and how to model non-gesture patterns effectively (Fig. 2). Each reference pattern for numbers (0-9) is modeled by LRB model with varying number of states ranging from 3 to 5 states based on its complexity (Fig. 1(b)). As, the excessive number of states can generate the over-fitting problem if the number of training samples is insufficient compared to the model parameters. It is not easy to obtain the set of non-gesture patterns because there are infinite varieties of meaningless motion. So, all other patterns rather than references pattern are modeled by a single HMM called a non-gesture model (garbage model) [2], [7]. The non-gesture model is constructed by collecting the states of all gesture models in the system [14]. The non-gesture model (Fig. 1(c)) is weak model for all trained gesture models and represents every possible pattern where its likelihood is smaller than the dedicated model for given gesture because of the reduced forward transition probabilities. Also, the likelihood of the non-gesture model provides a confidence limit for the calculated likelihood by other gesture models. Thereby, we can use confidence measures as an adaptive threshold for selecting the proper gesture model or gesture spotting. The number of states for non-gesture model increases as the number of gesture model increases. Moreover, there are many states in the non-gesture model with similar probability distribution, which in turn lead to waste time and space. To alleviate this problem, a relative entropy [15] is used. The relative entropy is the measure of the distance between two probability distributions. The proposed gesture spotting system contains two main modules; segmentation module and recognition module. In the gesture segmentation module, we use sliding window (Optimal size equal 5 empirically), which calculates the observation probability of all gesture models and non-gesture model for segmented parts. The start (end) point of gesture is spotted by competitive differential observation probability value between maximal gestures (λ_g) and non-gesture (Fig. 2). The maximal gesture model is

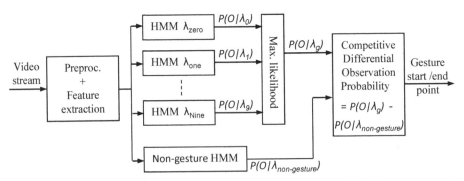

Fig. 2. Simplified structure showing the main computational modules for the proposed hand gesture spotting and recognition system. $p(O|\lambda_g)$ means the probability of the obsereved sequence O given the HMMs parameters of reference gesture g.

the gesture whose observation probability is the largest among all ten gesture $p(O|\lambda_g)$. When this value changes from negative to positive (Eq. 6, O can possibly as gesture g), the gesture starts. Similarly, the gesture ended around the time that this value changes from positive to negative (Eq. 7, O cannot be gesture).

$$\exists g : P(O|\lambda_g) > P(O|\lambda_{non-gesture}) \tag{6}$$

$$\forall g : P(O|\lambda_g) < P(O|\lambda_{non-gesture}) \tag{7}$$

After spotting start point in continuous image sequences, then it activates gesture recognition module, which performs the recognition task for the segmented part accumulatively until it receives the gesture end signal. At this point, the type of observed gesture is decided by Viterbi algorithm frame by frame.

5 Experimental Results

Our proposed system was capable for real-time implementation and showed good results to recognize numbers (0-9) from stereo color image sequences via the motion trajectory of single hand using HMMs. In our experimental results, the segmentation of the hand with complex background takes place using 3D depth map and color information over YC_bC_r color space, which is more robust to the disadvantageous lighting and partial occlusion. Gaussian Mixture Models (GMM) were considered where a large database of skin and non-skin pixel is used to train it. Moreover, morphological operations were used as a preprocessing, and Mean-shift algorithm in conjunction with Kalman filter is to track the hand to generate the hand motion trajectory. For more details, the reader can refer to [5], [14]. The input images were captured by Bumblebee stereo camera system that has 6 mm focal length at 15FPS with 240×320 pixels image resolution, Matlab implementation. Our experiments is carried out an isolated gesture recognition and meaningful gesture spotting test.

5.1 Isolated Gesture Recognition

In our experimental results, each isolated gesture number from 0 to 9 was based on 60 video sequences, which 42 video samples for training by BW algorithm and 18 video samples for testing (Totally, our database contains 420 video samples for training and 180 video sample for testing). The gesture recognition module match the tested gesture against the database of reference gestures, to classify which class it belongs to. The higher priority was computed by Viterbi algorithm to recognize the numbers in real-time frame by frame over Left-Right Banded topology with different number of states ranging from 3 to 5 based on gesture complexity. We evaluate the gesture recognition according to different clusters number from 28 to 37 to employee the extracted feature extraction for the HMMs codewords (Fig. 4(b)). Therefore, our experiments showed that the optimal number of clusters is equal to 33. The recognition ratio is the number

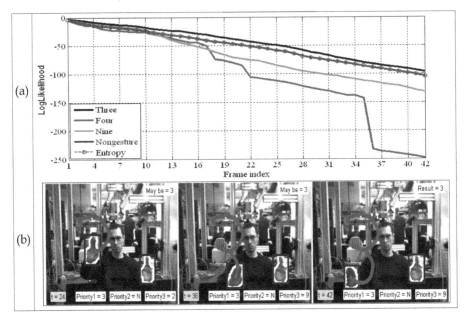

Fig. 3. In (a)&(b) isolated gesture '3' with high three priorities, where the probability of non-gesture before and after state reduction is the same (the no. of states of non-gesture model before reduction is 40 and after reduction is 28)

of correctly recognized gestures over the number of input gestures (Eq. 8). The recognition ratio of isolated gestures achieved best results with 98.33%.

$$Reco.\ ratio = \frac{no.\ of\ correctly\ recognized\ gestures}{no.\ of\ test\ gestures} \times 100\% \qquad (8)$$

5.2 Meaningful Gesture Spotting Test

Our database also contains 280 video samples for continuous hand motion. Each video sample either contains one or more than meaningful gestures. The number of states in the non-gesture model is equal to the sum of all states for all gesture models, except the two dummy states (ST and ET states in Fig. 1(c)). This means, the number of states for non-gesture model increases as the number of gesture model increases. Furthermore, an increase in the number of states is nothing but dues to waste time and space. To alleviate this problem, relative entropy is used to reduce the non-gesture model states because there are many states with similar probability distribution. In addition, we measure the gesture spotting accuracy according to different window size from 1 to 8 (Fig. 4(a)). We noted that, the gesture spotting accuracy is improved initially as the sliding window size increase, but degrades as sliding window size increase further. Therefore, the optimal size of sliding window is 5 empirically. Also, result of one meaningful gesture spotting '6' is shown in Fig. 4(a) where the start point detection at frame 15 and end point at frame 50. In automatic gesture spotting task, there are three types of errors, namely, insertion, substitution and deletion. The insertion error occurs when the spotter detects

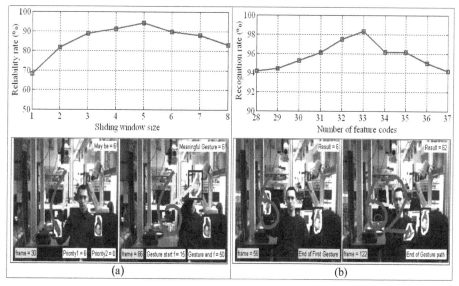

Fig. 4. (a) One meaningful gesture spotting '6' with spotting accuracy for different sliding window size from 1 to 8. (b) Gesture spotting '62'.

a nonexistent gesture. A substitution error occurs when the meaningful gesture is classified falsely. The deletion error occurs when the spotter fails to detect a meaningful gesture. Here, we note that some insertion errors cause the substitution errors or deletion errors where the insertion errors affect on the the gesture spotting ratio directly. Here, we note that some insertion errors cause the substitution errors or deletion errors where the insertion errors affect on the gesture spotting ratio directly. The reliability of automatic gesture spotting system is computed by Eq. 9 and achieved 94.35%. The reliability is the number of correctly recognized gestures over the number of tested gestures plus the number of insertion errors. Fig. 4(b) shows the results of continuous gesture path that contains within itself two meaningful gestures '6' and '2'.

$$Rel. = \frac{no.\ of\ correctly\ recognized\ gestures}{no.\ of\ test\ gestures\ +\ no.\ of\ insertion\ errors} \times 100\% \qquad (9)$$

6   Conclusion and Future Work

This paper proposes an automatic hand gesture spotting system for numbers from 0 to 9 in stereo color image sequences using Hidden Markov Models. Our system performs the hand gesture segmentation and recognition tasks simultaneously. Furthermore, it is suitable for real-time applications and solves the issues of time delay between the segmentation and the recognition tasks. The database contains 60 video sequences for each isolated gesture number (42 video sequences for training and 18 video sequences for testing) and 280 video sequences for continuous gestures. The results show that; the proposed system can successfully recognize isolated gestures with 98.33% and spotting meaningful gestures with

94.35% reliability. The future research will address the hand gesture spotting by fingertip trajectory instead of the hand centroid point to recognize American Sign Language over combined features using multi-camera system.

Acknowledgments. This work is supported by Forschungspraemie (BMBF-Förderung, FKZ: 03FPB00213) and Transregional Collaborative Research Centre SFB/TRR 62 "Companion-Technology for Cognitive Technical Systems" funded by DFG.

References

1. Mitra, S., Acharya, T.: Gesture Recognition: A Survey. IEEE Transactions on Systems, MAN, and Cybernetics, 311–324 (2007)
2. Yang, H., Park, A., Lee, S.: Gesture Spotting and Recognition for Human-Robot Interaction. IEEE Transaction on Robotics 23(2), 256–270 (2007)
3. Deyou, X.: A Network Approach for Hand Gesture Recognition in Virtual Reality Driving Training System of SPG. In: ICPR, pp. 519–522 (2006)
4. Kim, D., Song, J., Kim, D.: Simultaneous Gesture Segmentation and Recognition Based on Forward Spotting Accumlative HMMs. Journal of Pattern Recognition Society 40, 3012–3026 (2007)
5. Elmezain, M., Al-Hamadi, A., Michaelis, B.: Real-Time Capable System for Hand Gesture Recognition Using Hidden Markov Models in Stereo Color Image Sequences. Journal of WSCG 16(1), 65–72 (2008)
6. Takahashi, K., Sexi, S., Oka, R.: Spotting Recognition of Human Gestures From Motion Images. Technical Report IE92-134, pp. 9–16 (1992)
7. Lee, H., Kim, J.: An HMM-Based Threshold Model Approach for Gesture Recognition. IEEE Transactions on PAMI 21(10), 961–973 (1999)
8. Yang, H., Sclaroff, S., Lee, S.: Sign Language Spotting with a Threshold Model Based on Conditional Random Fields. IEEE Transactions on PAMI 31(7), 1264–1277 (2009)
9. Elmezain, M., Al-Hamadi, A., Michaelis, B.: Gesture Recognition for Alphabets from Hand Motion Trajectory Using HMM. In: IEEE International Symposium on Signal Processing and Information Technology (ISSPIT), pp. 1209–1214 (2007)
10. Elmezain, M., Al-Hamadi, A., Appenrodt, J., Michaelis, B.: A Hidden Markov Model-Based Continuous Gesture Recognition System for Hand Motion Trajectory. In: International Conference on Pattern Recognition (ICPR), pp. 519–522 (2008)
11. Yoon, H., Soh, J., Bae, Y.J., Yang, H.S.: Hand Gesture Recognition Using Combined Features of Location, Angle and Velocity. Journal of Pattern Recognition 34, 1491–1501 (2001)
12. Lawrence, R.R.: A Tutorial on Hidden Markov Models and Selected Applications in Speech Recognition. Proceeding of the IEEE 77(2), 257–286 (1989)
13. Nianjun, L., Brian, C.L., Peter, J.K., Richard, A.D.: Model Structure Selection & Training Algorithms for a HMM Gesture Recognition System. In: International Workshop in Frontiers of Handwriting Recognition, pp. 100–106 (2004)
14. Elmezain, M., Al-Hamadi, A., Michaelis, B.: A Novel System for Automatic Hand Gesture Spotting and Recognition in Stereo Color Image Sequences. Journal of WSCG 17(1), 89–96 (2009)
15. Cover, T.M., Thomas, J.A.: Entropy, Relative Entropy and Mutual Information. Elements of Information Theory, 12–49 (1991)

Objective Quality Evaluation of Laser Markings for Assembly Control

Jost Schnee[1], Norbert Bachfischer[2], Dirk Berndt[1], Matthias Hübner[2], and Christian Teutsch[1]

[1] Fraunhofer Institute for Factory Operation and Automation,
Sandtorstr. 22, 39106 Magdeburg, Germany
[2] Siemens AG, Industry Sector, I IA CD CC FTQ 31,
Werner-von-Siemens-Straße 48, 92220 Amberg, Germany

Abstract. This paper presents a novel method for the objective quality evaluation of letterings at the example of laser markings. The proposed method enables an automated quality evaluation during random inspection of small quantities of frequently changing items. The markings are segmented from the image within a first step and compared against a master piece in the second step. Each character and contour is evaluated with regard to completeness, contrast, edge width and homogeneity according to quality criteria based on the human visual perception. Thus our approach provides an user-independent evaluation method with high repeatability and reproducibility for its industrial application.

Keywords: letter quality, pattern analysis, template matching.

1   Introduction

Quality assurance has become a key instrument in the modern process of manufacturing. The field of quality evaluation of markings is still dominated by human inspection techniques. The motivation of our work is to objectify the quality evaluation of laser markings applied on plastic bodies for electric circuits to ensure a good readability. Actually, "good" is a subjective measure that varies between single persons. Our approach automates and objectifies this inspection using image processing techniques. Existing methods and measurement systems for the marking evaluation focus on specific tasks. Generally, variances between a golden sample and the current specimen are analyzed. [1] and [2] use neural networks and knowledge-based systems for quality evaluation while [3] compares a gray scale card against the current sample to evaluate the contrast. These methods are optimized for high clock frequencies and OK/NOK classifications of infrequently changing items. Compared to inspection systems humans have a larger dynamic range but lack the objectiveness of automated systems and do not achieve their high repeatability and reproducibility.

The purpose of our measurement system is to evaluate the marking quality in order to control and optimize the laser marking process according to the human visual perception. To allow rapid changes in the product appearance

D. Ślęzak et al. (Eds.): SIP 2009, CCIS 61, pp. 17–24, 2009.

a high flexibility is required as well. The considered markings are brought on ten materials with five different colors. The variety of the marking patterns is larger than one thousand, each of them consisting of several characters, symbols and contours. Thus, our approach avoids learning steps as it would be required for knowledge-based methods. It takes advantage of the fact that the master marking is stored as a vector graphic, which is used as a reference without teaching the system or searching for a faultless golden sample. The evaluation method includes completeness, contrast, edge width and homogeneity.

To identify the contours the master marking is aligned with the marking on the specimen first. Thereto a Canny edge detector [4] is deployed for segmentation. The mapping between the coordinates of the master marking and the examined marking uses an ICP algorithm [5] based on Horn's method to find a closed-form solution of absolute orientation [6]. The characteristic of the examined specimen allow for a histogram based peak-and-valley element segmentation algorithm comparable with the Gaussian smoothing procedure from Tsai [7]. Homogeneity measures are derived from co-occurrence matrices according to [8].

The optical resolution limit or point spread function of the human eye is 30 seconds of arc, and corresponds to the spacing of receptor cells on the fovea [9]. Therefore the theoretical minimum separabile of a sinusoidal pattern is 60 periods per degree. This equates to a visus of 2.0 [10] which is never reached in reality. At a visus of 2.0 the minimum separabile at a distance of 0.35 m is approximately 0.1 mm. Taking image sampling into account we chose a resolution for the camera system of 0.05 mm per pixel. Furthermore, the optical resolution limit of the human eye depends on the contrast. Thus the minimum separabile declines with descending contrast.

2   Marking Evaluation

In the following section we discuss a new method applied for objective quality evaluation of markings. In the first paragraph the marking selection and the template matching procedures are presented. Afterwards we propose an new method for distortion correction, which is needed to handle influences from optical systems. Finally a comparative study evaluates the results from a poll about human visual perception with the objective measures in the last paragraphs.

2.1   Marking Segmentation and Template Matching

The precondition for the evaluation is to segment the markings in the image and to align them with the master markings. This is done with a multistage approach as shown in Fig. 1. First the contours of the markings and the specimen are segmented using the Canny edge detector. Second the master markings are mapped to the segmented markings with an ICP algorithm. Third the master markings are finally aligned to the segmented markings with a new method for fractional template matching which reduces errors due to lens distortions. Finally each marking element is, again, segmented from its local environment.

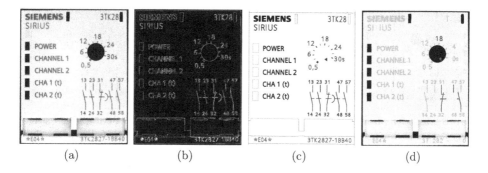

(a)   (b)   (c)   (d)

Fig. 1. Segmentation of the markings on a specimen (a), matching the edges (b) with a given vector graphic of the master marking (c), and final evaluation result from green (ok) to red (not ok) (d)

Thereafter the translation $\mathbf{t}$, the rotation $\mathbf{R}$ and the scaling $\mathbf{s}$ are calculated in order to map the master marking to the segmented marking. This is accomplished by an ICP algorithm supported by Kd-trees [11] with O(n) median partitioning [12]. The algorithm efficiently minimizes the euclidean distance $\mathbf{E}$ between the points of the segmented marking $\mathbf{m}_i$ and their nearest neighbor $\mathbf{d}_i$ from the transformed points of the master marking.

$$\mathbf{E}(\mathbf{s}, \mathbf{R}, \mathbf{t}) = \frac{1}{N} \sum_{i=1}^{N} \mathbf{m}_i - (\mathbf{s}\mathbf{R}\mathbf{d}_i + \mathbf{t}) \tag{1}$$

To reduce the influence of lens distortions from the optical systems the master marking contours are distorted instead of undistorting the taken image. This step preserves the input quality and is more efficient since it does not require an image distortion with time-consuming interpolations. Once translation, rotation and scaling are calculated, the whole master marking is mapped to the segmented marking by applying the global transformations.

The remaining distortions are corrected by the local adjustment as shown in section 2.2. In the last step the elements on the specimen are segmented in the local area of the elements of the master marking. This performed by a binarization algorithm benefiting from the bimodal histograms of the specimen [13]. After smoothing the histogram with a Gaussian, our algorithm computes the two peaks and the valley in between. The threshold is set to the point where the histogram falls below a value of 1% of the background peak or to the minimum.

2.2 Distortion Correction by Raster Alignment

There are several influences that affect the appearance of the taken image from the master. These are distortions due to the imaging and laser marking system (optical mapping through laser/camera optic, alignment of laser/camera

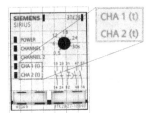 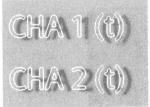 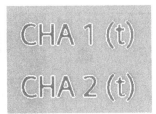

(a) Raster for fractional template matching.

(b) Master marking before template matching.

(c) Master marking after template matching.

Fig. 2. Raster approach for local template matching and detail at the beginning and the end of the whole matching process

and specimen). After having corrected the camera optics distortion the remaining distortions of the laser optics remain and are adjusted by a local template matching. Thereto the image is divided into sub-images in a raster of 100x100 pixels as shown in Fig. 2(a) which corresponds to a representative region of 0.5x0.5 mm. For every sub-image a template matching limited to translations between the equivalent part of the master marking and the segmented marking is accomplished. The translations are ordered in a grid positioned at the centers of the sub-images. The translation of every template pixel is calculated by bilinear interpolation from the translations of the adjacent grid positions. Compared to image correction based on the camera system this approach benefits from being able to adjust distortions due to the laser marking system without affecting the image quality as shown in Fig. 2(b)(c).

2.3 Poll about Human Perception

To ensure the results of the inquest about human perception a poll was accomplished. In this poll 15 test persons were interviewed twice. They had to evaluate four multiplied by ten specimen out of 4 different materials, which showed five markings in five different sizes and qualities. All test persons had to grade every marking into OK/partly OK/NOK respecting contrast, edge width (sharpness) and homogeneity. The analysis of the poll showed the following facts: There are strong correlations in the evaluations of specimen with good and bad quality but the evaluations of specimen with medium quality showed a lot of differences. For the overall impression contrast is more important than edge width (sharpness) and homogeneity. Furthermore, bigger markings (2.8 mm) are graded better than smaller ones (1.0 mm) and textures with a period length smaller than 0.2 mm do not affect the perception of homogeneity.

These facts show that humans generally have a good visual perception even for small differences, but each test person has its own threshold to distinguish between good, medium and bad quality. Furthermore minimum size and minimum contrast are most important. Although there was a certain variance between the

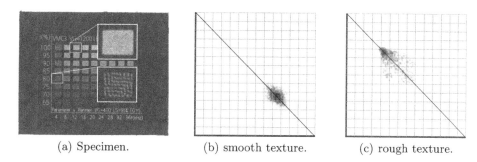

(a) Specimen.   (b) smooth texture.   (c) rough texture.

Fig. 3. Specimen and the different value distributions within the co-occurrence matrices for smooth and rough textured regions

subjective quality evaluations we could derive borders which were mapped to thresholds discussed in the following section.

2.4   Evaluation Methods

First the completeness of each object is computed by comparison with its equivalent in the master markings. If the completeness of one object does not reach the threshold it is discarded. Otherwise it is merged with its equivalent in the master markings to build a perfectly shaped model for the next calculations.

Afterwards, Michelson contrast [14] C of every object is derived from the intensities I_{max} and I_{min} of foreground and background. Subsequently the edge width E is calculated from the intensity difference I of each object and the average gradient magnitude along the object edges G.

$$C = (I_{max} - I_{min})/(I_{max} + I_{min}), \qquad I = I_{max} - I_{min} \qquad (2)$$

At last the homogeneity is calculated from 4 different criteria. These are covariance and contrast out of co-occurrence matrices, standard deviation and maximum intensity difference. The maximum intensity difference D is the difference between the average of the 100 highest intensity values I_{high} and the average of the 100 lowest intensity values I_{low} of the object normalized with the intensity difference I of object and background.

$$E = I/G, \qquad D = (I_{high} - I_{low})/I \qquad (3)$$

The co-occurrence matrices M consist of the occurrences $P_{x,y}$ of all intensity differences between adjacent points x and y along the coordinate axes and represent textures as shown in Fig. 3. The covariance COV along one axis a is derived from the differences to the local mean and normalized with the sum of the occurrences of all intensity differences p.

$$COV = \frac{1}{p}\sum_{x=0}^{255}\sum_{y=0}^{255}(x - \bar{x})(y - \bar{y})P_{x,y} \qquad COV = \sum_{a=0}^{1}COV^a\frac{p^a}{\sum\limits_{o=0}^{1}p^a} \qquad (4)$$

The local contrast C_L along one axis is calculated by the intensity difference between adjacent points and normalized with p.

$$C_L = \frac{1}{p}\sum_{x=0}^{255}\sum_{y=0}^{255}|x-y|\,P_{x,y} \qquad C_L = \sum_{a=0}^{1}C_L^a\,\frac{p^a}{\sum_{o=0}^{1}p^a} \qquad (5)$$

3   System Construction

To achieve a high repeatability and reproducibility a measuring system with constant characteristics was needed. Since the measuring system is based on a camera, the scene illumination is crucial. The main component is a hemispherical dome with a circular fluorescent lamp, an aperture and a circular opening for the camera at the crest. It ensures a homogeneous lighting and, thus, makes the measurement independent from external influences. The aperture prevents the object from being hit by light rays directly.

In order to handle different objects and heights an additional vertical positioning system with a controllable translatory stage has been integrated. It ensures the correct height above the specimen, which is necessary because of the small depth of field of 5 mm and to ensure the accurate ex ante scale of the master marking. The positioning step is additionally supported by showing the contours and the marking of the specimen on the display of the measurement software.

4   Verification

In order to introduce the system to the production process a Gage R&R analysis had to be performed. Therefore, we have verified the complete system and the proposed evaluation procedure with a set of 45 different test items, which have been previously classified by humans. To use the measuring system for OK/partly OK/NOK classification in the practical application, thresholds for the evaluation criteria according to the results of the survey about human perception have been ascertained (see Table 1). The feasibility of these thresholds was verified by comparative tests between human quality inspectors and the

Table 1. Thresholds for OK/partly OK/NOK classifications of ti-gray specimen

classifi-cation	completeness in %	contrast	edge width in mm	homogeneity covariance	contrast	max. diff.	σ
OK	>90	>0.25	<0.1	<100	<10	<0.9	<12
partly OK	80-90	0.18-0.25	0.1-0.15	100-220	10-14	0.9-1.2	12-18
NOK	<80	<0.18	>0.15	>220	>14	>1.2	>18

measuring system. For these tests the specimen used for the survey and other specimen from the same materials (marked with the same laser parameters) were used. The correlation of the results was higher than 90 % as shown in Table 2.

In order to determine the repeatability and reproducibility of the measuring system two test series were accomplished. The standard deviations (σ) were less than 5% of the mean value as shown in Table 3. The results in Table 4 show the high repeatability and reproducibility which was needed to successfully pass the verification procedure.

Table 2. Correlation between the mean evaluations of the survey about human perception and the results of the measuring system

quality classification	ti-gray	pale gray	yellow
OK	94.926	89.793	92.175
partly OK	87.038	86.305	89.259
NOK	93.084	90.372	91.583

Table 3. Results of 25 measurements of a ti-gray specimen to compute the repeatability using σ. The specimen has been re-positioned between the measurements.

evaluation criteria	mean	σ	σ/mean in ‰
completeness in %	99.998	0.047	0.470
contrast	0.375	0.003	8.00
edge width in mm	0.084	0.001	11.905
local covariance	32.478	1.363	41.967
local contrast	4.656	0.108	23,196
standard deviation	7.188	0.133	18.503
maximum difference	0.529	0.009	17.013

Table 4. Results of three times three measures of 10 specimen to collect the repeatability and reproducibility. The measures were performed triple by three persons.

evaluation criteria	person 1 mean	range	person 2 mean	range	person 3 mean	range
completeness in %	99.845	0.164	99.793	0.187	99.906	0.158
contrast	0.371	0.006	0.369	0.008	0.372	0.005
edge width in mm	0.093	0.004	0.092	0.004	0.087	0.006
local covariance	34.651	2.459	34.527	2.574	34.782	2.628
local contrast	5.048	0.184	4.983	0.198	5.179	0.212
standard deviation	7.863	0.265	7.738	0.300	8.073	0.309
maximum difference	0.627	0.023	0.664	0.035	0.668	0.031

5 Conclusion

We have presented a new method for objective quality evaluation of laser markings on plastic items. This method utilizes a template comparison of the markings which are to be evaluated only. The areas preset by the template are positioned via a two step ICP algorithm before the evaluation is accomplished. The chosen evaluation criteria completeness, contrast, edge width and homogeneity have been related to the human visual perception. To ensure a homogeneous lighting a measure dome that causes diffuse light and that is independent from external influences was built. In this work we have shown the high repeatability and reproducibility of the measuring system as well. Its ability to switch between different markings and to evaluate new markings without delay, as required when producing small quantities of frequently changing goods, makes it most useful for industrial applications.

References

1. Petry, J.P., Picard, L.L.: Mark quality inspection apparatus and method. US5859923. Cognex Corporation, Natick (1999)
2. Szatmari, I., Zarandy, A.: High-speed label inspection system for textile industry. In: 10th International Conference on IMEKO TC10 Technical Diagnostics, Budapest, vol. 1, pp. 99–106 (2005)
3. Peng, S.T.: Contrast measurement system for laser marks. US5969374. Hewlett-Packard Company, Palo Alto (1999)
4. Canny, J.: A computational approach to edge detection. IEEE Transactions on Pattern Analysis and Machine Intelligence (PAMI) 8, 679–698 (1986)
5. Besl, P., McKay, N.: A method for registration of 3-d shapes. IEEE Transactions on Pattern Analysis and Machine Intelligence (PAMI) 14, 239–256 (1992)
6. Horn, B.K.P., Hilden, H.M., Negahdaripour, S.: Closed-form solution of absolute orientation using orthonormal matrices. Journal of the Optical Society of America 5, 1127–1135 (1988)
7. Tsai, D.: A fast thresholding selection procedure for multimodal and unimodal histograms. Pattern Recognition Letters 16, 653–666 (1995)
8. Pietikäinen, M., Ojala, T., Xu, Z.: Rotation-invariant texture classification using feature distributions. Pattern Recognition Letters 33, 43–52 (2000)
9. Mallot, H.: Computational Vision. In: Information Processing in Perception and Visual Behavior, 2nd edn. The MIT Press, Cambridge (2000)
10. Graef, M.: Objektive pruefung der sehschaerfe. Zeitschrift der Ophthalmologie 97, 582–600 (2000)
11. Nuechter, A., Lingemann, K., Hertzberg, J.: Cached k-d tree search for icp algorithms. In: Sixth International Conference on 3-D Digital Imaging and Modeling, Montreal, vol. 1, pp. 419–426 (2007)
12. Press, W.H., Teukolsky, S.A., Vetterling, W.T., Flannery, B.P.: Numerical Recipes in C++: The Art of Scientific Computing, 2nd edn. Cambridge University Press, Cambridge (2002)
13. Rosin, P.L.: Unimodal thresholding. Pattern Recognition 34(11), 2083–2096 (2001)
14. Michelson, A.: Studies in Optics. U. of Chicago Press, Chicago (1927)

An Improved Object Detection and Contour Tracking Algorithm Based on Local Curvature

Jung-Ho Lee, Fang Hua, and Jong Whan Jang

PaiChai University, Daejeon, South Korea, 302-735
{style0428,fanghua,jangjw}@pcu.ac.kr

Abstract. Using the classical snake algorithm it is difficult to detect the contour of an object with complex concavities. Whereas the GVF (Gradient Vector Flow) method successfully detects the concavity of a contour, but consumes lots of time to compute the energy map. In this paper, we propose a fast snake algorithm to reduce computation time and to improve the performance of detecting and tracking the contour. In order to represent the object's contour accurately, a snake point inserting and deleting strategy is also proposed. Simulation results from a sequence of images show that our method performs well in detecting and tracking the object's contour.

Keyword: active contour model, snake algorithm, object detection, object contour tracking, local curvature.

1 Introduction

Various object tracking schemes for object extraction of 2D images have been developed over the last two decades [1-3, 6-7]. A very efficient method for tracking the object's contour is the active contour method, also commonly known as snake, which was first introduced by Kass[1]. The snake is an energy-minimizing spline guided by external constraint forces and is influenced by image forces that pull it toward features such as lines and edges. Object tracking utilizing the snake-based object segmentation approach has been explored in various research works in recent years. Unfortunately, this algorithm has a primary limitation. It is unable to detect concavities in the object's boundary because of the insufficiency of energy at the beginning of the concavities. To solve this, Xu[2] proposed an optical flow utilized method: "Gradient Vector Flow (GVF): a new external force for snakes". The resulting field has a large capture range and forces the active contour into concave regions. But it consumes a lot of time to compute the energy map. In addition this method is difficult to utilize for tracking object from a sequence of images [4-5, 8-11]. In order to improve these limitations of the snake algorithm, this paper proposes a new snake algorithm to reduce the computation time and promote performance in tracking the contour of an object from a sequence of images.

This paper is organized as follows. Chapter 2 covers an overview on active contour model, and chapter 3 presents our improved algorithm, and in chapter 4 we present the performance evaluation of our method. Conclusions are discussed in chapter 5.

D. Ślęzak et al. (Eds.): SIP 2009, CCIS 61, pp. 25–32, 2009.
© Springer-Verlag Berlin Heidelberg 2009

2 Active Contour Model

The active contour model, more widely known as the snake model, in the discrete formulation the contour is represented as a set of snake points $v_i = (x_i, y_i)$ for $i = 0,..., N-1$ where x_i and y_i are the x and y coordinates of the ith snake point, respectively and N is the total number of snake points. The energy function which minimizes the snake is expressed as follows.

$$E_{snake}(v) = \sum_{i=0}^{N-1}(E_{internal}(v_i) + E_{external}(v_i)) \tag{1}$$

where, $E_{internal}$ represents the internal energy term and $E_{external}$ represents the external energy term. The internal energy is composed of two terms.

$$E_{internal}(v_i) = E_{continuity}(v_i) + E_{curvature}(v_i) \tag{2}$$

The greedy snake method[3] maintains the points in the snake more evenly spaced than the original model of Kass.

$$E_{continuity}(v_i) = \frac{\left|\overline{d} - \|v_{i+1} - v_i\|\right|}{con_{max}} \tag{3}$$

where, con_{max} is the maximum value in the search neighborhood to which the point may move and $\overline{d} = \frac{1}{N}\sum_{i=0}^{N-1}\|v_{i+1} - v_i\|$. The main role of the continuity energy term is to create even spacing between the snake points.

$$E_{curvature}(v_i) = \frac{\|v_{i-1} - 2v_i + v_{i+1}\|}{cur_{max}} \tag{4}$$

where, cur_{max} is the maximum value in the search neighborhood to which the point may move. And the main role of the curvature energy term is to form a smooth contour between neighboring snake points.

$$E_{external}(v_i) = \frac{-|\nabla f(v_i)|}{ext_{max}} \tag{5}$$

where, $|\bullet|$ represents a scalar absolute value, ∇ is the gradient operator and ext_{max} is the maximum option in the search neighborhood. The external energy defines the gradient at a snake point. It helps the snake points to migrate toward the contour of the target object.

3 Proposed Algorithm

The curvature is the degree of a curve described by the change rate. To detect the details of a target object with complex contours, more snake points have to be initialized around this object. This paper proposes a new snake algorithm based on calculating the local curvature of the object's boundaries. We set a curvature threshold value first, and if the snake contour's curvature is higher than this value, new snake points are inserted. That way more details of the object's boundary could be detected. On the other hand, if the local curvature at a snake point is small enough, this kind of unnecessary points will be removed to better describe and track the object's contours.

3.1 Curvature Threshold and Inserting Additional Snake Points

Figure 1 shows the curvature of the object's boundaries. The curvature of the boundary in (b) is higher than the one in (a). The part of the object's boundary in (a) can almost be described by three snake points. But in case of (b), more snake points are of advantage. Additional points will be inserted at the snake contour (v_{a1}, v_{a2}) as initialization. Then a new convergence iteration starts to calculate the new local curvature of the snake contour. The iteration will stop when the local curvature does not exceed the threshold any more.

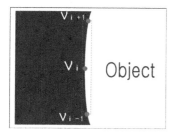

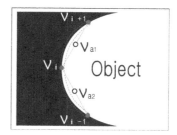

(a) Low-level Curvature   (b) High-level Curvature

Fig. 1. The curvature of the object's boundaries and the inserted new snake points (v_{a1}, v_{a2})

The curvature k, can be computed by the adjacent three snake points (v_{i-1}, v_i, and v_{i+1}).

Let two vectors

$$\vec{A} = v_{i-1} - v_i, \vec{B} = v_{i+1} - v_i \tag{6}$$

The distance of v_{i-1} and v_{i+1} and the angle between two vectors are d and θ respectively.

$$\cos \theta = \frac{\vec{A} \Box \vec{B}}{\left\| \vec{A} \right\| \left\| \vec{B} \right\|} \tag{7}$$

$$k = \frac{2\sin\theta}{d} \qquad (8)$$

where, $\sqcup$ represents the vectors inner product and $\|\ \|$ the represents the Euclidean distance.

High curvature implies that the object has complex contours. The threshold is determined by the measurement of the curvature value the adjacent three snake points made of.

$$k < k_{th} \qquad (9)$$

Once the value of the snake contour's curvature is above the threshold, two additional snake points, $v_{a1} = (v_{i-1} + v_i)/2$, $v_{a2} = (v_i + v_{i+1})/2$ are inserted so the detail of the contours can be detected, just like described in Figure 1 (b).

3.2 Object Contour Tracking

This paper proposes an object contour tracking method adapting to a video sequence. According to the characters of the object's shape in different images, the object is divided into fixed shape object and variable shape object. In this paper, we pay our attention to the variable shape object's contour tracking. From a continuous frame, the local curvature of the object's contour changes along with the object's movement. As Figure 2 shows, as the object rotated from shape (a) to shape (b), the local curvature of the contour also changed.

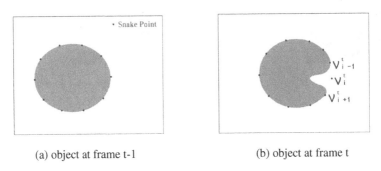

(a) object at frame t-1   (b) object at frame t

Fig. 2. Various object contours at different frames

Here, t-1 means the previous frame of t and v_i^t presents the ith snake point at time t.

3.3 Inserting and Deleting Snake Points

As described in 3.1, we first calculate the local curvature and then compare it with the threshold. If the curvature is above the threshold, two additional points (v_{a1}^t, v_{a2}^t) are inserted as Figure 3 illustrates.

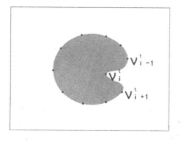 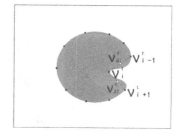

(a) snake point moved into concave region   (b) inserted snake points v^t_{a1}, v^t_{a2}

Fig. 3. Snake points insertion at frame t

At frame t, we substitute v^t_i for v_i to describe a snake point, then the neighboring points are v^t_{i-1} and v^t_{i+1}. Depending on the formulas from (6) to (9), we calculate the local curvature k^t. If k^t is not satisfying (9), additional points will be inserted and new convergence iteration starts, as presented in 3.1.

In the event that k^t calculated by continuous three snake points ($v^t_{i-1}, v^t_i, v^t_{i+1}$) is not satisfying (7), the three points are likely to be on a straight line. In this case, v^t_i can be safely removed as the two adjacent points can describe the boundary well enough on each side. And we use the remaining points to track the object's contour at frame t+1. Removing unnecessary points decreases the overall number of snake points and allows the tracking to be performed more efficiently.

4 Experiments

To verify the performance of the proposed algorithm, a set of experiments was performed. The algorithm was coded in Visual C++ 6.0 and the experiments were executed on an Intel Core2 machine running at 1.86GHz with DDR2 2GB RAM. The tracking experimental video was captured 15 Fps by a webcam (Logitech QuickCam) with the image size 320*240. The snake points were initialized at the first frame.

4.1 Performance of Detection Analysis and Comparison

We compared the performance of our proposed algorithm with the Greedy Snake algorithm and the GVF [2][4]. To measure the performance with a changing number of snake points, we define the Relative Shape Distortions as:

$$RSD(R) = \left(\sum_{(x,y) \in f} R_{original}(x,y) \oplus R_{estimation}(x,y) \right) \Bigg/ \sum_{(x,y) \in f} R_{original}(x,y) \qquad (10)$$

Where the $\oplus$ sign is the binary XOR operation. $R_{original}(x,y)$ are the coordinates of the real object's boundary and $R_{estimation}(x,y)$ are the estimated coordinates of the

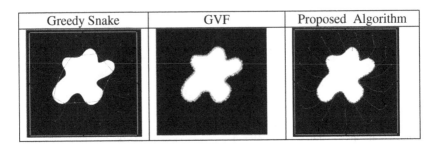

Fig. 4. Comparison with previous well known algorithms (k_{th} = 0.01)

Table 1. Comparison performances (N_i = 15)

	Greedy Snake	GVF	Proposed
T	0.011s	18.05s	0.079s
RSD	0.246	0.000	0.004

object's contour. The images of the experimental results are shown by three columns as follow:

In these results, T represents the processing time with 15 initialized snake points. Using Greedy Snake, the detection of the contour is inaccurate, as the initial snake points are not enough. The RSD value is too high to describe the correct edges. GVF reached a good result, but it took a long time. Therefore, this method can't be utilized into object tracking from a sequence of video frames. Our proposed algorithm needs little more time than Greedy Snake, but the results were more successful in all cases of the experiment.

4.2   Performance of Tracking Analysis and Comparison

First, we simulated the fixed shape object's contour tracking, we compare the block matching algorithm (BMA) based on Greedy snake with our proposed method [5]. The results for tracking the objects with boundary concavities are shown in Figure 5 and Table 2.

The first column shows the results for the traditional algorithm (BMA) could not detect the boundary of the OOI (object of interest) due to poor energy in the boundary concavities. Our proposed algorithm, as shown in the second column, the RSD value is small enough and the snake falls exactly on the edges of the object.

Figure 6 shows the results of an experiment on an object with variable shapes. We can see the closed snake contour tightly locks on the fist even though the fist is rotating.

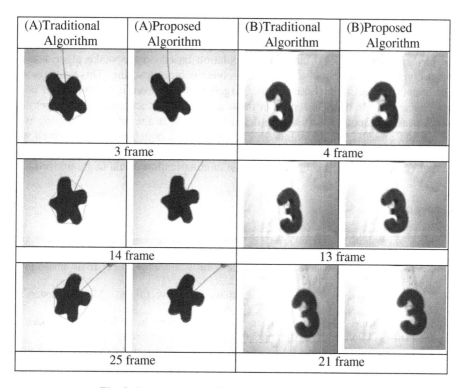

(A)Traditional Algorithm	(A)Proposed Algorithm	(B)Traditional Algorithm	(B)Proposed Algorithm
3 frame		4 frame	
14 frame		13 frame	
25 frame		21 frame	

Fig. 5. Contour tracking fixed shape object ($k_{th} = 0.01$)

Table 2. Performance evaluation on different frames

Frame	RSD		Frame	RSD	
	(A)Traditional	(A)Proposed		(B)Traditional	(B)Proposed
3	0.25	0.018	4	0.32	0.024
14	0.33	0.014	13	0.29	0.026
25	0.41	0.015	21	0.34	0.021

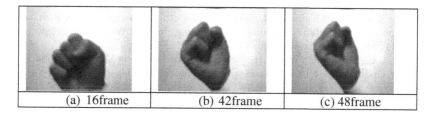

(a) 16frame	(b) 42frame	(c) 48frame

Fig. 6. Contour tracking variable shape object ($k_{th} = 0.01$)

5   Conclusions

We have presented a local curvature calculation based on the snake algorithm for object contour detection and a scheme for object contour tracking. This snake model is efficient in detecting and tracking the targeted contour even if the object has complex concavities. In order to represent the object's contour accurately, this paper also proposes a snake point inserting and deleting method. In the future, comparing with the method of setting snake points around the object's boundaries, our research will focused on how to automatically set snake points at the correct location where the contour could be detected better and the cost of computation would be minimized.

References

1. Kass, M., Witkin, A., Terzopoulos, D.: Snake: Active Contour Models. Int'l J. Computer Vision 1(4), 321–331 (1987)
2. Xu, C., Prince, J.L.: Snakes, Shapes, and Gradient Vector Flow. IEEE Trans. Image Processing 7(3), 359–369 (1998)
3. Lam, K.M., Yan, H.: Fast Greedy Algorithm for Active Contour. Electron Letters 30, 21–23 (1994)
4. Lin, Y.T., Chang, Y.L.: Tracking Deformable Objects with the Active Contour Model. In: International Conference on Multimedia Computing and System, pp. 608–609 (1997)
5. Pardas, M., Sayrol, E.: Motion Estimation Based Tracking of Active Contours. Pattern Recognition Letters 22, 1447–1456 (2001)
6. Kim, S.H., Alatter, A., Jang, J.W.: Accurate Contour Detection Based on Snake for Objects with Boundary Concavities. In: Campilho, A., Kamel, M.S. (eds.) ICIAR 2006. LNCS, vol. 4141, pp. 226–235. Springer, Heidelberg (2006)
7. Sum, K.W., Cheung, Y.S.: Paul: Boundary Vector Field for Parametric Active Contours. Pattern Recognition 40(6), 1635–1645 (2007)
8. Choi, J.J., Kim, J.S.: Modified Energy Function of the Active Contour Model for the Tracking of Deformable Objects. International Journal of Precision Engineering and Manufacturing 7(1), 47–50 (2007)
9. Shin, J.H., Kim, S.J., Kang, S.K., Lee, S.W., Park, J.K., Abidi, B., Abidi, M.: Optical Flow-Based Real-time Object Tracking Using Non-prior Training Active Feature Model. Elsevier Real-Time Imaging 11, 204–218 (2005)
10. Seo, K.H., Shin, J.H., Kim, W., Lee, J.J.: Real-Time Object Tracking and Segmentation Using Adaptive Color Snake Model. International Journal of Control, Automation, and Systems 4(2), 236–246 (2006)
11. Liu, H., Jiang, G., Wang, L.: Multiple Objects Tracking Based on Snake Model and Selective Attention Mechanism. International Journal of Information Technology. 12(2), 76–86 (2006)
12. Goumeidane, A.B., Khamadja, M., Odet, C.: Parametric Active Contour for Boundary Estimation of Weld Defects in Radiographic Testing. In: 9th International Symposium on Signal Processing and Its Applications, 2007. ISSPA 2007, pp. 1–4 (2007)
13. Tseng, C.C., Hsieh, J.G., Jeng, J.H.: Active Contour Model via Multi-population Particle Swarm Optimization. Expert Systems with Applications 36, 5348–5352 (2009)

An Efficient Method for Noisy Cell Image Segmentation Using Generalized α-Entropy

Samy Sadek[1], Ayoub Al-Hamadi[1], Bernd Michaelis[1], and Usama Sayed[2]

[1] Institute for Electronics, Signal Processing and Communications
Otto-von-Guericke-University Magdeburg, Germany
[2] Assiust University, Egypt
{Samy.Bakheet,Ayoub.Al-Hamadi}@ovgu.de

Abstract. In 1953, a functional extension by A. Rènyi to generalize traditional Shannon's entropy known as α-entropies was proposed. The functionalities of α-entropies share the major properties of Shannon's entropy. Moreover, these entropies can be easily estimated using a kernel estimate. This makes their use by many researchers in computer vision community highly appealing . In this paper, an efficient and fast entropic method for noisy cell image segmentation is presented. The method utilizes generalized α-entropy to measure the maximum structural information of image and to locate the optimal threshold desired by segmentation. To speed up the proposed method, computations are carried out on 1D histograms of image. Experimental results show that the proposed method is efficient and much more tolerant to noise than other state-of-the-art segmentation techniques.

Keywords: α-entropy, Cell image, Entropic image segmentation.

1 Introduction

The recent developments in Digital Mammography (DM), Magnetic Resonance Imaging (MRI), Computed Tomography (CT), and other diagnostic imaging techniques provide physicians with high resolution images which have significantly assisted the clinical diagnosis. These up-to-date technologies not only have a recognizably increased knowledge of normal and diseased anatomy for medical research but also become a significant part in diagnosis and treatment planning [1]. Due to the increasing number of medical images, taking advantage of computers to facilitate the processing and analyzing this huge number of images has become indispensable. Especially, algorithms for the delineation of anatomical structures and other regions of interest are a key component in assisting and automating specific radiological tasks. These algorithms, named image segmentation algorithms, play a fundamental role in many medical imaging applications such as the quantification of tissue volumes [2], diagnosis [3], localization of pathology [4], study of anatomical structure [5], treatment planning [6], partial volume correction of functional imaging data [7], and computer integrated surgery [8],[9]. Techniques for carrying out segmentations vary broadly

D. Ślęzak et al. (Eds.): SIP 2009, CCIS 61, pp. 33–40, 2009.
© Springer-Verlag Berlin Heidelberg 2009

depending on some factors such as specific application, imaging modality, etc. For instance, the segmentation of brain tissue has different requirements from the segmentation of the liver [10]. General imaging artifacts such as noise, partial volume effects, and motion can also have significant consequences on the performance of segmentation algorithms. Additionally, each imaging modality has its own idiosyncrasies with which to contend. There is currently no single segmentation technique that gives satisfactory results for each medical image.

Since the pioneering work by Shannon [11] in 1948 , entropy appears as an attention-grabbing tool in many areas of data processing. In 1953, Rènyi [14] introduced a wider class of entropies known as α-entropies. The functionalities of α-entropies share the major properties of Shannon's entropy. Moreover, the α-entropies can be easily estimated using a kernel estimate. This makes their use attractive in many areas of image processing [12]. In this paper we propose an efficient entropic technique for segmenting cell images which utilizes generalized Rènyi entropy. Our work for cell image segmentation has a relatively good performance in comparison to other related state-of-the-art techniques [13].

The outline of this paper is as follows. The next section discusses the generalized form of α-entropies especially generalized Rènyi entropy. The proposed entropic segmentation method is explained in section 3. Section 4 is to present the simulation results that validate the use of the proposed method. Advantages of our method and concluding remarks are outlined in Section 5.

2   Entropy of Generalized Distributions

Entropy has first appeared in thermodynamics as an information theoretical concept which is intimately related to the internal energy of the system. Then it has applied across physics, information theory, mathematics and other branches of science and engineering. When given a system whose exact description is not precisely known, the entropy is defined as the expected amount of information needed to exactly specify the state of the system, given what we know about the system.

Suppose $P = \{p_1, p_2, ..., p_n\}$ be a finite discrete probability distribution that satisfies these conditions $p_k \geq 0, k = 1, 2, ..., n$ and $\sum_{k=1}^{n} p_k = 1$. The amount of uncertainty of the distribution P, is called the entropy of the distribution, P. The Shannon entropy of the distribution, P, a measure of uncertainty and denoted by $H(P)$), is defined as

$$H(P) = -\sum_{k=1}^{n} p_k \log_2 p_k \qquad (1)$$

It should be noted that the Shannon entropy given by Eq. (1) is additive, i.e., it satisfies the following

$$H(A + B) = H(A) + H(B) \qquad (2)$$

for any two distributions A and B. Eq.(2) states one of the most important properties of entropy, namely, its additivity: the entropy of a combined experiment consisting of the performance of two independent experiments is equal to the sum of the entropies of these two experiments. The formalism defined by Eq. (1) has been shown to be restricted to the Boltzmann-Gibbs-Shannon (BGS) statistics. However, for nonextensive systems, some kind of extension appears to become necessary. Rènyi entropy, which is useful for describing the nonextensive systems, is defined as

$$H_\alpha(P) = \frac{1}{1-\alpha} \log_2 \sum_{k=1}^{n} p_k^\alpha \tag{3}$$

where $\alpha \geq 0$ and $\alpha \neq 1$. The real number α is called an entropic order that characterizes the degree of nonextensivity. This expression reduces to Shannon entropy in the limit $\alpha \longrightarrow 1$. We shall see that in order to get the fine characterization of Rànyi entropy, it is advantageous to extend the notion of a probability distribution, and define entropy for the generalized distributions. The characterization of measures of entropy (and information) becomes much simpler if we consider these quantities as defined on the set of generalized probability distributions.

Suppose $[\Omega, P]$ be a probability space that is, Ω an arbitrary nonempty set, called the set of elementary events, and P a probability measure, that is, a nonnegative and additive set function for which $P(\Omega)$. Let us call a function $\xi = \xi(\omega)$ which is defined for $\omega \in \Omega_1$, where $\Omega_1 \subset \Omega$. If $P(\Omega_1) = 1$ we call ξ an ordinary (or complete) random variable, while if $0 < P(\Omega_1) \leq 1$ we call ξ an incomplete random variable. Evidently, an incomplete random variable can be interpreted as a quantity describing the result of an experiment depending on chance which is not always observable, only with probability $P(\Omega_1) < 1$. The distribution of a generalized random variable is called a generalized probability distribution. Thus a finite discrete generalized probability distribution is simply a sequence $p_1, p_2, ..., p_n$ of nonnegative numbers such that setting $P = \{p_k\}_{k=1}^n$ and taking

$$\varpi(P) = \sum_{k=1}^{n} p_k \tag{4}$$

where $\varpi(P)$ is the weight of the distribution and $0 < \varpi(P) \leq 1$. A distribution that has a weight less than 1 will be called an incomplete distribution. Now, using Eq. (3) and Eq. (4), the Rànyi entropy for the generalized distribution can be written as

$$H_\alpha(P) = \frac{1}{1-\alpha} \log_2 [\frac{\sum_{k=1}^{n} p_k^\alpha}{\sum_{k=1}^{n} p_k}] \tag{5}$$

Note that Rànyi entropy has a nonextensive property for statistical independent systems, defined by the following pseudo additivity entropic formula

$$H_\alpha(A + B) = H_\alpha(A) + H_\alpha(B) + (\alpha - 1) \cdot H_\alpha(A) \cdot H_\alpha(B) \tag{6}$$

3   Proposed Segmentation Method

Image segmentation problem is considered to be one of the most holy grail challenges of computer vision field especially when done for noisy images. Consequently it has received considerable attention by many researchers in computer vision community. There are many approach for image segmentation, however, these approach are still inadequate. In this work, we propose an entropic method that achieves the task of segmentation in a novel way. This method not only overcomes image noise, but also utilizes time and memory optimally. This wisely happens by the advantage of using the Rànyi entropy of generalized distributions to measure the structural information of image and then locate the optimal threshold depending on the postulation that the optimal threshold corresponds to the segmentation with maximum structure (i.e., maximum information content of the distribution). In the following subsections, we show how our segmentation method is realized.

3.1   Preprocessing

Preprocessing ultimately aims at improving the image in ways that increase the opportunity for success of the other ulterior processes. In this step, we apply a Gaussian filter to the input image prior to any process in order to reduce the amount of noise in an image.

3.2   Entropies Calculation

Suppose $\{p_i\}_{i=1}^n$ be the probability distribution for the image. At the threshold, t this distribution is divided into two sub distributions; one for the foreground (class f) and the other for the background (class b) given by $P^f = \{p_i\}_{i=1}^t$ and $P^b = \{p_i\}_{i=t+1}^n$ respectively. Thus, the generalized Rànyi entropies for the two distributions as functions of t are given as

$$H_\alpha^f(t) = \frac{1}{\alpha - 1} \log_2\left[\frac{\sum_{k=1}^t p_k^\alpha}{\sum_{k=1}^t p_k}\right] \tag{7}$$

$$H_\alpha^b(t) = \frac{1}{\alpha - 1} \log_2\left[\frac{\sum_{k=t+1}^n p_k^\alpha}{\sum_{k=t+1}^n p_k}\right] \tag{8}$$

3.3   Image Thresholding

Thresholding is the most often used technique to distinguish objects from background. In this step an input image is converted by threshed into a binary image so that the objects in the input image can be easily separated from the background. To get the desired optimum threshold value t^*, we have to maximize the total entropy, $H_\alpha^{f+b}(t)$. When the function $H_\alpha^{f+b}(t)$ is maximized, the value of parameter t that maximizes the function is believed to be the optimum threshold value[18]. Mathematically, the problem can be formulated as

$$t^* = argmax[H_\alpha^{f+b}(t)] = argmax[H_\alpha^f(t) + H_\alpha^b(t) + (1 - \alpha) \cdot H_\alpha^f(t) \cdot H_\alpha^b(t)] \tag{9}$$

3.4 Morphology-Based Operations

In image processing, dilation, erosion, closing and opening are all well-known as morphological operations. In this step we aim at improving the results of the previous thresholding step. Due to the inconsistency within the color of objects, the resulting binary image perhaps includes some holes inside. By applying the closing morphological operation, we can get rid of the holes form the binary image. Furthermore Opening operation with small structure element can be used to separate some objects that are still connected in small number of pixels [19].

3.5 Overlapping Cancelation

In this step we attempt to remove the overlapping between objects that perhaps happened through extensively applying the previous morphological operations. To perform this, we first get the Euclidean Distance Transform (EDT) of the binary image. Then we apply the well-known watershed algorithm [20] on the resulting EDT image. The EDT ultimately converts the binary image into one where each pixel has a value equal to its distance to the nearest foreground pixel. The distances are measured in Euclidean distance metric. The peaks of the distance transform are assumed to be in the centers of the objects. Then the overlapping objects can be yet easily separated.

3.6 Non-objects Removal

This step helps in removing incorrect objects according to the object size. Sizes of objects are measured in comparison to the total size of image. Each tiny noise object of size less than a predefined minimum threshold can be discarded. Also each object whose size is greater than the maximum threshold size can be removed as well. Note that thresholds of size used herein are often dependent on the application, and so they are considered as user-defined data. The preceding steps of the proposed entropic segmentation approach are depicted by the block diagram shown in Fig. 1.

4 Experimental Results

In this section, the results of the proposed approach are presented. First to investigate the proposed approach for image segmentation we began by different image histograms. Each of these histograms describes the "objects" and the "background". Additionally, to verify the benefit of using the generalized Rènyi entropy, we have tried using another formula of entropy(e.g. Tsallis entropy) which is given by

$$H_\alpha = \frac{1 - \sum_{k=1}^{n} p_k^\alpha}{\alpha - 1} \tag{10}$$

The results of segmentation have testified to the higher efficiency of our entropic segmentation approach especially when generalized Rènyi entropy is used.

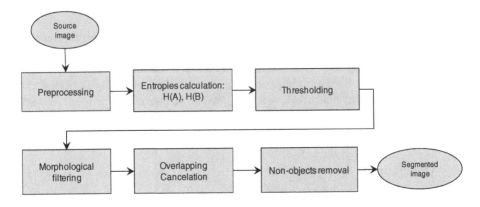

Fig. 1. Block diagram of the proposed segmentation method

In Fig. 2, an image of a mammogram showing breast cancer with a bright region (tumefaction) surrounded by a noisy region. The histogram roughly exemplifies an unimodal distribution of the graylevel values. The proposed entropic method will look for regions with uniform distribution in order to find the maximum entropy. This will regularly take place at the peak limit. It is well-known that segmenting this type of images is typically a challenging task. However the proposed method could performed well when applied on this type of images. Additionally, segmentation results in the figure show that using generalized Rènyi entropy is better than using Tsallis entropy.

Fig. 3 shows another example of our segmentation method. We present an image of a medical domain with a spatial background scattering noise; a stained brain cell that shows branching of cell dendrites-fibers that receive input from other brain cells. Several values of α are experimented. But the superior segmentation results has been obtained at $\alpha = 0.9$.

In Fig. 4, we show the segmentation results of the proposed method on a sample of color medical images. In this example the images are segmented with α equal to 0.8.

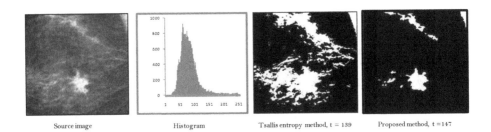

Source image   Histogram   Tsallis entropy method, t = 139   Proposed method, t = 147

Fig. 2. Entropic segmentation for noisy mammography image

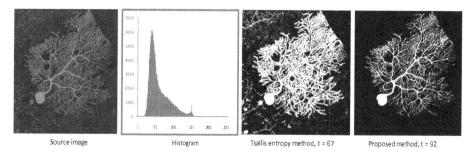

Source image | Histogram | Tsallis entropy method, t = 67 | Proposed method, t = 92

Fig. 3. Entropic segmentation for a brain cell image with a spatial noise around

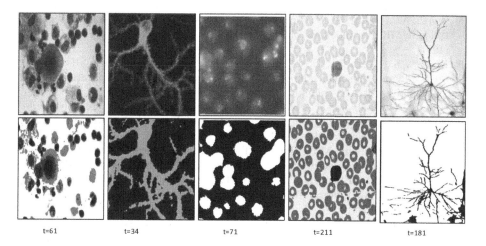

t=61   t=34   t=71   t=211   t=181

Fig. 4. Results of the proposed segmentation method for a sample of test images

5   Conclusion

In this paper, an efficient method for cell image segmentation based on generalized α-entropy has been presented . The proposed method has achieved the task of segmentation in a novel way. This method could give good results in many cases and perform well when applied to noisy cell images. The simulation results of the proposed method illustrate that using generalized Rènyi formalism of entropy is more viable than using Tsallis counterpart in segmentating cell image. The chief advantages of the proposed method are its high rapidity, and its tolerance to image noise.

Acknowledgments. This work is supported by Forschungspraemie(BMBF-Förderung, FKZ: 03FPB00213) and Transregional Collaborative Research Centre SFB/TRR 62 "Companion-Technology for Cognitive Technical Systems" funded by the German Research Foundation(DFG).

References

1. Bazin, P.-L., Pham, D.L.: Topology Correction of Segmented Medical Images using a Fast Marching Algorithm. Computer Methods and Programs in Biomedicine 88(2), 182–290 (2007)
2. Larie, S.M., Abukmeil, S.S.: Brain abnormality in schizophrenia: a systematic and quantitative review of volumetric magnetic resonance imaging studies. J. Psych. 172, 110–120 (1998)
3. Taylor, P.: Invited review: Computer aids for decision-making in diagnostic radiology- a literature review. Brit. J. Radiol. 68, 945–957 (1995)
4. Resnick, S.M., Pham, D.L., Kraut, M.A., Zonderman, A.B., Davatzikos, C.: Longitudinal MRI Studies of Older Adults: A Shrinking Brain. Journal of Neuroscience 23(8), 3295–3301 (2003)
5. Worth, A.J., Makris, N., Caviness, V.S., Kennedy, D.N.: Neuroanatomical segmentation in MRI: technological objectives. Int. J. Patt. Rec. Art. Intel. 11, 1161–1187 (1997)
6. Khoo, V.S., Dearnaley, D.P., Finnigan, D.J., Padhani, A., Tanner, S.F., Leach, M.O.: Magnetic resonance imaging (MRI): considerations and applications in radiotheraphy treatment planning. Radiother. Oncol. 42, 1–15 (1997)
7. Carter, J.C., Lanham, D.C., Bibat, G., Naidu, S., Kaufmann, W.E.: Selective Cerebral Volume Reduction in Rett Syndrome: A multiple approach MRI study. American Journal of Neuroradiology 29(3), 436–441 (2008)
8. Tosun, D., Rettmann, M.E., Han, X., Tao, X., Xu, C., Resnick, S.M., Prince, J.L.: Cortical Surface Segmentation and Mapping. NeuroImage 23, S108–S118 (2004)
9. Grimson, W.E.L., Ettinger, G.J., Kapur, T., Leventon, M.E., Wells, W.M.: Utilizing segmented MRI data in image-guided surgery. Int. J. Patt. Rec. Art. Intel. 11, 1367–1397 (1997)
10. Bazin, P.-L., Pham, D.L.: Homeomorphic Brain Image Segmentation with Topological and Statistical Atlases. Medical Image Analysis 12(5), 616–625 (2008)
11. Shannon, C.E., Weaver, W.: The Mathematical Theory of Communication. University of Illinois Press, Urbana (1949)
12. Tsallis, C., Abe, S., Okamoto, Y.: Nonextensive Statistical Mechanics and its Applications. Lecture Notes in Physics. Springer, Berlin (2001)
13. Albuquerque, M., Esquef, I.A., Gesualdi Mello, A.R.: Image thresholding using Tsallis entropy. Pattern Recognition Letters 25, 1059–1065 (2004)
14. Rényi, A.: On a theorem of P. Erdős and its application in information theory, Mathematica 1, 341–344 (1959)
15. Strzałka, D., Grabowski, F.: Towards possible q-generalizations of the Malthus and Verhulst growth models. Physica A 387(11), 2511–2518 (2008)
16. Singh, B., Partap, A.: Edge Detection in Gray Level Images based on the Shannon Entropy. Journal. of Computer Sci. 4(3), 186–191 (2008)
17. Tsallis, C., Albuquerque, M.P.: Are citations of scientific paper a case of nonextensivity? Euro. Phys. J. B 13, 777–780 (2000)
18. Tatsuaki, W., Takeshi, S.: When nonextensive entropy becomes extensive. Physica A 301, 284–290 (2001)
19. Gonzalez, R.C., Woods, R.E.: Digital Image Processing Using Matlab, 2nd edn. Prentice Hall, Inc., Upper Saddle River (2003)
20. Levner, l., Zhang, H.: Classification-Driven Watershed Segmentation. IEEE Transactions on Image Processing 16(5), 1437–1445 (2007)

An Algorithm for Moving Multi-target Prediction in a Celestial Background

Lu Zhang[1,2,*], Bingliang Hu[1], Yun Li[1,2], and Weiwei Yu[1,2]

[1] Xi'an Institute of Optics and Precision Mechanics of Chinese Academy Sciences,
Xi'an 710119, China
[2] Graduate University of Chinese Academy Sciences,
Beijing 100049, China

Abstract. Nowadays, there are thousands of in-orbit spacecrafts. It is of strategic significance to identify, track the space targets, and calculate their orbits and coverage area. In the celestial space observation of space targets on a micro spacecraft detector, it is difficult to observe and track targets because of their small size, unobvious geometric shape, and the complex movements relative to observed targets, background and detectors. The prediction algorithm of space targets must satisfy the demands of high reliability, low computation, real-time processing and so on. Based on multi-thread image processing and the Kalman algorithm, a new algorithm is proposed for multi-target prediction. Using this approach, the coordinates of targets can be predicted in real-time and precisely. Through theoretical calculation and computer simulation, this algorithm has shown that it is simple, effective and can satisfy the requirement of working conditions of the micro spacecraft detector well.

Keywords: Multi-target prediction; Kalman filter; Image block processing.

1 Introduction

Generally, based on a space platform, the optical observation and other methods are used to monitor space targets on space target monitoring. However, the images, obtained from the remote star field observation, are often small, with low luminance and unobvious geometric shape. Besides, there are a great number of stars in the observation background, the size and appearance of targets and these stars are similar, they can be hardly distinguished from each other. In addition, space targets as well as micro spacecraft are moving all the time, especially to the micro spacecraft, which is whirling itself; these two kinds of movement mingled together really add to the difficulties of multi-target capture and prediction [1] [2].

The prediction algorithm of space targets must satisfy the demands of high reliability, low computation, real-time processing and so on. To achieve this goal, an algorithm has been proposed and discussed in this paper.

---

* Corresponding author. Tel: +86-29-88887704, zl830608@gmail.com

D. Ślęzak et al. (Eds.): SIP 2009, CCIS 61, pp. 41–47, 2009.

2 Prediction of Moving Multi-targets

A new algorithm is introduced for multi-target prediction, which includes targets capture, coordinates calculation, image block processing and targets prediction.

2.1 Targets Capture

In this paper, the targets are captured by three image differences method, the flow chart is as Fig. 1. Fig. 2 shows the image of multi-target capture.

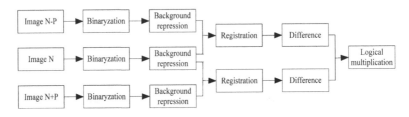

Fig. 1. The algorithm of multi-target capture

(a)   (b)

Fig. 2. The image of multi-target capture: (a) Original image of frame No.9; (b) Final image of frame No.9 after processing

Here, it is important to consider the impact of noise. The noise in celestial space mainly comes from three aspects: 1. Space radiation noise: high-energy particles from outer space may hit on the focal plane array, and a particle can affect several pixels in this situation. 2. Dark current noise from detector: usually there is single-pixel location of a fixed point in image, and it can be removed by determine if there is a fixed point in the same location from several frames. 3. Deep space background noise; usually, this kind of noise is negligible.

2.2 Calculation of Targets Number and Coordinates

As the paper mentioned before, the targets size, that ranges from several to more than twenty pixels, are so small in the images, and some of them are not even a connected domain. Based on these characters, a method is proposed to obtain the number and coordinates of the targets; its steps are as follows:

Step 1. To search the left feature points (x_{11}, y_{11}), (x_{12}, y_{12})... (x_{1n}, y_{1n}) of all the targets in the image, where n is the number of targets. The values of left lower pixels of them are 0. The distances of these points are more than a certain threshold value (40 pixels in this paper).

Step 2. To search the right feature points of all the targets in the image, their coordinates (x_{21}, y_{21}), (x_{22}, y_{22})... (x_{2n}, y_{2n}) will be recorded.

Step 3. To calculate the feature points of the targets (x_1, y_1), (x_2, y_2)... (x_n, y_n). For the first target, the coordinates of its feature pixel are as follows: $x_1=0.5*(x_{11}+x_{2n})$, $y_1=0.5*(y_{11}+y_{2n})$.

Using this method, all the targets can be identified and calculated simply and stably, and the miscounting of the targets number can be avoided in this method.

2.3 Image Block Processing

In order to reduce the process time, every image is divided into several image blocks for separate calculation in the same time. Image block processing can be achieved by four steps as follows:

Step 1. To calculate the targets numbers n and their coordinates $P_i(x, y)$ in an image, where x, $y \in [0, 1023]$, $i \in [0,n]$.

Step 2. To find dividing points in the X axis. List x value of the targets coordinates by small to large order in every image like $(x_1, x_2, ..., x_i)$, if the distance of two targets are larger than threshold value, find their dividing points, for example: $x'_1=(x_1 + x_2)/2$. Then we get the dividing points array $(x'_1, x'_2, ..., x'_p)$, where m is the number of the dividing points in X axis and p $<n$.

Step 3. To find dividing points in the Y axis in the same way, and get the dividing points array $(y'_1, y'_2, ..., y'_q)$, where q is the number of the dividing points in Y axis and q $<n$.

Step 4. To divide the image into several blocks. Based on the X axis and the Y axis dividing points, generate horizontal dividing lines x= x'_j, $j \in [1, p]$, and vertical lines y= y'_k, $k \in [1, q]$.

Fig. 3 is an example of the division of the image, dividing points are x'_1 and y'_1 so there are two lines x= x'_1 and y= y'_1, that divide this image into four image blocks.

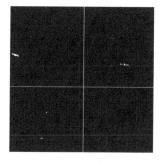

Fig. 3. The block division of image

Experiments have shown that we can process the next twenty or thirty images based on the division of the present image. In practice, there may be some image blocks which do not contain a target, thus there is no need to deal with these blocks and the quantity of calculation will be cut down directly. The rest of the image blocks can be processed in multi-threads, and the time of calculation will be reduced greatly.

2.4 Target Prediction

In this paper, the Kalman filter was used to predict the coordinate position of targets in the next image by processing the former five frames of adjacent images.

Kalman Filter

Kalman filters are the statistically optimal sequential estimation procedure for dynamic systems. Observations are recursively combined with recent forecasts with weights that minimize the corresponding biases [3]. The system of Kalman filter can be described by using the dynamic system equation and observation equation, which are as follows:

$$x_k = \Phi_{k, k-1} \, x_{k-1} + W_{k-1} \, . \tag{1}$$

$$Y_k = H_k \, x_k + V_k \, . \tag{2}$$

In the application of the Kalman filter, a first estimate of x_k and P_k based on the previous time step values is given by:

$$x_{k,k-1} = \Phi_{k, k-1} \, x_{k-1} \, . \tag{3}$$

$$P_{k,k-1} = \Phi_{k, k-1} \, P_{k-1} \Phi_{k, k-1}{}' + Q_k \, . \tag{4}$$

After knowing the new observation value Y_k, the estimate of x at time k becomes

$$x_k = \Phi_{k, k-1} \, x_{k-1} + K_k \, (\, Y_k - H_k \Phi_{k, k-1} \, x_{k-1} \,) \, . \tag{5}$$

Where

$$K_k = P_{k, k-1} \, H_k{}' \, (\, H_k \, P_{k, k-1} \, H_k{}' + V_k \,)^{-1} \, . \tag{6}$$

K_k, gain coefficient matrix of Kalman, determines how easily the filter adjusts to possible new conditions.

The final estimate of P_k is

$$P_k = (\, 1 - K_k \, H_k \,) \, P_{k,k-1} \, . \tag{7}$$

The status of system from time k-1 to k is updates by Eqs. (3)–(7).

Parameter Definition

The system status of the Kalman filter that is defined as x_k. x_k is a four-dimensional array (x_1, x_2, x_3, x_4), where x_1 stands for X coordinate component, x_2 stands for X velocity component, x_3 stands for Y coordinate component, and x_4 stands for Y velocity component. Because the measured value of the target is a two-dimensional coordinate position in the image, the observational status vector Y_k is a two-dimensional coordinate vector (y_1, y_2).

Due to the related short time interval, the movement of target can be regarded as uniform motion and its state transition matrix. $\Phi_{k, k-1}$ is as follows:

$$\Phi_{k,k-1} = \begin{bmatrix} 1 & \Delta t & 0 & 0 \\ 0 & 1 & 0 & 0 \\ 0 & 0 & 1 & \Delta t \\ 0 & 0 & 0 & 1 \end{bmatrix} \tag{8}$$

Where, Δt is the time interval between t_{k-1} and t_k from the relationship between system state and observation state, the observation matrix is as follows:

$$H_k = \begin{bmatrix} 1 & 0 & 0 & 0 \\ 0 & 0 & 1 & 0 \end{bmatrix} \tag{9}$$

Based on the approach above for the single target prediction, the multi-target movement prediction is implemented separately by using the Kalman filter, while there are some problems need be settled in the prediction of multi-target movement:

1. Dynamic change of targets. When the new targets flies into the field of view, the position and velocity of the new target should be calculated in real time, and when the target flies out of the field of view, its record should be deleted immediately.

2. The possible blank of the observation target. It requires that the prediction of the moving target can be continued until the end of the blank.

3. Targets re-capturing after blank time. It is a problem to ensure that the moving target can be predicted and re-captured accurately after the blank.

The target number and coordinates play a very important role in the multi-target movement prediction. As we know, the coordinates of a target in two adjacent frames are much closed; usually the distance of them is usually shorter than 5 pixels, and they can be judged as a pair. Then the problems mentioned above can be settled by the following analysis.

1. If a target in the present frame is not closed enough without any pair in the previous frame, or the total number of the targets in the present frame is more than that of the last frame, it is evident that new targets appear in this situation. At this time the system will predict these new ones immediately and record the coordinates of them at the same time.

2. If a target in the last frame cannot make a pair with anyone in the present frame, it is evident that this target has moved out of the image. In this case, there is no need to track it any more, unless you want to keep tracking.

3. If two or more pairs of targets are closed enough, and meanwhile the target number in this area is less than last frame, it is considered that some targets has been blanked by other targets. In this case, the predicted coordinates of these blanked targets can seem to be the real coordinates of them until they appear again.

3   Algorithm Simulation

To simulate and test this algorithm, the Satellite Tool Kit (STK) was used to get the video flow of multi targets movement in the simulated star background. Fig 4 shows the

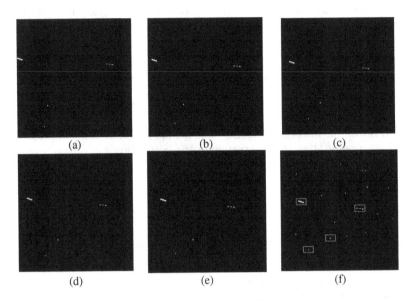

Fig. 4. The image of multi-target capture: (a) raw frame 9;(b) raw frame 10; (c) raw frame 11; (d) raw frame 12; (e) raw frame 13; (f) prediction of raw frame 14

prediction results of multi movement targets using multi-thread image processing and the Kalman filter algorithm, and targets acquisition were shown in Fig 4(a) – Fig 4(e). The algorithm introduced in the paper is shown to be reliable and precise.

Compared with the algorithm of single thread image processing, multi thread image processing can greatly reduce the processing time, and by about 60% in a simulation of three groups of 1024*1024 video flows, which is shown in Table 1.

Table 1. Process time comparison of multi thread and single thread

| | | Process time (ms) | |
		Multi Thread	Single Thread
Group 1	Capture	38	93
	Prediction	43	109
Group 2	Capture	40	102
	Prediction	51	115
Group 3	Capture	37	110
	Prediction	44	109

4   Conclusion

To satisfy the special requirements of space targets observation in the micro spacecraft, a prediction algorithm for moving multi-targets is proposed based on multi-thread image processing and the Kalman filter method. This algorithm has shown the capabilities of background suppression, precise prediction, high reliability, low complexities.

References

1. Marco, D.E., Michele, G., Sergio, V.: A bistatic SAR mission for earth observation based on a small satellite. J. Acta Astronautica 39, 837–846 (1996)
2. Zhukov, B., Lorenz, E., Oertel, D., Wooster, M., Roberts, G.: Spaceborne detection and characterization of fires during the bi-spectral infrared detection (BIRD) experimental small satellite mission (2001–2004). J. Remote Sensing of Environment 100, 29–51 (2006)
3. Kalman, R.E.: A new approach to linear filtering and prediction problems. Transactions of the ASME Series D, Journal of Basic Engineering 82, 35–45 (1960)

Automatic Control Signal Generation of a CCD Camera for Object Tracking

Jin-Tae Kim[1], Yong-In Yun[2], and Jong-Soo Choi[2]

[1] Dept. Computer anf Information Eng., Hanseo University,
Chungnam 356-756, Korea
jtkim@hanseo.ac.kr
[2] Dept. Image Eng., Graduate School of Advanced Imaging Science,
Multimedia, and Film, Chung-Ang University, Seoul 156-756, Korea
{yoonyi,jschoi}@imagelab.cau.ac.kr

Abstract. CCD cameras monitoring a moving object generally operate under fixed view point or a defined pattern. So a free moving object goes out of the CCD camera's sight in a short time and automatic control is required to observe it continuously. This paper proposes a signal generation algorithm for the automatic control of the CCD camera. The proposed algorithm computes horizontal and vertical displacements between the detected object position and the image center and coverts them into angles. Finally pan/tilt data are generated from the angles. To estimate the proposed algorithm, we compare the generated data in the automatic control with the measured data in manual control and carry out the object tracking using a simple object. The experiment results show that the difference between the generated and measured data is a negligible quantity and the object, which is moving in $\pm 52^\circ/\pm 40^\circ$ of pan/tilt, is kept in about $\pm 13^\circ/\pm 10^\circ$.

Keywords: CCD camera, moving object, pan/tilt, automatic control.

1 Introduction

As science, technologies and networks develop, CCD cameras are actively being utilized in diverse areas as a core element of systems intended to monitor moving objects. Since the objects that are monitored are moving whereas CCD cameras' fields of view (FOV) are limited, the objects can go out of the FOV of CCD cameras in a short time. Therefore, most CCD cameras move along defined patterns but that is not enough. Methods to solve this problem would be installing more CCD cameras or controlling CCD cameras when the object of interest is found to continuously track the objects thereby keeping the object of interest located always in the center of the image. In general, the former case will required additional manpower and equipment and the latter case will require monitoring persons to permanently stay in order to manually control the cameras following the movements of the object. If the latter case can be handled by just processing images, that must be a more efficient method.

D. Ślęzak et al. (Eds.): SIP 2009, CCIS 61, pp. 48–55, 2009.

Monitoring systems using image processing techniques should obtain images through CCD cameras, detect the positions of the moving object of interest and create camera control signals based on the positions of the object in order to track the object [1,2]. This paper suggests a method to appropriately create CCD camera controlling signals including pan and tilt information based on the positions of objects using image processing techniques in order to continuously monitor moving objects. In general, pan/tilt modules for CCD camera control provide information necessary to control cameras but when the modules are applied to actual systems, the information is often insufficient or not quite accurate. Therefore, this paper describes methods to calculate information necessary to control given controllable CCD cameras through image processing techniques and compares the results of the calculation with results of actual measurements thereby demonstrating that the information is useful for CCD camera control. In addition, a tracking experiment is conducted with a simple object using the control signals created by the method suggested in this paper to show that the created control signals are valid.

2 Signal to Control CCD Cameras

In general, signals to control CCD cameras include pan, tilt and zoom signals and pan and tilt signals are mainly used to track objects. Pan signals are to move cameras horizontally as shown in Fig. 1(a). They are the camera control signals that make it possible to effectively track objects moving horizontally in images. Tilt signals are to move cameras vertically as shown in Fig. 1(b). They are the camera control signals that make it possible to effectively track objects moving vertically in images. The dome cameras DMP23 from Daiwa used in this study provides pan signals ranging from $0°$ to $360°$ and tilt signals ranging from $-5°$ to $90°$.

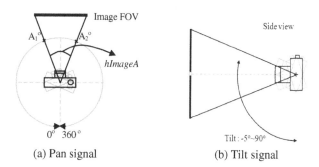

(a) Pan signal   (b) Tilt signal

Fig. 1. CCD camera control signals

3 Creation of CCD Camera Control Signals

In this paper, the purpose of CCD camera control is to detect moving objects from inputted images and make the objects located at the center of the images. This chapter presents methods to create pan/tilt signals based on object positions.

3.1 Creation of Pan Signals

To create pan signals, the object of interest should be detected from the inputted images, the horizontal movements should be shown as positions of picture elements and the positions should be converted into relative angles to the horizontal center and then changed into pan data suitable to the characteristics of the camera. To this end, the horizontal image angle ($hImageA$) should be calculated using Equation (1). As shown in Fig. 1(a), this is the ratio of the pan data required to move the object from one end of the horizontal axis of the image to the other end to the pan data ($PanD_x$) required to pan the camera by $360°$.

$$hImageA = \frac{|PanD_{A_1} - PanD_{A_2}|}{|PanD_0 - PanD_{360}|} \times 360[°] \tag{1}$$

As shown in Fig. 2, by applying the horizontal image angle and the image size ($hImSize$) to the trigonometric function, the virtual distance between the camera and the image ($CamtoImL$) can be obtained using Equation (2).

$$CamtoImL = \frac{hImSize/2}{\tan(hImageA/2)} \tag{2}$$

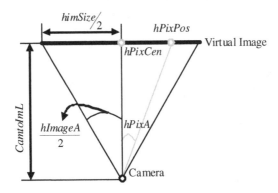

Fig. 2. Structural relationships between the camera and the image in the top view

The horizontal angle ($hPixA$) of an object based on its horizontal position ($hPixPos$) can be calculated using the virtual distance between the camera and the image utilizing Equation (3).

$$hPixA = \tan^{-1}\left(\frac{hPixPos - hPixCen}{CamtoImL}\right) \tag{3}$$

The pan signal ($PanD$) of the CCD camera based on the horizontal position of the object can be calculated using the horizontal angle and the pan signal ($PanDp1$) intended to pan the CCD camera by $1°$ in the horizontal direction by Equation (4).

$$PanD_{hPixA} = PanDp1 \times hPixA \tag{4}$$

3.2 Creation of Tilt Signals

The creation of tilt signals is similar to the creation of pan signals except that it considers the horizontal movements of objects instead of vertical movements. Therefore, the movements of the object of interest in the inputted image should be shown as the positions of picture elements, the positions should be expressed as vertical angles and then converted into tilt data suitable to the characteristics of the camera. First, under the assumption that picture elements consist squares, the virtual distance between the camera and the image as calculated in Equation (2) can be applied as it is and the relative vertical angle (*vPixA*) based on the vertical position (*vPixPos*) of an object as shown in Fig. 3 can be calculated using Equation (5).

$$vPixA = \tan^{-1}\left(\frac{vPixPos - vPixCen}{CamtoImL}\right) \qquad (5)$$

The tilt signal (*TiltD*) of the CCD camera based on the vertical position of the object can be calculated using the horizontal angle and the tilt signal (*TiltDp1*) intended to pan the camera by $1°$ in the vertical direction by Equation (6).

$$\qquad\qquad\qquad\qquad\qquad\qquad\qquad (6)$$
$$TiltD_{vPixA} = TiltDp1 \times vPixA$$

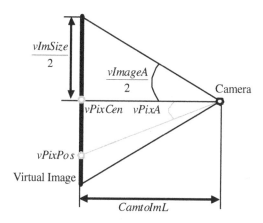

Fig. 3. Structural relationships between the camera and the image in the side view

4 Object Tracking Test of the CCD Camera

In this Section, it performs object tracking tests in order to evaluate the performance of the control signal generation algorithm of the proposed CCD camera. Prior to the tracking test, appropriate frame frequency and pan/tilt data are selected to track objects in the test environment, with the dome camera DMP23, RS 232C communication and the PC for image processing having P4 2.4GHz.

4.1 Initial Starting and Acceleration of CCD Camera Pan/Tilt

For the CCD camera to track an object, control signals generated based on the location of the object shall be provided to the camera at regular intervals. Therefore, for smooth object tracking, the motions of the camera responding to the control signals must be analyzed. Fig. 4 shows the location of an object in an image after movements of the camera when control signals commanding the camera to move the object located away from the center of the image to the center had been sent to the camera. From Fig. 4, it can be seen that initial movement distances are very short, movement distances are different among different frames and time is required for the speed to increase and reach a certain level. Therefore, when controlling the camera by the unit of frame, control signals with appropriate frequencies and sizes must be generated for normal operation of the camera.

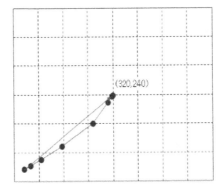

Fig. 4. Changes in the location of an object made through single control signals

4.2 Effect of Frame Frequency

Since a certain time is required for a CCD camera to receive control signals and move to the designated location and the control signals are produced in the unit of frame, frame frequencies must be considered. Under high frame frequencies, smooth object movements may be shown but the camera may not perceive control signals due to the short time intervals. Under low frame frequencies, the camera will normally receive control signals but unnatural object movements may be shown. Fig. 5 shows the effect of frame frequencies on camera control when signals are generated to move an object located in an outer area of an image to the center of the image. In the case of 15Hz, the movements of the object are progressing smoothly and continuously and the movements maintain the image quality not repellent to the eyes. On the other hand, in the case of 20Hz, one of every two control signals is not received thus there is no movement in one of every two frames and the data are accumulated. The accumulated data make the next movement larger. Consequently, these are the same movements as those made when moving the camera using two times larger data with the frame frequency of 10Hz thus the result is that the low frequency produces movements that are repellent to the eyes. In this test, 15Hz is used as a basic frequency to control the CCD camera.

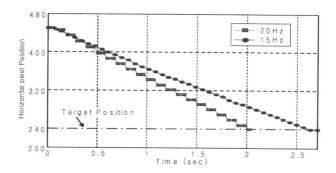

Fig. 5. Object movements depending on frame frequencies

4.3   Effect of the Size of Pan/Tilt Signals

As mentioned above, the distance to be moved at a time is limited when moving the camera by the unit of frame due to the limited initial movements of the CCD, thus this should be analyzed. If signal data are small, the movements of the camera are not affected but the speed to track objects is reduced thus objects may disappear from the field of vision of the camera if the objects move fast. On the other hand, if control signals are large, objects may be tracked fast but data that are not moved within a cycle are generated and accumulated. When the accumulated data grow to a size larger than a certain size, then the camera will perform abnormal movements.

In object tracking, control signals are generated sent to the camera in the unit of frame and the camera moves only one time per cycle unless the frame frequency is very low. These are important data to generate effective pan/tilt data for continuous object tracking. Although there are some differences depending on the location, the average numbers of pixels of the movements are maintained at the similar values of 12.2/7.6 and if these are converted into pan/tilt data, the values are 39.9/42.2. Based on this, the pan/tilt data were assumed to be 40/40 in order to track objects. Fig. 6 shows the movement distances in the unit of frame when moving an object located away from the center to the center of an image with pan data. The movements are stable to some extent with the pan data of 30 and 40 but with the pan data 50, it is shown that the data that could not be moved in a frame are accumulated and produce movements at a moment. The pan data 30 is limited in tracking fast objects thus the pan data 40 is more effective. Similar results were identified in tilt data too.

Pan/tilt data 40/40 are also limited in tracking fast objects. In this paper, to track fast objects, the frame frequency will be maintained and control signals will be generated once per 3 frames. If one control signal is generated per two frames, two opportunities for movements will be available but these are limited in enhancing the speed at the beginning and end of the movement. However, if one control signal is generated per three frames, three opportunities are given and the speed may be enhanced at the second movement. Consequently, the speed can be doubled in the test and this enables fast object tracking.

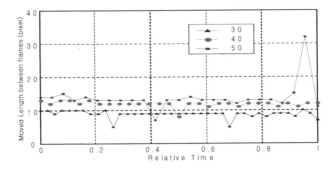

Fig. 6. Movement distances during a frame by pan data

4.4 Object Tracking of CCD Camera

This test will detect an object from images obtained through the CCD camera to identify the location and will generate pan/tilt data with the proposed algorithm based on the location of the object and control the camera to perform object tracking. Since this test is interested in generating control signals for the CCD camera for object tracking, a simple red square was used as the object. Fig. 7 shows the result of tracking of the freely moving object using the frame frequency of 15Hz determined as above and the pan/tilt data of 40/40 for maximum movements. With the object tracking using the proposed algorithm, even if the object moves in a square consisting of (-640,-480) and (640,480), it becomes to exist in the center of the image consisting of (-320,-240) and (320,240). On the basis of pan/tilt, this gives the effect of holding an object moving in the field of vision of around ±52°/±40° within the field of vision of around ±13°/±10°.

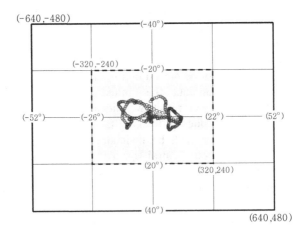

Fig. 7. Object tracking test

5 Conclusion

CCD cameras that monitor/observe moving objects are fixed or move in certain patterns. Therefore, moving objects disappear from the fields of vision of CCD cameras within a short time. This paper suggested a method to generate signals to automatically control CCD cameras in order to continuously monitor/observe moving objects, determined important variables for object tracking and performed tracking tests with an object thereby identifying the effectiveness of the proposed algorithm. The control signal generation tests were performed by marking a location within an image and moving the location to the center of the image by the proposed method and by a manual method in order to compare between the methods for evaluation. The difference between the two methods was negligibly small thus the effectiveness of the proposed method was confirmed. Also, through the object tracking test, it was identified that the method was holding the object moving in the field of vision of around $\pm 52°/\pm 40°$ on the basis of pan/tilt within the field of vision of around $\pm 13°/\pm 10°$ thus it was confirmed that the proposed algorithm was effective.

References

1. Suryadarma, S., Adiono, T., Machbub, C., Mengko, T.: Camera Object Tracking System. In: Han, Y., Quing, S. (eds.) ICICS 1997. LNCS, vol. 1334, pp. 1557–1561. Springer, Heidelberg (1997)
2. Collins, R.T., Lipton, A.J., Kanade, T.: A System for Video Surveillance and Monitoring. In: ANS 8 International Meeting on Robotics and Remote Systems (1999)
3. Gonzalez, R.C., Woods, R.E.: Digital Image Processing. Addison Wesley, New York (1993)
4. Nakamura, S.: Applied Numerical Methods in C. Prentice Hall, London (1995)
5. Standard Installation Hardware Reference. Meteor II Manual, Matrox (2001)
6. http://www.matrox.com/imaging/products/mil-lite/home.cfm
7. Speed Dome Camera (DMP23-H2) Command Manual. Dure Optronix (2003)
8. Ahn, S.-H., Koo, H., Kim, J.-T., Oh, J.-S., Willis, A.: Automatic Control of a Camera Signal for Monitoring Systems. In: Proc. International Conference on Imaging Processing, Computer Vision, and Pattern Recognition, pp. 321–324 (2008)
9. Kim, J.-T., Ahn, S.-H., Oh, J.-S.: Signal Generation for Automatic Control of a CCD Camera. J. Korean Society for Imaging Science and Technology 15(2), 1–8 (2009)

An Efficient Approach for Region-Based Image Classification and Retrieval

Samy Sadek[1], Ayoub Al-Hamadi[1], Bernd Michaelis[1], and Usama Sayed[2]

[1] Institute for Electronics, Signal Processing and Communications
Otto-von-Guericke-University Magdeburg, Germany
[2] Assiust University, Egypt
{Samy.Bakheet,Ayoub.Al-Hamadi}@ovgu.de

Abstract. In this paper, a fast and efficient approach for region-based image classification and retrieval using multi-level neural network model is proposed. The advantages of this particular model in image classification and retrieval domain will be highlighted. The proposed approach accomplishes its goal in two main steps. First, by aid of a mean-shift based segmentation algorithm, significant regions of the image are isolated. Then, features of these regions are extracted and then classified by the multi-level model into five categories, i.e., "Sky", "Building", "Sand\Rock", "Grass" and "Water". Features extraction is done by using color moments and 2D wavelets decomposition technique. Experimental results show that the proposed approach can achieve precision of better than 93% that justifies the viability of the proposed approach compared with other state-of-the-art classification and retrieval approaches.

Keywords: Multi-level neural networks, Content-based image retrieval, Feature extraction, Wavelets decomposition.

1 Introduction

With the advent of high powerful digital imaging hardware and software along with the accessibility of the internet, databases of billions of images are now available and constitute a dense sampling of the visual world. As a result, efficient approaches for image databases classification and retrieval are highly needed. Image classification and retrieval have been addressed by many researchers based on low-level image features such as color, texture, and simple shape properties. Recently a great deal of work has been done to investigate and develop techniques for classification and retrieval of large image databases [1],[2],[3]. While the work of many researchers within image classification and retrieval [4],[5],[6] has focused on more general, consumer-level, semantic classes such as indoor/outdoor, people/non-people, city, landscape, sunset, forest, etc , there is another work done by other researchers that is centered on the separation of computer-generated graphic images, such as presentation slides, from photographic images [7],[8],[9]. Yet image classifcation and retrieval techniques are still far from achieving the goal of being a complete, requiring additional research to improve their effectiveness.

D. Ślęzak et al. (Eds.): SIP 2009, CCIS 61, pp. 56–64, 2009.
© Springer-Verlag Berlin Heidelberg 2009

Machine learning particularly artificial neural networks are increasingly employed to deal with many tasks of image processing especially image classification and retrieval. The neural classifier has the advantages of being effortlessly trainable, highly rapid, and capable to create arbitrary partitions of feature space [10]. However a neural model, in the standard form, is incompetent to correctly classify images into more than two classes [11]. This comes out of the fact that each of the component single neuron makes use of the standard bi-level activation function. As the bi-level activation function produces only binary responses, the neurons can generate only binary outputs. So, in order to produce multiple color responses either an architectural or a functional extension to the existing neural model is required.

The remainder of the paper is organized as follows. Section 2 highlights multi-level activation functions used by the multi-level neural model. In section 3, a fast segmentation technique based on a mean shift algorithm is presented. Color moments and multi-level wavelet decomposition are discussed in section 4. In section 5, the proposed image classification and retrieval approach is introduced. Section 6 presents the simulation results of the proposed approach and Section 7 closes the paper with some concluding remarks.

2   Multi-level Neural Networks

Multi-level neural model utilizes an activation function, named multi-level activation function. A multi-level activation function is originally a functional extension of the standard activation functions. Several multi-level forms belonging to several standard activation functions can be defined. This section is to show how to obtain the multilevel form of an activation function from its standard form. Suppose the standard sigmoidal activation function is given by

$$f(x) = \frac{1}{1 + e^{-\beta x}} \qquad (1)$$

where, β is the steepness factor of the function. Thus the multi-level form of the sigmoidal functions can be derived from the previous standard form given by Eq. (1) as follows:

$$\varphi(x) = f(x) + (\lambda - 1)f(c) \qquad (2)$$

where $(\lambda - 1)c \leq x \leq \lambda c$ and $1 \leq \lambda \leq n$.

In Eq. (2), λ represents the color index, n is the number of categories, and c represents the color scale contribution. Note that the learning method of the MSNN model does not considerably differ from the other learning methods used in training artificial neural networks. It is employing some form of gradient descent. This is done by taking the derivative of the cost function with respect to the network parameters and then altering those parameters in a gradient-related direction [12].

3   Image Segmentation

In this section, a fast segmentation technique based on a mean shift algorithm; a simple nonparametric procedure for estimating density gradients is used to recover significant image features (for more details see [13],[14]). Mean shift algorithm is really a tool required for feature space analysis. We randomly choose an initial location of the search window to allow the unimodality condition to be settled down. The algorithm then converges to the closest high density region. The steps of the color image segmentation method are outlined as follows

1. Initially, define the segmentation parameters (e.g. radius of the search window, smallest number of elements required for a significant color, and smallest number of contiguous pixels required for significant image regions).
2. Map the image domain into the feature space.
3. Define an appropriate number of search windows at random locations in the feature space.
4. Apply the mean shift algorithm to each window to find the high density regions centers.
5. Verify the centers with image domain constraints to get the feature palette.
6. Assign all the feature vectors to the feature palette using the information of image domain.
7. Finally, remove small connected components of size < predefined threshold.

It should be noticed that the preceding procedure is universal and valid for applying with any feature space. Furthermore, all feature space computations mentioned above are performed in HSV space. An example of image segmentation by the previous mean shift based algorithm is shown in Fig. 1.

Fig. 1. Example of image segmentation based on mean shift algorithm

4   Feature Extraction

Image classification and retrieval are regularly using some image features that characterize the image. In the existing content-based image classification and retrieval systems the most common features are color, shape, and texture. Color histograms are commonly used in image classification and retrieval. In this paper, we use both color moments and approximation coefficients of multi-level wavelet decomposition to extract features from each image region.

4.1   Wavelet Decomposition

Discrete Wavelet Transform (DWT) captures image features and localizes them in both time and frequency content accurately. DWT employs two sets of functions called scaling functions and wavelet functions, which are related to low-pass and high-pass filters, respectively. The decomposition of the signal into the different frequency bands is merely obtained by consecutive high-pass and low-pass filtering of the time domain signal. The procedure of multi-resolution decomposition of a signal x[n] is schematically. Each stage of this scheme consists of two digital filters and two down-samplers by 2. The first filter H0 is the discrete mother wavelet; high pass in nature, and the second, H1 is its mirror version, low-pass in nature. The down-sampled outputs of first high-pass and low-pass filters provide the detail, D1 and the approximation, A1, respectively. The first approximation, A1 is further decomposed and this process is continued as shown in Fig. 2.

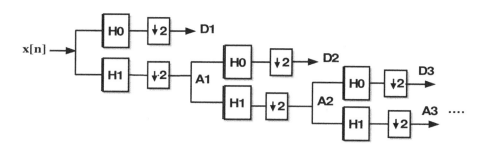

Fig. 2. Multi-level wavelets decomposition

4.2   Color Moments

The basis of color moments lays in the assumption that the distribution of color in an image can be interpreted as a probability distribution [15]. Probability distributions are characterized by a number of unique moments (e.g. normal distributions are differentiated by their mean and variance). It therefore follows that if the color in an image follows a certain probability distribution, the moments of that distribution can then be used as features to identify that image based on color. The three central moments (Mean, Standard deviation, and Skewness) of an image's color distribution can be defined as

$$\mu_k = \frac{1}{n}\sum_{i=1}^{n}p_i^k, \sigma_k = \sqrt{\frac{1}{n}\sum_{i=1}^{n}(p_i^k - \mu_k)^2}, s_k = \sqrt[3]{\frac{1}{n}\sum_{i=1}^{n}(p_i^k - \mu_k)^3} \qquad (3)$$

where p_i^k is the value of the k'th color channel for the i'th pixel, and n is the size of the image.

5   Proposed Approach

The prime difficulty with any image retrieval process is that the unit of information in image is the pixel and each pixel has properties of position and color value; however, by itself, the knowledge of the position and value of a particular pixel should generally convey all information related to the image contents [16],[17]. To surmount this difficulty, features are extracted using two-way. The extracted features consist of two folds: color moments and approximate coefficients of multi-level wavelet decomposition. This allows us to extract from an image a set of numerical features, expressed as coded characteristics of the selected object, and used to differentiate one class of objects from another. The main steps of the proposed approach are depicted in Fig. 3.

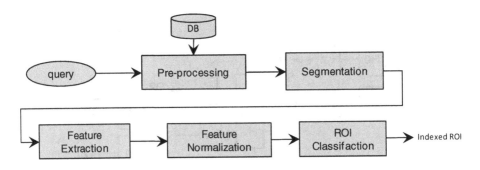

Fig. 3. Block diagram of the proposed approach

In the following subsections, the main steps of the proposed approach depicted in Fig. 3 are described.

5.1   Preprocessing

In image processing, preprocessing mainly purpose to enhance the image in ways that raise the opportunity for success of the other succeeding processes (i.e. segmentation, features extraction, classification, etc). Preprocessing characteristically deals with techniques for enhancing contrast, segregating regions, and eliminating or suppressing noise. Preprocessing herein includes normalizing the images by bringing them to a common resolution, performing histogram equalization and applying the Gaussian filter to remove small distortions without reducing the sharpness of the image.

5.2 Segmentation

In this step the fast mean shift based segmentation technique described above in section 2 is used to segment the image into distinct regions. To get rid of the segmentation errors, regions of small area (i.e. less than predefined threshold,t=0.05) are discarded. The significant regions (i.e. regions of areas greater than or equal 0.05 of the image area) are the candidates where the feature vectors are extracted from.

5.3 Feature Extraction

In this step, we utilize 2D multi-level wavelets transform to decompose image regions. Each level of decomposition gives two categories of coefficients, i.e., approximate coefficients and details coefficients. Both approximate coefficients and color moments are considered as the features for our retrieval problem.

5.4 Feature Normalization

To prevent singular features from dominating the others and to obtain comparable value ranges, we do feature normalization by transforming the feature component, x to a random variable with zero mean and one variance as follows

$$\bar{x} = \frac{x - \mu}{\sigma} \tag{4}$$

where μ and σ are the mean and the standard deviation of the sample respectively. If we assume that each feature is normally distributed, the probability of $\bar{x}$ to be in the [-1,1] is 0.68.

5.5 Classification of Image Regions

As a matter of fact, it should be stated that the neural classifier can accomplish better classification if each region belongs to only one of the predefined categories. Therefore, it is hard to build up a full trustworthy classifier due to the truth that different categories may have similar visual features (such as Water and Sky categories). Before doing any classification process, categories that reflect the semantics in the image regions are first defined. Then multi-level neural classifier has to learn the semantics of each category via the "training" process. So it is possible now to classify a specific region into one of the predefined semantic categories which humans easily understand. To do so, extracted features of the region are fed into the trained multi-level classifier and then it directly predicts the category of that region.

6   Experimental Results

In this section classification and retrieval results of the proposed approach are presented. First, to train the multi-level classifier, we have manually prepared a training set comprising of 200 regions; on the average, 40 training samples per category. o verify the ability of the proposed approach in image classification and retrieval, we have used a test set containing about 500 regions covering 5 categories, "Sky", "Building", "Sand \Rock", "Grass" and "Water. Table 1 tabulates the classification results done by the proposed approach.

Table 1. Classification results of image regions of dataset collections

Category	Precision
Sky	96%
Building	91%
Sand\Rock	89%
Water	98%
Grass	95%
Average	93.8%

It should be noted that the raw figures tabulated in table 1 are considered as a quite quantitative measure for the performance of the proposed approach, indicating to the high performance of the proposed approach compared with other classification approaches, specifically that has been proposed in [18].

Once the semantic classification of image regions is successfully done, in this case, an image can be represented by the categories in which the image regions are classified. That is, each image can be characterized by a set of keywords (i.e., categories' indices) which allows for things such as a highly intuitive query to be possible. Therefore, by using one or more keywords, image databases can easily be searched. In a query of such type, all images in the database that contain the selected keywords will be retrieved. For instance, if the keyword "Sky" is selected.

Fig. 4. Result of image retrieval for the query: keyword ="sky"

The purpose of this query is to retrieve all images that include a region of sky. The retrieval results of such query are shown in Fig.4.

7   Conclusion

In this paper, we have proposed a fast and efficient approach for region-based image classification and retrieval. The proposed approach makes use of a neural model called multi-level neural Network. The main property of this model is the low computational complexity as well as the easiness of implementation. The simulation results on image classification and retrieval show that the multi-level model is very effective in terms of learning capabilities and classification accuracies. This allows the proposed approach to be comparable to the other state-of-the-art classification approaches. Although our approach has dealt with still images, it can be straightforwardly extended to handle motion scenes.

Acknowledgments. This work is supported by Forschungspraemie (BMBF-Förderung, FKZ: 03FPB00213) and Transregional Collaborative Research Centre SFB/TRR 62 ŤCompanion-Technology for Cognitive Technical SystemsŤ funded by the German Research Foundation (DFG).

References

1. Deng, H., Clausi, D.A.: Gaussian MRF Rotation-Invariant Features for SAR Sea Ice Classification. IEEE PAMI 26(7), 951–955 (2004)
2. Goodrum: Image Information Retrieval: An Overview of Current Research. Special Issue on Information Science Research 3(2) (2000)
3. O'Connor, N.E., Cooke, E., Borgne, H., Blighe, M., Adamek, T.: The aceToolbox: Lowe-Level Audiovisual Feature Extraction for Retrieval and Classification. In: Proc. of EWIMT 2005 (November 2005)
4. Vailaya, A., Jain, K., Zhang, H.-J.: On Image Classification: City Images vs. Landscapes. Pattern Recognition Journal 31, 1921–1936 (1998)
5. Zhao, R., Grosky, W.I.: Bridging the Semantic Gap in Image Retrieval. In: Shih, T.K. (ed.) Distributed Multimedia Databases: Techniques and Applications, pp. 14–36. Idea Group Publishing, Hershey (2001)
6. Luo, J., Savakis, A.: Indoor vs. Outdoor Classification of Consumer Photographs using Low-level and Semantic Features. In: Proc. of ICIP, pp. 745–748 (2001)
7. Hartmann, L.R.: Automatic Classification of Images on the Web. In: Proc of SPIE Storage and Retrieval for Media Databases, pp. 31–40 (2002)
8. Wang, J.Z., Li, G., Wiederhold, G.: SIMPLIcity: Semantics-sensitive Integrated Matching for Picture Libraries. IEEE Trans. on Pattern Analysis and Machine Intelligence 23, 947–963 (2001)
9. Prabhakar, S., Cheng, H., Handley, J.C., Fan, Z., Lin, Y.W.: Picture-graphics Color Image Classification. In: Proc. of ICIP, pp. 785–788 (2002)
10. Kuffler, S.W., Nicholls, J.G.: From Neuron to Brain. Sinauer Associates, Sunderland (1976); Mir, Moscow (1979)
11. Bhattacharyya, S., Dutta, P.: Multi-scale Object Extraction with MUSIG and MUBET with CONSENT: A Comparative Study. In: Proceedings of KBCS 2004, pp. 100–109 (2004)

12. Escudero, G., Escudero, G., Marquez, L., Rigau, G.: A Comparison between Supervised Learning Algorithms for Word Sense Disambiguation. In: Proc. of CoNLL 2000, pp. 31–36. ACL (2000)
13. Beveridge, J.R., Grith, J.R., Kohler, R., Hanson, A.R., Riseman, E.M.: Segmenting images using localized histograms and region merging. Int'l J. of Comp. Vis. 2, 311–347 (1989)
14. Cheng, Y.: Mean shift, mode seeking, and clustering. IEEE Trans. Pattern Anal. Machine Intell. 17, 790–799 (1995)
15. Yu, H., Li, M., Zhang, H.-J., Feng, J.: Color texture moments for content-based image retrieval. In: Internat. Conf. on Image Processing, vol. 3, pp. 929–932 (2002)
16. Li, D.-C., Fang, Y.-H.: An algorithm to cluster data for efficient classification of support vector machines. Expert Systems with Applications 34, 2013–2018 (2008)
17. Marmo, R., et al.: Textural identification of carbonate rocks by image processing and neural network: Methodology proposal and examples. Computers and Geosciences 31, 649–659 (2005)
18. Ohashi, T., Aghbari, Z., Makinouchi, A.: Semantic Approach to Image Database Classification and Retrieval. NII. Journal 7 (September 2003)

An Improved Design for Digital Audio Effect of Flanging

Jianping Chen, Xiaodong Ji, Xiang Gu, and Jinjie Zhou

School of Computer Science and Technology, Nantong University
Nantong, Jiangsu, 226019, P.R. China
chen.jp@ntu.edu.cn

Abstract. Digital audio effect is a digital signal processing technique used to modify or modulate an audio signal to make it have some special hearing effect such as reverberation and flanging. Starting from a discussion of the general model for the generation of digital flanging effect, this paper presents an improved model for flanging effect generation. Simulation experiments are done and the performances of the two models are compared. The results show that the new model can enrich the harmonic components and enhance the flanging effect. The new flanging model is also examined for different modulation waveforms.

Keywords: Digital Audio Processing; Audio Effect; Flanging.

1 Introduction

Audio effects, such as echo, reverberation, flanging, chorusing and pitching are indispensable in movie, television, broadcast and music production and performance. They play an important role for the tone beautification of music, song and voice, and the generation of special sound effects. Traditionally, audio effects are produced using artificial methods or analogy technology. With the development of digital signal processing (DSP) technology, they are now mostly realized using digital signal processing systems. In a digital audio effect system, the sampled audio signal is processed or modified with a specific DSP algorithm which usually involves the basic operations of delay, filtering and modulation. The understanding of the generation mechanism and the building of the basic model for a specific digital audio effect is very important for the design of the corresponding DSP algorithm. This paper focuses on the generation of digital audio effect of flanging, the one of the most often used audio effects. Based on the basic principle and the general model of digital flanging effect, a modified model is proposed, which proves to be with an enhanced tone inflection effect.

2 Generation of Flanging Effect

Flanging is an audio effect that is created by smoothly modulating the duration of a delay line with a very low frequency. This effect can circularly and repeatedly exaggerate the odd or even harmonic components of an audio signal, making its

D. Ślęzak et al. (Eds.): SIP 2009, CCIS 61, pp. 65–71, 2009.
© Springer-Verlag Berlin Heidelberg 2009

frequency harmonics vary periodically and producing the sound effects with hallucinogenic coloring like the amphoric, spouting, and alternating, as if the original sound is mounted with a peculiar sound edge.

2.1 The General Model

The general model of digital flanging effect that is often used in practice is shown in Figure 1.

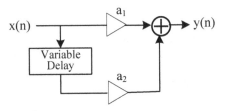

Fig. 1. General model of flanging effect

The input audio signal $x(n)$ goes through a variable or dynamic delay unit and generate a dynamically delayed signal $s(n)=x(n-d(n))$, where $d(n)$ is a dynamic delay modulation waveform. Controlled by an attenuation coefficient a_2, the dynamically delayed signal is mixed with the direct signal $a_1 x(n)$ (a_1 is an attenuation coefficient of $x(n)$ with $0 < a_1 < 1$) to form the output signal:

$$y(n) = a_1 x(n) + a_2 x(n-d(n)) . \qquad (1)$$

When the dynamic delay modulation waveform $d(n)$ is a low-frequency signal with a frequency of less than 1 Hz and the maximum delay of $d(n)$ is less than 10 msec, the output signal will produce a flanging effect.

The generation of flanging effect involves the waveform of modulation signal, the frequency of modulation signal, and the attenuation coefficients. Different modulation waveforms, different modulation frequencies and different attenuation coefficients would produce different flanging effects.

2.2 The Modified Model

The general model in Figure 1 is a simple addition of the dynamically delayed signal to the original signal. Such a structure acts as a comb filter. The slight changes of the dynamic delay produce different comb filtering effects, which hence form the flanging effect. By increasing the depth of the notches and the selectivity of the comb filter, the flanging effect can be enhanced. The depth of the notches of a comb filter can be increased with a feedback. Introducing a feedback link to the general model in Figure 1, we propose a modified model as shown in Figure 2.

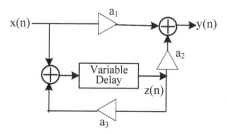

Fig. 2. Modified model for flanging effect

In this new model, the dynamic delay unit has the same function as in the original model in Figure 1, which performs a dynamic delay operation to the input signal. A feedback unit is added to increase the depth of the notches of comb filtering. The output signal of the modified model is as follows.

$$y(n) = a_1 x(n) + a_2 z(n) .$$ (2)

$$z(n) = x(n\text{-}d(n)) + a_3 z(n\text{-}d(n)) .$$ (3)

The generation of flanging effect with the new model in Figure 2 also involves the waveform and the frequency of modulation signal, and the attenuation coefficients. Generally, the values of the attenuation coefficients a_1, a_2 and a_3 are between 0 and 1. The frequency of the modulation waveform should also be less than 1 Hz. The modulation waveform may be of four kinds: (a) Sinusoidal, (b) Triangle, (c) Logarithmic, and (d) Exponential.

3 Simulation Experiments and Result Analysis

Simulation experiments of the two models are conducted using Matlab. The digitized sampling rate is 44.1 KHz. The precision of sampled data is 16 bits.

3.1 Comparison of Modified Model with General Model

For both the modified model and the original model in Figure 2 and Figure 1, use a sinusoidal wave as the modulation waveform:

$$d(n) = \frac{D}{2}(1 - \cos(2\pi F_d n)) .$$ (4)

Here F_d =0.01 cycles/sample, and the maximum delay D=20 samples. A single-frequency sinusoidal wave is used as the input signal. Its frequency is F=0.05 cycles/sample. For the general model in Figure 1, select a_1=a_2=0.5. For the modified model in Figure 2, select a_1=0.5 and a_2=a_3=0.8. Let $y_1(n)$ and $y_2(n)$ represent the output signals of the two models in Figure 1 and Figure 2, respectively. Figure 3 shows the waveforms of input signal *x(n)* and output signals $y_1(n)$ and $y_2(n)$. It can be seen from Figure 3 that the time domain output waveform of the modified model has a greater amplitude variation.

Figure 4 shows the frequency spectrum diagrams of the input *x(n)* and outputs $y_1(n)$ and $y_2(n)$. As shown in Figure 4, the output signal of the modified model has richer harmonic components, which indicates an enhancement of the flanging effect.

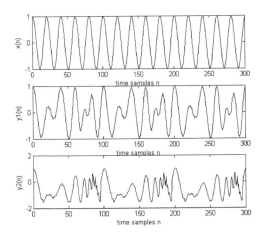

Fig. 3. Comparison of output waveforms of two models

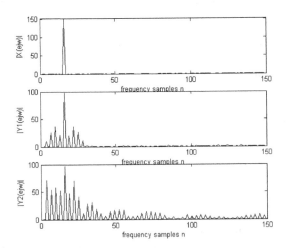

Fig. 4. Comparison of frequency spectrums of output waveforms

Using the WAVPLAY function of Matlab, we can obtain and listen to the audio outputs of the two models. An enhanced flanging effect can be perceived for the modified model.

3.2   Performances of Modified Model for Different Modulation Waveforms

For the modified model, we use four different modulation waveforms to compare the output flanging effects. Figure 5 shows the waveforms of four kinds of modulation signals which are sinusoidal wave, triangle wave, exponential wave and logarithmic wave, respectively. The modulation frequency is set at $F_d=0.01$ cycles/sample, and the maximum delay is $D=20$ samples. The input $x(n)$ is a single-frequency sinusoidal signal with a frequency of $F=0.05$ cycles/sample. The attenuation coefficients are $a_1=0.5$, $a_2=a_3=0.8$.

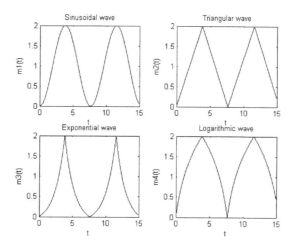

Fig. 5. Waveforms of four kinds of modulation signals

Figure 6(a), (b), (c) and (d) show the waveforms of output signals $y_1(n)$, $y_2(n)$, $y_3(n)$ and $y_4(n)$, which correspond to the four modulation signals of the sinusoidal, triangle, exponential and logarithmic, respectively. As can be seen from Figure 6, the amplitude of the output signal varies slowly and smoothly with the sinusoidal modulation waveform. The amplitude varies faster with the triangle modulation waveform. For the exponential and logarithmic waveforms, the output amplitude varies the fastest.

Similarly, using the WAVPLAY function of Matlab, we can obtain and listen to the audio outputs of the above four kinds of flanging effects. Assessing by the subjective hearing, for the sinusoidal modulation waveform the output has a smooth tone inflection effect. The triangle modulation waveform produces a flanging effect of linear tone inflection that has a rapid tonic rise and fall in the vicinity of the turning point. The tone inflection effects produced by the exponential and the logarithmic waveforms are similar and sound the best.

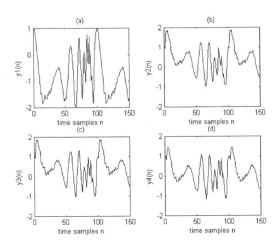

Fig. 6. Output waveforms corresponding to four modulation signals

4 Conclusion

The generation of digital audio effect of flanging is studied. The general model of flanging effect is modified to create a new model, in which a feedback unit is added to enhance the tone inflection effect. Simulation experiments and comparison are made for the two models. The results show that the modified model enriches the harmonic components and improves the flanging effect. For the modified model, further simulation experiments are conducted using different modulation waveforms including the sinusoidal, the triangle, the exponential and the logarithmic. The sinusoidal modulation wave produces a smooth tone inflection. The tone inflection produced by the triangle modulation wave has a rapid tonic rise and fall in the vicinity of the turning point. The exponential and the logarithmic modulation waves produce similar flanging effects that sound the best.

Acknowledgments. Financial supports from the Application Research Program of Nantong City (No. K2007007) of P. R. China are acknowledged. The author also gratefully acknowledges the support of K. C. Wong Education Foundation, Hong Kong.

References

1. Orfanidis, S.: Introduction to Signal Processing. Tsinghua University Press/Prentice Hall, Beijing (1998)
2. Mitra, S.: Digital Signal Processing: A Computer Based Approach. Tsinghua University Press/McGraw-hill, Beijing (2001)
3. Dattorro, J.: Effect Design: Part 1 Reverberator and Other Filters. Audio Engineering Society 45(10), 660–684 (1997)

4. Dattorro, J.: Effect Design: Part 2 Delay-Line Modulation and Chorus. Audio Engineering Society 45(10), 764–788 (1997)
5. Verfaille, V., Zolzer, U., Arfib, D.: Adaptive Digital Audio Effects (A-DAFx): A New Class of Sound Transformations. IEEE Transactions on Audio, Speech and Language Processing 14(5), 1817–1831 (2006)
6. Caputa, M.: Developing Real-Time Digital Audio Effects for Electric Guitar in an Introductory Digital Signal Processing Class. IEEE Transactions on Education 41(4), 341–346 (1998)
7. Martens, W., Atsushi, M.: Categories of Perception for Vibrato, Flange, and Stereo Chorus: Mapping out the Musically Useful Ranges of Modulation Rate and Depth for Delay-Based Effects. In: 9th Int. Conference on Digital Audio Effects (DAFx 2006), Montreal, pp. 149–152 (2006)

Robust Speech Enhancement Using Two-Stage Filtered Minima Controlled Recursive Averaging

Negar Ghourchian[1], Sid-Ahmed Selouani[2], and Douglas O'Shaughnessy[3]

[1] INRS-EMT, University of Quebec, Montreal, Canada
[2] University of Moncton, Campus de Shippagan, Canada

Abstract. In this paper we propose an algorithm for estimating noise in highly non-stationary noisy environments, which is a challenging problem in speech enhancement. This method is based on minima-controlled recursive averaging (MCRA) whereby an accurate, robust and efficient noise power spectrum estimation is demonstrated. We propose a two-stage technique to prevent the appearance of musical noise after enhancement. This algorithm filters the noisy speech to achieve a robust signal with minimum distortion in the first stage. Subsequently, it estimates the residual noise using MCRA and removes it with spectral subtraction. The proposed Filtered MCRA (FMCRA) performance is evaluated using objective tests on the Aurora database under various noisy environments. These measures indicate the higher output SNR and lower output residual noise and distortion.

Keywords: speech enhancement, noise estimation, musical noise, spectral subtraction.

1 Introduction

In all speech enhancement systems, noise power spectrum estimation has been a principal component, as it has a major impact on the quality of the enhanced signal. Overestimation causes distortion in intelligibility and underestimation yields residual noise audibility. A common approach is to average the noisy signal during the silent intervals using voice activity detection (VAD), which makes a binary decision on a frame of the captured signal to represent silence or speech regions. This decision mostly depends on the energy level and causes a weak performance in high-level noises. Additionally, in the presence of nonstationary noises wherein we have a rapid change in the spectral characteristics of the noise, insufficient silent sections prevent having an accurate and adaptive noise estimation. Alternatively, some algorithms estimate and update the noise spectrum continuously even during the speech sections [1]-[3]. These noise estimation algorithms are based on the fact that, during the speech-presence periods, there are some segments wherein the spectral energy approaches zero. These silent segments not only can be detected in the stop closures, but also they can occur during unvoiced fricatives at low frequencies and during vowels or generally voiced sounds, at high frequencies. Thus, due to this nature of

D. Ślęzak et al. (Eds.): SIP 2009, CCIS 61, pp. 72–81, 2009.
© Springer-Verlag Berlin Heidelberg 2009

speech, different frequency bands in the spectrum will be corrupted nonuniformly and will have different signal-to-noise ratios (SNR). This remark has led to recursive averaging-based noise estimation algorithms in which the noise spectrum is estimated and updated more reliably due to the extra information [4] (chapter 9). In [1] Martin has presented a noise estimation algorithm based on tracking the minimum of the noisy signal power spectral density (PSD). This method is sensitive to outliers with a large variance due to its tardy update rate.

In [2] Cohen introduces a minima-controlled recursive averaging (MCRA) approach in which the noise spectrum is updated recursively. The choice of using MCRA is motivated by its performance that combines the robustness of the minimum tracking with the simplicity of the recursive averaging. However, the domination of the minima over the weak speech components results in speech leakage in the noise spectrum. This overestimation of the noise spectrum introduces speech distortion to the system after the denoising process. Furthermore, in the original MCRA, misclassification of noise-only intervals in the within-speech and inter-speech/silent segments results in underestimation of the noise spectrum. Therefore, in the enhanced signal, some artifacts appear randomly in the time and frequency domains and have some audible dominant peaks in spectral sub-bands. These fluctuations in the residual noise are called musical noise.

In this work, we have proposed a two-stage noise enhancement method to address the inefficiency of VAD and to remedy the inadequacies of MCRA. The motivation is to jointly reduce the distortion and residual noise caused by spectral filtering.

The layout of this paper is as follows. In section 2 we introduce the pre-processing step. In Section 3 we review the general concept behind original MCRA and its application as the main part of the second stage. Evaluations are depicted in Section 4, and in Section 5 the work is concluded.

2 Overview of the Filtering Stage

The overall block diagram of the proposed method is depicted in Fig. 1. The first stage of this speech enhancement system is composed of the following blocks:

- A VAD is applied to estimate the noise power spectrum during speech pauses.
- The *a priori* SNR, which depends on the *a posterior* SNR, and the speech spectrum estimated from the previous frame.
- The gain function of our proposed filter produces the first-level enhanced speech.

The output of this stage then will be the input of MCRA in the second stage instead of the original noisy speech.

2.1 Concept of Pre-processing

We consider a speech signal $x(m)$ corrupted by uncorrelated additive noise $n(m)$, where m is a discrete-time index. The observed signal,

$$y(m) = x(m) + n(m),\tag{1}$$

is divided into K overlapping frames by windowing $y(m)$ and then computing the F-point FFT of the windowed signal. The distorted signal in the transformed domain is

$$Y(k,f) = X(k,f) + N(k,f),\qquad(2)$$

where $Y \in R^{K \times F}, X \in R^{K \times F}$ and $N \in R^{K \times F}$ represent the spectra of $y(m), x(m)$ and $n(m)$, respectively. An estimation of the clean speech is obtained as follows:

$$\hat{X}(k,f) = G_{opt}(k,f)Y(k,f),\qquad(3)$$

where $G_{opt}(k,f)$ is the gain function of pre-processing filter.

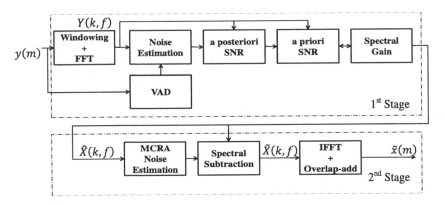

Fig. 1. Block diagram of the overall proposed two-stage enhancement algorithm

This gain function and other steps of the first stage are designed in order to ameliorate the drawbacks of the original MCRA, owing to the following observations:

- One of the drawbacks of MCRA is in classifying the transition frames between speech and non-speech segments. Specifically, when shifting from speech to silence, the momentum inherited from the recursive averaging noise estimation will influence the performance of the estimator in a few frames following the speech transition. Since this transition happens in some stop closures, unvoiced fricatives and voiced sounds as well, addressing the problem will probably have a big effect on the results. The suggested refinement has been done applying a VAD in first stage and eliminating the power of noise during the noise-only intervals.
- The pre-processing filter needs to smooth the input of the MCRA by excluding the relatively strong speech components with a view to lowering the voice component influence. This smoothing prevents the information from leaking out of noisy intervals to clean parts of the speech, where the noise magnitude is erroneously produced from the masking voice components. The required attenuation was applied on high SNR regions by a smoothing factor, τ, in (4) and the lower SNR regions were amplified in the pre-processing filter.

- The filtered speech signal has a lower noise level compared to the original noisy speech and this improvement will be reflected in the accuracy of the noise estimator in the second stage, which yields more residual noise reduction and less distortion.

2.2 Spectral Gain Calculation

Let $G_{opt}(k, f)$ be defined as below,

$$
G_{opt}(k, f) = \begin{cases} G(k, f) + \zeta, & G(k, f) + \zeta \leq 1 \\ \\ \tau & otherwise \end{cases} \tag{4}
$$

where τ, the smoothing/damping factor in higher SNRs, is set to 1, and ζ (the shifting factor) is set experimentally in order to minimize the speech leakage. We have used a modified version of the Wiener principle [5] [6] to filter the noisy speech in the first stage. $G(k, f)$ in (4) is a function of *a priori* SNR and *a posteriori* SNR given by

$$
G(k, f) = \frac{\Lambda(k,f)}{\Lambda(k,f)+1}, \tag{5}
$$

where $\Lambda(\lambda, f)$ is the *a priori* SNR which is estimated as follows,

$$
\Lambda(k, f) = (1 - \beta)D(\Gamma(k, f)) + \beta \frac{|G(k-1,f)Y(k-1,f)|^2}{\eta_n(k,f)}, \tag{6}
$$

where β is a real constant and $D(x) = \frac{x+|x|}{2}$. In equation (6), the $\eta_n(k, f)$ is the estimated noise PSD in the fth sub-band,

$$
\eta_n(k, f) = E\{|N(k, f)|^2\}, \tag{7}
$$

wherein $N(k, f)$ is the noise power spectrum estimated during the speech pauses. As we just demonstrated in (5) and (6) the $G(k, f)$ is a function of the *a posteriori* SNR, $\Gamma(k, f)$, as well that represented by

$$
\Gamma(k, f) = \frac{|Y(k,f)|^2}{\eta_n(k,f)} - 1. \tag{8}
$$

After computing the filter's parameters, the gain function can be simply derived from (4) and then we have the output of our first stage enhancement, i.e., $\hat{X}(k, f)$.

3 Enhancement Using FMCRA

In the recursive averaging algorithms not only the individual frequency bands of the noise spectrum are estimated and updated from the non-speech intervals, they also can be extracted whenever the effective SNR or the probability of speech being present is extremely low during the speech activity. Therefore, in the second stage of the proposed method we used MCRA noise estimation to cover the VAD problems in lower SNRs, particularly when the noise is nonstationary and changes abruptly estimation.

3.1 Validation of the Two-Stage Progress

In order to be able to feed the output signal of the proposed filter, $\hat{X}(k,f)$, to the MCRA block, we need to satisfy the condition of having an uncorrelated additive noise. From what we had in (3) we can derive that

$$\hat{X}(k,f) = G_{opt}(k,f)Y(mk,f) = G_{opt}(k,f)X(k,f) + G_{opt}(k,f)N(k,f). \quad (9)$$

Thus, the first term in (9), i.e., $G_{opt}(k,f)X(k,f)$ must be uncorrelated with the second term $G_{opt}(k,f)N(k,f)$. Therefore, in Equation (4) we define the G_{opt} to be just a function of noise in silence regions and be a constant value in the presence of both clean and noise signals.

3.2 Noise Estimation

Let $\hat{X}(k,f)$ be the noisy input to the MCRA as illustrated in Fig. 1. The presence or absence of speech in the kth frame of the fth sub-band is defined using the following two hypotheses:

$$\begin{aligned} H_0^f &: speech\ absent:\ \hat{X}(k,f) = D(k,f), \\ H_1^f &: speech\ present: \hat{X}(k,f) = \tilde{X}(k,f) + D(k,f), \end{aligned} \quad (10)$$

where $D \in R^{K \times F}$ and $X \in R^{K \times F}$ represent the spectra of the residual noise and the final clean signal, respectively. The noise PSD in the fth sub-band is given by $\sigma_d(k,f) = E\{|D(k,f)|^2\}$ and $\tilde{\sigma}_d(k,f)$ is the optimum PSD in terms of minimum mean square error.

The original MCRA [2] uses the temporal modified hypotheses to obtain an estimate for $\tilde{\sigma}_d(k,f)$ as follows:

$$\begin{aligned} H_0^f &: \tilde{\sigma}_d(k,f) = \alpha_d \tilde{\sigma}_d(k-1,f) + (1 - \alpha_d)|\hat{X}(k,f)|^2, \\ H_1^f &: \tilde{\sigma}_d(k,f) = \tilde{\sigma}_d(k-1,f), \end{aligned} \quad (11)$$

where α_d ($0 < \alpha_d < 1$) is the smoothing factor that is updated based on the conditional probability of speech being absent. Let $p(k,f) = P(H_1^f|\hat{X}(k,f))$ indicate the conditional probability of speech presence. Then using [3] we can express $\tilde{\sigma}_d(k,f)$ as follows:

$$\begin{aligned} \tilde{\sigma}_d(k,f) &= \tilde{\sigma}_d(k-1,f)p(k,f) + \left[\alpha_d \tilde{\sigma}_d(k-1,f) + (1-\alpha_d)|\hat{X}(k,f)|^2\right](1 - p(k,f)) \\ &= \hat{\alpha}_d(k,f)\tilde{\sigma}_d(k-1,f) + [1 - \hat{\alpha}_d(k,f)]|\hat{X}(k,f)|^2, \end{aligned} \quad (12)$$

where

$$\hat{\alpha}_d(k,f) \triangleq \alpha_d + (1 - \alpha_d)p(k,f). \tag{13}$$

The required probability in (13) can be calculated by the ratio between the instantaneous and minimum power spectra of the noisy speech. The above ratio is then formed as:

$$S_r(k,f) = \frac{S(k,f)}{S_{min}(k,f)}, \tag{14}$$

where the smoothed noisy speech PSD, $S(k,f)$, is estimated as

$$S(k,f) = \alpha_s S(k-1,f) + (1 - \alpha_s)S_f(k,f), \tag{15}$$

and the local minimum $S_{min}(k,f)$ is found over a fixed window by sample-wise comparison of past values of $S(k,f)$. In (15) α_s is a smoothing factor and $S_f(k,f)$ is the smoothed (over frequency) noisy speech PSD.

The ratio obtained in (14) is compared against a threshold to verify the speech presence/absence regions in the spectrum; thus, when the ratio exceeds this threshold, the period is labeled as speech presence. Afterwards, a temporal smoothing is required on the presence probability over time to lower the fluctuations between speech and non-speech segments. The smoothed speech presence probability $\hat{p}(k,f)$ is calculated using a smoothing factor, α_p, i.e.,

$$\hat{p}(k,f) = \alpha_p\hat{p}(k-1,f) + (1 - \alpha_p)p(k,f). \tag{16}$$

In the end, using the obtained $\hat{p}(k,f)$, a time-frequency dependent smoothing factor, $\hat{\alpha}_d(k,f)$, can be determined from (13) and subsequently we will have the updated noise PSD, $\tilde{\sigma}_n(k,f)$, using (12).

The process of obtaining the noise spectrum is usually followed by a denoising phase, in order to extract the clean signal from the degraded one. Here a typical spectral subtraction (SS) [9] is used.

4 Experiments

In this section we analyze the performance of the FMCRA (Filtered-MCRA), comparing to other algorithms in terms of extensive objective quality tests under various noisy environments. These measurements are used in standard forms from [4] (chapter 10) as listed below:

- SegSNR: Segmental SNR
- WSS: Weighted Spectral Slope distance, (smaller reflects less distortion).
- PESQ: Perceptual Evaluation of Speech Quality [7], (the larger the better).

To evaluate and compare the performance of different methods, we carried out computer simulations with the Aurora 2 database [8] ($f_s = 8\,kHz$). The speech signals in TESTA and TESTB of the Aurora database were corrupted by eight types of noise at different global SNR levels, from $-5dB$ to $15dB$:

Test A: N_1 Subway noise, N_2 Babble noise, N_3 Car noise, N_4 Exhibition hall noise

Test B: N_5 Restaurant noise, N_6 Street noise, N_7 Airport noise, N_8 Train station noise

In the segmentation process, the frame length was chosen to be 30 ms with 50% overlap and a hamming window was applied. The shifting factor, ζ, is set to 0.35, and β is heuristically set to 0.98. Also in the MCRA implementation the ratio threshold of 5, $\hat{\alpha}_n$ of 0.95, α_s of 0.8 and α_p of 0.2 are used.

We have proved the efficiency of our method (FMCRA), as opposed to

- (MCRA): The original MCRA [2] noise estimation
- (IMCRA): The Improved MCRA [10], in which Cohen proposed a different formula to improve the estimation of the speech absence/presence probability.
- (SS-VAD): A VAD-based spectral subtraction [9].

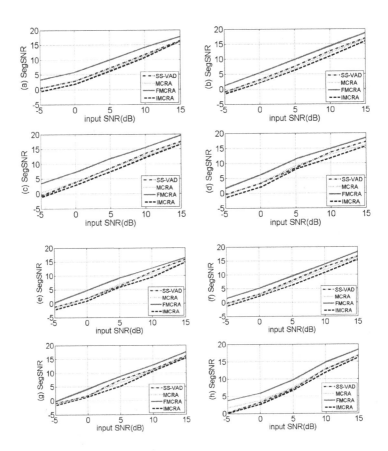

Fig. 2. SegSNR improvements in different noisy conditions: (a) N_1: Subway noise, (b) N_2: Babble noise, (c) N_3: Car noise, (d) N_4: Exhibition hall noise: (e) N_5: Restaurant noise, (f) N_6: Street noise, (g) N_7: Airport noise, (h) N_8: Train station noise

In Fig. 2 and Fig. 3, respectively, the SegSNRs and PESQ scores achieved by each method are shown and as we can see, FMCRA outperforms the other estimators. Fig. 4 depicts the WSS of all methods and the results report the minimum distance from the original speech in FMCRA, among all other algorithms.

5 Conclusions

In this paper we have developed an algorithm to enhance noisy speech in highly noisy environments and non-stationary conditions. Our approach is a two-stage enhancing method to improve the accuracy of the estimated noise spectrum with a view to reduce the speech leakage and also to remedy the musical noise effects and speech distortion in the enhanced signal. The drawbacks of MCRA were addressed and reduced in this work. The FMCRA seems to be a very effective algorithm based on subjective evaluations such as SegSNR, PESQ score and WSS distance.

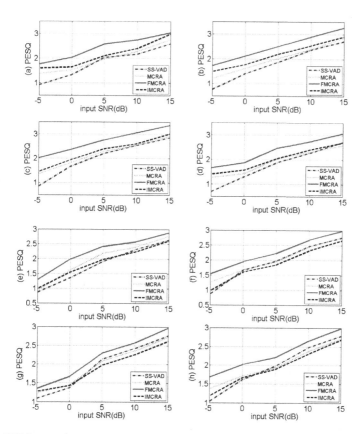

Fig. 3. PESQ score improvements in different noisy conditions: (a) N_1: Subway noise, (b) N_2: Babble noise, (c) N_3: Car noise, (d) N_4: Exhibition hall noise, (e) N_5: Restaurant noise, (f) N_6: Street noise, (g) N_7: Airport noise, (h) N_8: Train station noise

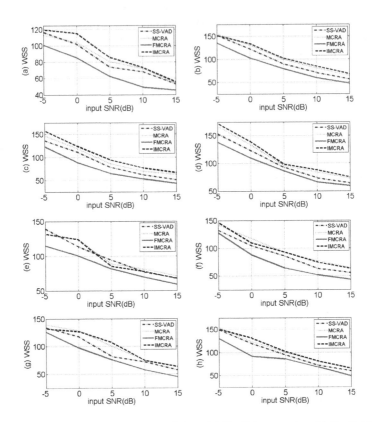

Fig. 4. WSS distance in different noisy conditions: (a) N_1: Subway noise, (b) N_2: Babble noise, (c) N_3: Car noise, (d) N_4: Exhibition hall noise, (e) N_5: Restaurant noise, (f) N_6: Street noise, (g) N_7: Airport noise, (h) N_8: Train station noise

References

[1] Martin, R.: Noise Power Spectral Density Estimation Based on Optimal Smoothing and Minimum Statistics. IEEE Trans. Speech Audio Processing 9(5), 504–512 (2001)

[2] Cohen, I.: Noise Estimation by Minima Controlled Recursive Averaging for Robust Speech Enhancement. IEEE Signal Processing Lett. 9(1), 12–15 (2002)

[3] Sohn, J., Sung, W.: A Voice Activity Detector Employing Soft Decision Based Noise Spectrum Adaptation. In: IEEE ICASSP, pp. 365–368 (1998)

[4] Loizou, P.: Speech Enhancement Theory and Practice, 1st edn. CRC Press, Boca Raton (2007)

[5] Ben Aicha, A., Ben Jebara, S.: Perceptual Musical Noise Reduction using Critical Band Tonality Coefficients. In: INTERSPEECH Conf., Antwerp, Belgium, pp. 822–825 (2007)

[6] Ephraim, Y., Mallah, D.: Speech Enhancement using Optimal Non-Linear Spectral Amplitude Estimation. In: IEEE ICASSP, pp. 1118–1121 (1983)

[7] Rix, A., Beerends, J., Hollier, M., Hekstra, A.: Perceptual Evaluation of Speech Quality (PESQ)- a New Method for Speech Quality Assessment of Telephone Networks and Codecs. In: IEEE ICASSP, pp. 749–752 (2001)

[8] Hirsch, H., Pearce, D.: The Aurora Experimental Framework for the Performance Evaluation of Speech Recognition Systems under Noisy Conditions. In: ISCA ITRW ASR 2000 (2000)

[9] Berouti, M., Schwartz, M., Makhoul, J.: Enhancement of Speech Corrupted by Acoustic Noise. In: IEEE ICASSP, pp. 208–211 (1979)

[10] Cohen, I.: Noise Estimation in Adverse Environments: Improved Minima Controlled Recursive Averaging. IEEE Trans. Speech Audio Processing 11(5), 466–475 (2003)

Automatic Colon Cleansing in CTC Image Using Gradient Magnitude and Similarity Measure

Krisorn Chunhapongpipat[1], Laddawan Vajragupta[2], Bundit Chaopathomkul[2], Nagul Cooharojananone[1], and Rajalida Lipikorn[1,*]

[1] Department of Mathematics, Faculty of Science, Chulalongkorn University
Bangkok, Thailand
Rajalida.L@chula.ac.th
[2] Department of Radiology, Faculty of Medicine, Chulalongkorn University
Bangkok, Thailand

Abstract. Electronic colon cleansing (ECC) is an alternative method that can be used to remove remaining tagged fecal material from colon during polyp detection in virtual colonoscopy (VC). This paper presents a new method for automatic electronic colon cleansing using gradient magnitude and similarity measure to detect and remove tagged material from colon in CTC images. First, Canny edge detection and 8-adjacency are used to generate closed boundary for low density (air) and high density regions (tagged material and bone). Then similarity measure is used to classify pixels into high density regions, and all the pixels that are classified as high density materials are removed. Finally, gradient magnitude and thresholding are used to detect AT and ATT layers. The proposed method was evaluated on four pilot dadasets from two patients and the experimental results reveal that the proposed method can perform colon cleansing effectively.

Keywords: Electronic colon cleansing, virtual colonoscopy, similarity measure, edge detection.

1 Introduction

Colorectal cancer is the forth leading cause of cancer-related deaths in the world, affecting approximately 665,000 deaths per year. Colonoscopy is an endoscopic technique that allows for inspection and therapeutic operations of the entire colon through insertion of the endoscope (with the camera at the tip of the scope) via the anus into the colon, however, it is invasive, time consuming, and expensive. Computer Tomography Colonography (CTC) or virtual colonoscopy is a rapidly emerging technique for polyp detection. Patients undergo a bowel preparation process and take oral contrast agent prior to coming for CTC. The patient is then placed in the supine or prone position and CT scanning is performed. The effectiveness of CTC for polyp detection depends on several factors such as quality of bowel preparation, bowel distention, and CTC interpretation

---

* Corresponding author.

D. Ślęzak et al. (Eds.): SIP 2009, CCIS 61, pp. 82–89, 2009.

by radiologists. Computer-aided detection of polyps along with new 3D viewing methods beyond the current fly-through and fly-back may significantly reduce the time taken for radiologists to view and interpret CTC data. Several methods have been proposed [1]-[9] but improvements of these methods are still possible. Most of the existing techniques cannot determine whether a pixel is on an AT layer or on an ATT layer. This is considered to be a crucial problem because a pixel on an ATT layer should be identified as soft-tissue and must be kept while a pixel on an AT layer should be identified as an artifact and should be removed. Thus, the main objective of this paper is to introduce a new method for electronic colon cleansing without prior-knowledge of intensity value and differentiate AT layer from ATT layer [9]. The method was evaluated on CTC image datasets in supine and prone positions obtained from Chulalongkorn hospital.

The rest of the paper is organized as follows: section 2 presents the proposed method. Section 3 describes the experimental results, and section 4 presents discussion and conclusions.

2   Proposed Method

The proposed method consists of four steps: preprocessing, edge detection, region classification using histogram peak and artifact removal, and reconstruction.

2.1   Preprocessing

The first step in colon cleansing is to prepare image data for electronic colon cleansing by finding the boundary of a patient's body using Canny edge detection and then filling in the outer boundary of the body to generate a body mask as shown in Fig. 1(b). Next, the body mask is eroded in order to remove the boundary of the body as shown in Fig. 1(c). Then we perform logical AND operation between the original image and the mask. The result image is shown in Fig. 1(d).

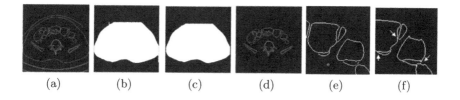

(a)   (b)   (c)   (d)   (e)   (f)

Fig. 1. (a) An edge image, (b) a body mask (white position), (c) an eroded image of (b), (d) a result image of (a)AND(c). (e) Part of Canny edge image, (f) an edge image after interpolation (white arrow points to successful connection, whereas, yellow arrow shows incorrect interpolation).

The next problem that must be solved before electronic cleansing can be performed is to close the boundary of colons. This problem is solved by using

8-adjecency to connect between two end points of any opened-connected components as shown in Fig. 1(e)and Fig.1(f). The process starts from generating a mask of size 3×3 pixels and placing the center of the mask at one end of the edge. Then any pixels that are 8-adjecency is assumed to be another end and they are connected together to form a closed boundary.

The last step in preprocessing is to fill the regions within all closed boundaries as shown in Fig. 2.

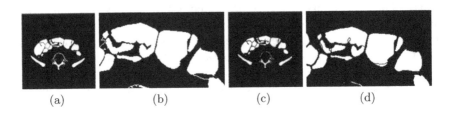

(a)   (b)   (c)   (d)

Fig. 2. (a) An image with filled regions,(b) a close-up image of (a),(c) a result image from subtraction of a filled-region image from an edge image,(d) a close-up of (c).

2.2 Histogram Peak Classification

In this step, each region is classified as either air region or tagged material region. First, we select the highest peak from the histogram of a body as shown in Fig. 3(b) (without background intensity). This peak represents tissue intensities and it is called the max_frequency_intensity. Thus any region that its average intensity is higher than the max_frequency_intensity is classified as a high density region(tagged material and bone). Any region that its average intensity is lower than the max_frequency_intensity is classified as a low density region(air).

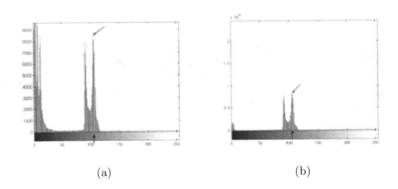

(a)   (b)

Fig. 3. (a) The histogram of all pixels in a CT scan image, (b) the histogram of pixels within the body of CT scan image: red arrow points to the peak and black arrow points to intensity value corresponding to the histogram peak, called **max_frequency_intensity**.

2.3   Similarity Measure Classification

The purpose of similarity measure is to classify any pixel, Ir_i, whose intensity is higher than the max_frequency_intensity and is located outside of high density regions as shown in Fig. 4(a). The similarity can be measured by calculating the sum of the differences between Ir_i and all pixels in a group of high density regions, s_1, and the sum of the differences between Ir_i and a group of tissue regions, s_2, whose intensities are higher than or equal to the max_frequency_intensity and do not belong to high density regions. For the second measurement of s_2, we first initialize a region of tissue pixels to contain only pixels with the max_frequency_intensity. if s_1 is greater than s_2 then Ir_i is classified as a pixel in a high density region, otherwise, it is classified as a pixel in a tissue region. The process can be expressed as follows:

$$s_1 = \frac{1}{m} \sum_{j=1}^{m} \mid (Ir_i - e_j) \mid \tag{1}$$

$$s_2 = \frac{1}{n} \sum_{j=1}^{n} \mid (Ir_i - t_j) \mid \tag{2}$$

where Ir_i is the ith pixel whose intensity value is higher than the max_frequency _intensity and is not in any region, s_1 is the difference between Ir_i and all pixels in a group of high density regions, s_2 is the difference between Ir_i and all pixels in a group of tissue regions, m is the number of pixels in a high density group, n is the number of pixels in a tissue group, e_j is the jth pixel in a group of high density regions, and t_j is the jth pixel in a group of tissue regions.

After the classification, all pixels that belong to high density regions are removed from an image.

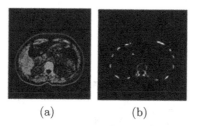

(a)   (b)

Fig. 4. (a) An image shows pixels whose intensity values are higher than max_frequency_intensity and are located outside of high density regions, (b) an image shows a high density group which are tagged material and bone

2.4   AT vs. ATT Layers

In order to distinguish AT from ATT, we use the magnitude of gradient which corresponds to the rate of change in intensities to detect the layers. These layers are located between air and tagged material which have high intensity change rate.

The magnitude of gradient is normalized to [0,1] and the location of edges in an image is shown in Fig. 5(c). Then we perform logical AND operation between the magnitude image in Fig. 5(b) and an eroded body mask (Fig. 1(c)) in order to extract the boundary of the body from an image as shown in Fig. 5(d).

AT and ATT are the layers that lie between air and high density components. This layer could be the layer between air and tagged material, which is called AT, or could be the tissue layer between air and tagged material, which is called ATT. The AT and ATT layers are extracted from gradient magnitude image by using thresholding with the threshold value, $threshold\_mag$, that can be computed from:

$$threshold\_mag = mean + \frac{3}{4}median \qquad (3)$$

where $mean$ is the average of the gradient magnitude of edges(Fig. 5(d)) and $median$ is the median of the gradient magnitude of edges(Fig. 5(d)). Any edge pixels whose gradient magnitudes are higher than $threshold\_mag$, are identified as pixels in AT or ATT layer as shown in Fig.5 (e) where yellow arrows point to AT layer and white arrow points to ATT layer.

Next, we have to seperate AT from ATT layer by using a combination of two techniques: thresholding and patient's posture. Firstly, the threshold value, $threshold\_AT$, is computed from:

$$threshold\_AT = \frac{4}{3}mean + median \qquad (4)$$

where $mean$ and $median$ are defined as above. Any layer that one of its pixels has gradient magnitude higher than the $threshold\_AT$ is identified as AT layer(yellow arrows in Fig. 5(e)).

Secondly, we also use patient's posture to identify AT layer. If the patient is scanned with supine position, the AT layer must lie between air on the top and tagged material on the bottom, otherwise, this layer is not an AT layer. If the

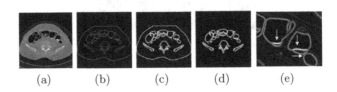

(a)   (b)   (c)   (d)   (e)

Fig. 5. (a) An original image, (b) a gradient magnitude image,(c) edge positions,(d) a result image of (c) AND Fig. 1(c), (e) the gray-scale gradient magnitude of AT (yellow arrow) and ATT (white arrow) layers

patient is scanned with prone position, the AT layer must lie between tagged material on the top and air on the bottom.

Finally, the results from two techniques are unioned in order to idetify all AT layers in an image. Fig. 6 shows a case when the image was taken in prone position but the tagged material is located in up-side-down position as pointed by the down arrow on the left image, thus our method has to use both the thresholding and the patient's posture to remove tagged material.

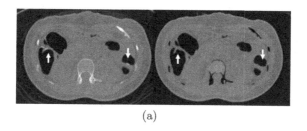

(a)

Fig. 6. ((Left) An image with tagged material (white arrow) in prone position, (right) an image after tagged materials were remvoed (white arrow)

.

2.5   Reconstruction

After all high density regions are removed, the reconstruction is the process that enhances colon border by blurring the sharp edges(Fig. 7(b)) along the border using Gaussian filter and average filter.

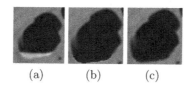

(a)   (b)   (c)

Fig. 7. (a) An image with tagged material.(b) an image without tagged material. (c) reconstruction of edges.

3   Experimental Results

The experimental results were evaluated by the physician under two criteria: percentage of cleansing and confidence score on accuracy. The percentage of cleansing is divided into 5 levels of completeness where level 1 means 0-25%, level 2 means 26-50%, level 3 means 51-75%, level 4 means 75-99%, and level 5 means 100%. The confidence score on accuracy is divided into 4 levels of confidence on the results where level 0 means no confidence, level 1 means low confidence, level

2 means moderate confidence, and level 3 means high confidence. Tabel 1 shows percentage of cleansing and table 2 shows confidence score on accuracy on four datasets of two patients. Fig. 8(b) and (d) illustrate how complete our method can clean tagged material from colon regardless of its location and size.

Table 1. Percentage of cleansing

case	level 1	level 2	level 3	level 4	level 5
case#1 prone	9.94 %	0 %	0 %	1.84 %	88.21 %
case#1 supine	12.68 %	1.87 %	0.72 %	0.43 %	84.29 %
case#2 prone	17.91 %	2.09 %	2.39 %	4.78 %	72.84 %
case#2 supine	11.07 %	0 %	0 %	6.56 %	82.38 %
average	12.9 %	0.99 %	0.78 %	3.4 %	81.93 %

Table 2. Confidence score on accuracy

case	level 0	level 1	level 2	level 3
case#1 prone	7 %	0 %	0.74 %	92.27 %
case#1 supine	14.84 %	0.29 %	0.58 %	84.44 %
case#2 prone	14.93 %	7.46 %	5.07 %	72.54 %
case#2 supine	11.46 %	0 %	11.07 %	77.45 %
average	12.06 %	1.94 %	4.37 %	81.68 %

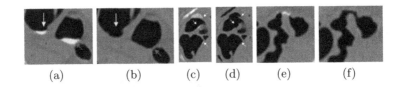

(a)   (b)   (c)   (d)   (e)   (f)

Fig. 8. (a) The area of tagged or enhanced material region. (b) the removed tagged area and red arrow points to ATT. (c) shows the small tagged area of prone patient pose. (d) shows the removed small tagged area.(e) The area of tagged material region.(f) the removed tagged material area.

4   Discussion and Conclusions

From the experimental results on four datasets from two patients in supine and prone positions where each dataset consists of CT-scan images of size 512x512x400. Fig.8(b) shows that tagged materials located at normal position can be cleaned efficiently, especially at the position pointed by the red arrow at the lower right corner where tagged materials are removed while colonic walls are kept untouched. As a summary, we propose a new method for electronic colon cleansing that can effectively remove tagged materials and AT layers while preserving the thin tissue layer that is located next to the tagged regions. Fig. 8(e)and Fig. 8(f) shows that removal of tagged materials allows virtual colonoscopy to fly through this section of colon lumen.

Acknowledgement

This research has been partially supported by grant under the A1B1 Scholarships from Faculty of Science, Chulalongkorn University, Thailand.

References

1. Liang, Z., Chen, D., Chiou, R., Kaufman, A., Wax, M., Viswambharan, A.: On segmentation of colon lumen for virtual colonoscopy. In: The SPIE Medical Imaging Conference, vol. 3660, pp. 270–278. SPIE, San Diego (1999)
2. Chen, D., Liang, Z., Wax, M.R., Li, L., Li, B., Kaufman, A.E.: A novel approach to extract colon lumen from CT images for virtual colonoscopy. IEEE transactions on medical imaging 19(12), 1220–1226 (2000)
3. Yoshida, H., Näppi, J.: Three-dimensional computer-aided diagnosis scheme for detection of colonic polyps. IEEE transactions on medical imaging 20(12), 1261–1274 (2001)
4. Li, L., Chen, D., Lakare, S., Kreeger, K., Bitter, I., Kaufman, A., Wax, M., Djuruc, P., Liang, Z.: An image segmentation approach to extract colon lumen through colonic material tagging and hidden Markov random field model for virtual colonoscopy. In: SPIE Medical Imaging Conference, vol. 4683, pp. 406–411. SPIE, San Diego (2002)
5. Serlie, I., Truoyen, R., Florie, J., Post, F., van Vliet, L.J., Vos, F.: Computed cleansing for virtual colonscopy using a three-material transaction model. In: Ellis, R.E., Peters, T.M. (eds.) MICCAI 2003. LNCS, vol. 2879, pp. 175–183. Springer, Heidelberg (2003)
6. Zalis, M.E., Perumpilichira, J., Hahn, P.F.: Digital substraction bowel cleansing for CT colonography using morphological and linear filtration methods. IEEE transactions on medical imaging 23(11), 1335–1343 (2004)
7. Wang, Z., Liang, Z., Li, X., Li, L., Li, B., Eremina, D., Lu, H.: An Improved Electronic Colon Cleansing Method for Detection of Colonic Polyps by Virtual Colonoscopy. IEEE transactions on biomedical engineering 53(8), 1635–1646 (2006)
8. Serlie, I.W.O., Vos, F.M., Truyen, R., Post, F.H., van Vliet, L.J.: Classifying CT Image Data Into Material Fractions by a Scale and Rotation Invariant Edge Model. IEEE transactions on image processing 16(12), 2891–2904 (2007)
9. Cai, W., Zalis, M.E., Näppi, J., Harris, G.J., Yoshida, H.: Structure-analysis method for electronic cleansing in cathartic and noncathartic CT coloscopy. Medical Physics 35(7), 3259–3277 (2008)
10. Ouadfel, S., Batouche, M.: An Efficient Ant Algorithm for Swarm-Based Image Clustering. Journal of Computer Science 3(3), 162–167 (2007)

A Scale and Rotation Invariant Interest Points Detector Based on Gabor Filters

Wanying Xu[1], Xinsheng Huang[1], Yongbin Zheng[1],
Yuzhuang Yan[1], and Wei Zhang[2]

[1] College of Mechatronics Engineering and Automation,
National University of Defence Technology, Changsha, China
[2] College of Computer, National University of Defence Technology, Changsha, China
{wanying_xu,zybnudt,yuzhuangyan,wadezhang}@nudt.edu.cn
huangxinsheng@163.com

Abstract. This paper presents a new method for detecting scale and rotation invariant interest points. The method is based on a representation of the image that involves both spatial and spatial-frequency variables in its description. The method is based on two main conclusions: 1) Interest points can be extracted based on the local maxima of the normalized local energy maps. 2) Local extrema over scale of a new established Gabor scale-space indicate the presence of characteristic local structures. Our method first extract interest points at multi-scales from the local energy map constructed by Gabor filter responses, and then select points at which a local measure is maximal over scales. This allows a selection of distinctive points for which the characteristic scale is known. The interest points are invariant to scale and rotation and give repeatable results (geometric stable). Comparative evaluation using the repeatability criteria shows the good performance of our approach.

Keywords: interest points, scale invariance, rotation invariance, Gabor filter, Gabor scale-space.

1 Introduction

Interest points are characteristic points in an image where the signal changes bidimensionally. Interest points characterized with local descriptors can be well adapted to geometric and photometric deformations, therefore have been widely used in obtaining image to image correspondence for 3D reconstruction [1-2] , searching databases of photographs [3] and as an early stage in object recognition [4] and image retrieval [5]. In a typical scenario, an interest point detector is used to select matchable points in an image. The output of a detector is a bunch of interesting locations which are expected to be stable under various image distortions.

There are numerous approaches for interest point detection and the points extracted by these algorithms differ in localization, scale and structure (corners, blobs, multi-junctions). In the literature, blob-like structures have been presented by [6-8],

D. Ślęzak et al. (Eds.): SIP 2009, CCIS 61, pp. 90–97, 2009.

corner-like structures by [3, 9-11], and edge-based structures by [8, 12], among which rotation invariants have been presented by [11-13], rotation and scale invariants by [3, 6, 14-16], and affine invariants by [8, 10].

Lindeberg [7] adopt different expression of Gaussian derivatives to capture different types of image structures such as blobs, corners, ridges and edges, and then searches for 3D maxima of the scale normalized Laplacian-of-Gaussian (LoG) function for automatic scale selection. Lowe [6, 14] proposed an efficient algorithm for detecting blob-like features and its characteristic scale based on local 3D extrema in the scale-space pyramid built with Difference-of-Gaussian (DoG) filters. This method has achieved great success in object recognition and image registration. Since then a lot of work has been done based on the method [15-16]. Mikolajczyk and Schmid [3] detect corner-like features based on Harris corner detector and obtain the characteristic scale using the Laplacian measure, which is referred as the Harris-Laplacian method. Schmid and Mohr [11] extract a set of interest points and characterize each of the points by rotational invariant descriptors which are combination of Gaussian derivative. Robustness to scale change is obtained by computing Gaussian derivatives at several scales. Furthermore, affine invariant features are presented in [8, 10], which start with various kinds of interest points like Harris corner [10] or intensity extrema [8]etc.

In this paper we proposed a new method which detects perceptually significant features based on the local energy map constructed by Gabor filter responses, and together with our proposal of detecting the local characteristic scale based on Gabor scale-space, it can achieves scale invariance as well as rotation invariance. The detected points are invariant to scale, rotation and translation as well as robust to illumination changes and limited changes of viewpoint. Experiment results show that its repeatability is better than some of the approaches proposed in the literature.

This paper is organized as follows. Section 2 reviews the basic properties of Gabor function. In section 3 the rotation invariant interest point detection method based on Gabor filter responses is introduced, and in section 4 the scale selection technique based on Gabor scale-space is presented. Section 5 presents the scale and rotation interest point detection algorithm integrally, and experiment results are given in section 6. Conclusion and future work are presented in section 7.

2   Gabor Functions

The 2D Gabor functions are local spatial bandpass filters that achieve the theoretical limit for conjoint resolution of information in the 2D spatial and 2D Fourier domains. They now play an important role in image representation because of their increasing importance in many computer vision applications and also in modeling biological vision, since recent neurophysiological evidence from the visual cortex of mammalian brains suggests that the filter response profiles of the main class of linearly-responding cortical neurons (called simple cells) are best modeled as a family of self-similar 2D Gabor wavelets [17].

Gabor functions are Gaussians modulated by complex sinusoids. A 2D complex Gabor function that models the responses of simple cells is denoted by [18]:

$$g_{\sigma,\theta}(x,y) = \frac{1}{2\pi\sigma^2} e^{-\frac{x^2+y^2}{2\sigma^2}} e^{i\frac{\pi}{\sigma}(x\cos\theta + y\sin\theta)} \tag{1}$$

where σ denotes the width of the Gaussian envelope in space domain, and θ is the preferred orientation. Gabor wavelets have the optimal energy concentration in the time and frequency plane. Furthermore, they provide multi-scale and multi-orientation information of the input image.

3   Rotation Invariant Interest Points Detection

What kinds of features in an image seem to draw the observer's attention and thus become interesting? A great deal of biological vision research has addressed such a problem [19-20]. Fdez-Valdivia et al. [19] present a method for identifying features based on the local maxima of energy maps for a few active sensors for a multichannel partition of the reference image. The local energy map, for each active channel, is defined as the sum of the squared response of the even-symmetric and odd-symmetric Gabor filters, which defines the reasonable candidates for locations where the visual system perceives something of interest.

Given an image $I(x, y)$, let $W_{\sigma,\theta}(x, y)$ be the image filtered by a complex 2D Gabor filter, as given in equation:

$$W_{\sigma,\theta}(x,y) = I(x,y) * g_{\sigma,\theta}(x,y) \tag{2}$$

where * denotes the convolution operator.

Then the local energy map from viewpoint of sensor (σ,θ), noted as $Lem_{\sigma,\theta}(x, y)$, is defined by:

$$Lem_{\sigma,\theta}(x,y) = \left\| W_{\sigma,\theta}(x,y) \right\| = \left\| I(x,y) * g_{\sigma,\theta}(x,y) \right\| \tag{3}$$

where $\|\cdot\|$ denotes modulus and is equal to the sum of the squared quadrature pair of sine and cosine part of the complex value.

To detect features at all orientations, a bank of filters is designed so that they tile the frequency plane uniformly. The important issue here is to ensure that features at all possible orientations are treated equally and all possible conjunctions of features are treated uniformly. That is, to obtain rotation invariance in feature detection. Therefore, instead of treating the local energy map from different channels, we eliminate the influence of parameter θ by summing energies over all orientations to obtain rotation invariance.

The approach that has been adopted is as follows. In the frequency plane, the orientation range $[0, \pi]$ is divided into N intervals. At each scale σ, we calculate the energy at each location of the image, denoted as $Lem_{\sigma,\theta}(x, y)$, in each orientation, and then sum the values over all orientations. This sum of energies is then normalized by dividing the number of orientations. So the normalized local energy map at scale σ, denoted as $Q_\sigma(x, y)$, is obtained by the following equation:

$$Q_\sigma(x, y) = \frac{\pi}{N} \sum_{k=1}^{N} Lem_{\sigma,\theta_k}(x, y) = \frac{\pi}{N} \sum_{k=1}^{N} \left\| W_{\sigma,\theta_k}(x, y) \right\| \tag{4}$$

where $k = \{1,...N\}$, $\theta_k = k\pi / N$.

To detect the peaks of the local energy map, non-maximum suppression is applied to the result obtained above. And to suppress false detection, a threshold T is applied to the results. Then a point at location (x^Δ, y^Δ) in the image is identified as a candidate interest point if:

$$Q_\sigma(x^\Delta, y^\Delta) = \max_{(x,y) \in N_{xy}} (Q_\sigma(x, y)) \quad and \quad Q_\sigma(x^\Delta, y^\Delta) \geq T \tag{5}$$

where N_{xy} represents a local neighborhood of (x,y) within which the search is conducted.

4 Scale Selection in Gabor Scale-Space

In this section, we establish a new Gabor scale-space and show how to select the characteristic scale in the Gabor scale-space.

The scale-space representation of an image is a set of representations created by convolving with the scale-space kernel at different levels of scale. Here we define a Gabor scale-space kernel as:

$$GSS_{ker nel}(x, y, \sigma) = \frac{\pi}{N} \sum_{k=1}^{N} g_{\sigma,\theta_k}(x, y) \tag{6}$$

where N has the same meaning as in formula (4), and $k = \{1,...N\}$, $\theta_k = k\pi / N$.

Then the Gabor scale-space representation of image $I(x, y)$ is denoted as:

$$GSS(x, y, \sigma) = I(x, y) * GSS_{ker nel}(x, y, \sigma) \tag{7}$$

The characteristic scale (denoted as σ_{char}) of a point (x_0, y_0) in the image can be obtained by searching for the local extrema over scales in Gabor scale-space, which can be expressed as:

$$\sigma_{char} = \arg \max_{\sigma} \left\| GSS(x_0, y_0, \sigma) \right\| \tag{8}$$

Given a point in an image, we compute the Gabor scale-space responses for a lot of scale factor σ, see Figure 1. The characteristic scale is the one at which the function achieves the local maximum (peaks of the curves), which is also indicated by the size of the circles in the left column images. It can be seen that the characteristic scale indicates the size of the image structure around the interest point and is independent of the image resolution. Furthermore, the ratio of the scales, at which the extrema were found for corresponding points in two rescaled images, is equal to the scale factor between the images.

The new established scale-space kernel is built on very well localized time-frequency Gabor atoms, so it has better frequency characteristics and nice selectivity of the energy spectrum than the LoG kernel. That makes it more applicable and reliable than the LoG method in characteristic scale selection.

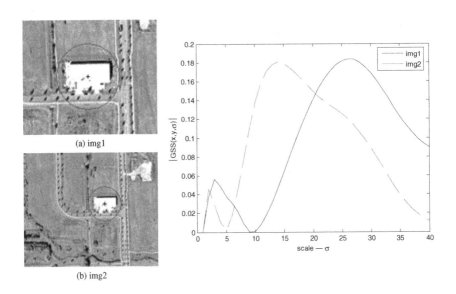

Fig. 1. The left column shows two images taken from different distance. The right column shows the response of GSS(x,y,σ) over scales. The extrema of the two curves are at 26.5 and 14 respectively, which corresponds to the scale factor (2.0) between the two images.

5 Scale Invariant Interest Points Detection

In the following section we present our scale invariant interest points in detail. Our detection algorithm works as follows:

Step 1. First we build a scale-space representation for the input image.

$$W_{\sigma_n,\theta_k}(x,y) = I(x,y) * g_{\sigma_n,\theta_k}(x,y) \tag{9}$$

where $n = \{3,...17\}$, $k = \{1,...8\}$, $\theta_k = k\pi/8$, $\sigma_n = \alpha^n \sigma_0$ with $\alpha = 1.2$ and $\sigma_0 = 1.6$.

Step 2. At each level of the scale-space we detect candidate points using the method proposed in section 3.

Step 3. In order to select the characteristic scale of the points, we verify for each of the candidate points found on different levels if it forms a maximum in the scale dimension. We reject the points for which the $GSS(x^\Delta, y^\Delta, \sigma_n)$ function attains no extrema. So a candidate is kept as an interest point if:

$$\left\| GSS(x^\Delta, y^\Delta, \sigma_n) \right\| > \left\| GSS(x^\Delta, y^\Delta, \sigma_{n-1}) \right\| and$$
$$\left\| GSS(x^\Delta, y^\Delta, \sigma_n) \right\| > \left\| GSS(x^\Delta, y^\Delta, \sigma_{n+1}) \right\| \tag{10}$$

and its characteristic scale is recorded as σ_n. Else the candidate point is rejected.

Figure 2 shows the interest points detected by our approach for two scaled images. Each point is circled with the radius equal to its characteristic scale. It can be seen that there are many point-to-point correspondences between the two images, and the characteristic scale indicates the size of the local structure of the feature. In two images, corresponding points cover the same surface content.

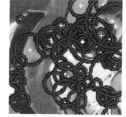

Fig. 2. Detected points of image pair at different resolutions with their characteristic scale. Circles are drawn with radius equal to σ_{char} of the detected point.

6 Comparative Evaluation

In this section, we first compare our approach with the traditional Harris detector for rotation invariance, since the Harris detector was evaluated as superior to other methods under rotation changes [13]. Then we compare our approach with several scale invariant interest points detectors in the literature, including the approaches of Lindeberg (Laplacian and gradient) [7], Lowe (SIFT) [6] as well as the

Harris-Laplace detector [3]. The stability of detectors is evaluated using the repeatability criteria introduced in [13]. The repeatability score is computed as a ratio between the number of point-to-point correspondences that can be established for detected points and the mean number of points detected in two images. Two points are taken to correspond if the error in relative location does not exceed 1.5 pixels in the coarse image and the ratio of detected scales for these points does not differ from the real scale ratio by more than 20%.

Figure 3 presents the repeatability score for the compared methods. The experiments were done on two sequences of real images. One sequence consist of rotated images for which the rotation degrees varies from 0 to 180 at interval of 15 degrees, and the other sequence consist of scaled images for which the scale factor varies from 1.2 up to 4.5. It can be seen that the proposed method has better rotation repeatability than the original Harris detector. The scale repeatability is a little smaller than Harris-Laplacian [3] method but better than the Laplacian method and SIFT method proposed by Lindeberg and Lowe respectively. And it should be cleared that our approach detected more features than the Harris-Laplacian method.

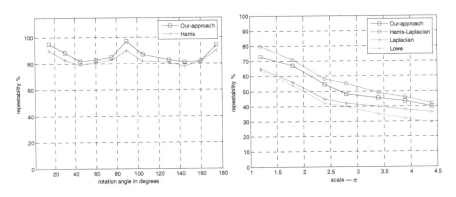

Fig. 3. Repeatability of interest point detectors with respect to rotation changes (left) and scale changes (right)

7 Conclusion and Future Work

In this paper, we have presented a new algorithm for interest points detection which is invariant to rotation and scale changes based on Gabor filters. The main content of the paper includes: 1) A rotation invariant interest point detector based on local energy map is presented. 2) A new Gabor scale-space is established, and the characteristic scale can be obtained by searching for the extrema in the scale-space. 3) A multi-scale interest points detector which is invariant to rotation and scale changes is developed base on the two results above.

The evaluation using repeatable criteria shows that our detector gives good stability under rotation and scale changes. Furthermore, our approach is robust to illumination changes as well. A comparison with some interest points detectors shows that our detector gives better results. In our future research, we intend to focus on the problem of local descriptor based on the Gabor representation.

References

1. Ohbuchi, R., Osada, K., Furuya, T., Banno, T., et al.: Salient Local Visual Features for Shape-Based 3d Model Retrieval. In: IEEE International Conference on Shape Modeling and Applications, Stony Brook, NY, pp. 93–102 (2008)
2. Snavely, N., Seitz, S.M., Szeliski, R.: Photo Tourism: Exploring Photo Collections in 3d, pp. 835–846. ACM, New York (2006)
3. Mikolajczyk, K., Schmid, C.: Indexing Based On Scale Invariant Interest Points. In: Eighth IEEE International Conference on Computer Vision, Vancouver, Canada, pp. 525–531 (2001)
4. Nister, D., Stewenius, H.: Scalable Recognition with a Vocabulary Tree. In: IEEE Computer Society Conference on Computer Vision and Pattern Recognition, pp. 2161–2168. IEEE Computer Society, Los Alamitos (2006)
5. Deselaers, T., Keysers, D., Ney, H.: Features for Image Retrieval: An Experimental Comparison. Information Retrieval 11, 77–107 (2008)
6. Lowe, D.G.: Distinctive Image Features From Scale-Invariant Keypoints. International Journal of Computer Vision 60, 91–110 (2004)
7. Lindeberg, T.: Feature Detection with Automatic Scale Selection. International Journal of Computer Vision 30, 79–116 (1998)
8. Tuytelaars, T., Van, G.L.: Matching Widely Separated Views Based On Affine Invariant Regions. International Journal of Computer Vision 59, 61–85 (2004)
9. Brown, M., Lowe, D.G.: Invariant Features From Interest Point Groups. In: British Machine Vision Conference, Cardiff, Wales, pp. 656–665 (2002)
10. Mikolajczyk, K., Schmid, C.: Scale & Affine Invariant Interest Point Detectors. International Journal of Computer Vision 60, 63–86 (2004)
11. Schmid, C., Mohr, R.: Local Grayvalue Invariants for Image Retrieval. IEEE Transactions on Pattern Analysis and Machine Intelligence 19, 530–535 (1997)
12. Sebe, N., Lew, M.S.: Comparing Salient Point Detectors. Pattern recognition letters 24, 89–96 (2003)
13. Schmid, C., Mohr, R., Bauckhage, C.: Evaluation of Interest Point Detectors. International Journal of computer vision 37, 151–172 (2000)
14. Lowe, D.G.: Object Recognition From Local Scale-Invariant Features. In: Seventh IEEE International Conference on Computer Vision, Kerkyra, Greece, pp. 1150–1157 (1999)
15. Bay, H., Ess, A., Tuytelaars, T., Van, G.L., et al.: Speeded-Up Robust Features (Surf). Computer Vision and Image Understanding 110, 346–359 (2008)
16. Grabner, M., Grabner, H., Bischof, H.: Fast Approximated Sift. In: 7th Asian Conference of Computer Vision, Hyderabad, India, pp. 918–927 (2006)
17. Lee, T.S.: Image Representation Using 2d Gabor Wavelets. IEEE Transaction on pattern analysis and machine intelligence 18, 959–971 (1996)
18. Wuertz, R.P., Lourens, T.: Corner Detection in Color Images through a Multiscale Combination of End-Stopped Cortical Cells. Image and Vision Computing 18, 531–541 (2000)
19. Fdez-Vidal, X.R., Garc, J.A., Fdez-Valdivia, J.: Using Models of Feature Perception in Distortion Measure Guidance. Pattern Recognition Letters 19, 77–88 (1998)
20. Kovesi, P.: Image Features From Phase Congruency. Videre: Journal of Computer Vision Research 1, 1–26 (1999)

Signature Verification Based on Handwritten Text Recognition

Serestina Viriri and Jules-R. Tapamo

School of Computer Science
University of KwaZulu-Natal
Durban, South Africa
{viriris,tapamoj}@ukzn.ac.za

Abstract. Signatures continue to be an important biometric trait because it remains widely used primarily for authenticating the identity of human beings. This paper presents an efficient text-based directional signature recognition algorithm which verifies signatures, even when they are composed of special unconstrained cursive characters which are superimposed and embellished. This algorithm extends the character-based signature verification technique. The experiments carried out on the GPDS signature database and an additional database created from signatures captured using the ePadInk tablet, show that the approach is effective and efficient, with a positive verification rate of 94.95%.

Keywords: Signature Recognition, Direction Feature, Feature Normalization.

1 Introduction

Signature verification has received an extensive research attention in recent years. Signatures are still widely used as the primary method for the identification and verification of human beings, and for authentication and authorization in legal transaction. In fact, it is the most widely accepted biometric trait that has been used for enforcing binding contracts in paper based documents for centuries. Handwriting recognition has reached its maturity level; especially for the recognition of isolated characters recognition, hand printed words recognition, automatic address processing, etc. Signatures are composed of special unconstrained cursive characters. In most cases they are superimposed and embellished, as a result are not readable.

The intrapersonal variations and interpersonal differences in signing make it necessary to analyse signatures absolutely as images instead of applying character and word recognition [3]. The other challenge of signature verification is that a signature is termed a behavioral biometric trait so it may change depending on behavioral characteristics such as mood, fatigue, etc [2]. Hence innovative and accurate approaches need to be employed to improve the recognition and verification rates.

D. Ślęzak et al. (Eds.): SIP 2009, CCIS 61, pp. 98–105, 2009.

1.1   Related Work

There has been substantial research work carried out in the area of signature recognition and verification. Justino et al [6], proposed an off-line signature verification system using Hidden Markov Model. In [7] a handwritten signature verification system, based on Neural 'Gas' Vector Quantization is proposed. Other recent approaches to signature recognition and verification include: the use of Modified Direction Feature (MDF) which generated encouraging results, reaching an accuracy rate of 81.58% for cursive handwritten character recognition [1]; a Support Vector Machine approach based on the geometrical properties of the signature is proposed by Ozgunduz et al [3]; and an Enhanced Modified Direction Feature with Single and Multi-classifier approaches [2]. Various classifiers have been successful in off-line signature verification, with Support Vector Machines (SVMs) providing an overall better result than all others such as Hidden Markov Models. SVMs have achieved a true classification ratio of 0.95 [2].

In this paper, a new approach that improves of the Modified Direction Feature is proposed. It adopts the whole text recognition rather character or word recognition for signature recognition and verification.

The rest of the paper is structured as follows: *Section 2* describes an overall signature recognition system, including image preprocessing, feature extraction, feature normalization and signature verification; *Section 3* presents experimental results; and *Section 4* draws the conclusions and future work.

2   Signature Recognition System

The basic modules for a signature recognition system: *capture the signature, preprocessing, feature extraction, feature vector*, and *classification* are depicted in **Fig.1**, and presented in the following subsections.

2.1   Preprocessing

The preprocessing module is applied to both training and testing phases. It includes noise reduction and skeletonization. Hilditch's thinning algorithm was adopted as the skeletonization technique [9].

2.2   Feature Extraction

The choice of a meaningful set of features is crucial in any signature verification system. Different feature methods have been studied to capture various structural and geometric properties of the signatures. The extracted features for this system are mask features and grid features. Mask features provide information about directions of the segments of the signature, while grid features give an overall signature appearance [3]. The extracted mask features are direction and transition features, and the grid features are trifold and sixfold features.

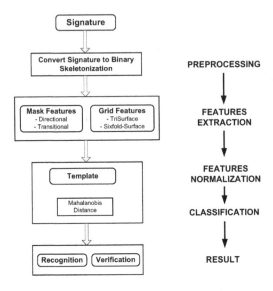

Fig. 1. Signature Recognition System

Direction Features

The direction feature technique was originally designed to simplify each character's boundary or thinned representation through identification of individual stroke or line segments in the image [5]. Signatures are composed of special unconstrained cursive characters which are in most of the cases superimposed and embellished. This makes it impossible to apply character recognition to all the signatures. For the latter reason, we have chosen to use direction feature technique for handwritten text recognition, instead of the segmented character recognition. The proposed text-based signature recognition approach explained below, is an extension of the character-based direction feature technique.

The line segments which give the direction transitions are determined for the whole text image rather than on each character image, and categorized as *Vertical, Horizontal, Right diagonal* and *Left diagonal*. The intersection points between types of lines are located. The techniques for locating the starting point and intersection points, and labelling the line segments are discussed in [2], [5], [11]. The direction transitions are coded as shown in Table 1.

Line Segments Categorization. As mentioned above, four types of line segments are distinguished by following neighbouring pixels along the thinned signature from the starting point. A semi 8-neighbourhood[1] based algorithm determines the beginning and end of individual line segments. This algorithm is described below.

---

[1] In our context semi 8-neighbourhood of a pixel (i,j) is made of (i-1,j-1), (i,j-1), (i+1,j-1) and (i+1,j).

Table 1. Direction Transition Codes

Direction Transition	Code
Vertical	2
Right Diagonal	3
Horizontal	4
Left Diagonal	5
Intersection	9

Algorithm 1. Line Segments Categorization

1: **if** $pixel[i,j] == INTERSECTION$ **then**
2:    $pixel[i,j] \leftarrow 9$
3: **else**
4:    **if** $SOUTH(pixel[i,j]) == FOREGROUND$ **then**
5:       $pixel[i,j] \leftarrow 2$
6:    **end if**
7: **else**
8:    **if** $SOUTHWEST(pixel[i,j]) == FOREGROUND$ **then**
9:       $pixel[i,j] \leftarrow 3$
10:    **end if**
11: **else**
12:    **if** $WEST(pixel[i,j]) == FOREGROUND$ **then**
13:       $pixel[i,j] \leftarrow 4$
14:    **end if**
15: **else**
16:    **if** $NORTHWEST(pixel[i,j]) == FOREGROUND$ **then**
17:       $pixel[i,j] \leftarrow 5$
18:    **end if**
19: **end if**

In Algorithm 1, NORTHWEST, WEST, SOUTHWEST and SOUTH give semi 8-neighbourhood values. After labelling the whole thinned signature, the algorithm for extracting and storing line segment information first locates the starting point and then extracts the number of intersection points (see equation (3)), the number and lengths of line segments resulting in a set of nine feature values. The number and lengths features of line segments are extracted using equations (1) and (2) respectively.

$$Number = \frac{Number\ of\ segments\ in\ a\ particular\ direction}{Area\ of\ signature\ bounding\ box} \qquad (1)$$

$$Length = \frac{Number\ of\ pixels\ in\ a\ particular\ direction}{Area\ of\ signature\ bounding\ box} \qquad (2)$$

$$Intersection = \frac{Number\ of\ intersection\ points}{Area\ of\ signature\ bounding\ box} \qquad (3)$$

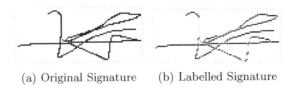

(a) Original Signature   (b) Labelled Signature

Fig. 2.   Original   Signature   and   its   corresponding   Labelled   Signature.
(Black=Horizontal,   Yellow=Vertical,   Red=Intersection,   Blue=Right   Diagonal,
Cyan=Left Diagonal).

Transition Features

Transition feature (TF) extraction technique records the locations of the transitions between foreground and background pixels in the vertical and horizontal directions of binary images. The image is traversed in the following four directions: left to right, right to left, top to bottom and bottom to top. Each time there is a change from background to foreground or vice versa, then the ratio between the locations of the transition and the length/width of the image traversed is recorded as a feature [1], [2], [4]. The average ratios of the x and y positions of the transitions for each of the four directions are recorded, resulting in eight features. In addition, the total number of transitions is also recorded, resulting in two more features.

Grid Features

1. **TriSurface Feature** -The trisurface area was implemented in an attempt to increase the accuracy of a feature describing the surface area of a signature [2],[11]. This is achieved by splitting up vertically the signature into three equal parts and calculating the proportion of the signature surface of each part over the total surface area of the image using equation (4). This results in a set of three feature values.

2. **Sixfold** -The concept of sixfold feature is similar to the concept of the trisurface feature. The signature image is split up vertically into three equal parts. Then the center of gravity is calculated for each of the three parts, and the signature surface above and below the horizontal line of the center of gravity is calculated [2], [11]. The result is a set of six feature values corresponding to the proportion of the signature surface of each of the six parts over the total surface area of the entire signature image.

$$Ratio = \frac{Area\ of\ signature}{Area\ of\ bounding\ box} \qquad (4)$$

2.3   Feature Normalization

The extracted feature vector $\vec{v}$ of 28 dimensions is defined as follows:

$$\vec{v} = (v_0, \ v_1, \ ..., \ v_{27})\tag{5}$$

where $\vec{v}$ is composed of 3 trisurface, 6 sixfold-surface, 10 transitional, and 9 directional features. The scales of these individual features extracted are completely different. Firstly, this disparity can be due to the fact that each feature is computed using a formula that can produce various range of values. Secondly, the features may have the same approximate scale, but the distribution of their values has different means and standard deviations [12]. A statistical normalization technique is used to transform each feature in such a way that each transformed feature distribution has means equal to 0 and variance of 1.

Let n be the number of features and m the size of the distribution, features matrix X is defined in (6):

$$X = \begin{bmatrix} x_{11} & x_{12} & ... & x_{1n} \\ x_{21} & x_{22} & ... & x_{2n} \\ \vdots & & \ddots & \vdots \\ x_{m1} & x_{m2} & ... & x_{mn} \end{bmatrix}\tag{6}$$

where X_{ij} is the jth feature of the ith candidate for $(i = 1, 2, ..., m)$ and $(j = 1, 2, ..., n)$, the corresponding value is $newX_{ij}$ and is defined as (7):

$$newX_{ij} = \frac{X_{ij} - \overline{X}_j}{\sigma_j}\tag{7}$$

where $\overline{X}_j$ is the mean defined in equation (8) and σ_j is the standard deviation defined in equation (9)

$$\overline{X}_j = \frac{1}{m}\sum_{i=1}^{m} X_{ij}\tag{8}$$

$$\sigma_j = \sqrt{\frac{1}{m-1}\sum_{i=1}^{m}(X_{ij} - \overline{X}_j)^2}\tag{9}$$

2.4   Signature Verification

To verify the similarity of two signatures, Mahalanobis Distance (MD) based on correlations between signatures is used. It differs from Euclidean distance in that it takes into account the correlations of the data set and is scale-invariant. The smaller the MD, the higher the similarity of the compared signatures, and is defined in equation (10).

$$MD(\overrightarrow{x}, \overrightarrow{y}) = \sqrt{(\overrightarrow{x} - \overrightarrow{y})^T S^{-1} (\overrightarrow{x} - \overrightarrow{y})} \qquad (10)$$

where $\overrightarrow{x}$ and $\overrightarrow{y}$ denote the enrolled feature vector and the new signature feature vector to be verified, with the covariance matrix S.

3   Experimental Results and Discussions

Our experiments were carried out using two data sets: the GPDS signature database [10], and a database created from signatures captured using the ePadInk tablet. The results obtained were based on a total of 2106 signatures from GPDS database, and 400 signatures captured using ePadInk tablet. The 2106 signatures comprise of 39 individuals, 24 genuine signatures for each individual, and 39 samples of forgeries per each genuine signature.

Table 2 shows the results obtained by directional features alone, and the combination of directional features with one or more of the following features: transitional, sixfold and trifold. The overall verification rate is exceptionally positive.

Table 3 shows that the proposed approach using the text-based directional signature recognition algorithm outperforms the character-based directional signature recognition. The statistical normalization applied to the features as described in *section 2.3*, also enhanced the verification rate since it takes into account the correlations of the data set verified by Mahalanobis Distance.

Table 2. Verification Rates. *(D=Directional, T=Transitional, S=Sixfold, Tr=Trifold)*

	Verification Rate (%)		
	GPDS Database	ePadInk Signatures	Avg.
D	94.76	95.14	94.95
DT	95.28	97.43	96.36
DTS	93.77	98.01	95.89
DTSTr	90.79	94.14	92.47

Table 3. Comparison of Directional Feature Extraction Algorithms. *(O=Original, D=Directional, F=Feature, M=Modified, E=Enhanced)*

	Verification Rate (%)		
	GPDS Database	ePadInk Signatures	Avg.
ODF	83.65	80.02	81.84
MDF	84.57	85.50	85.04
EMDF	89.61	86.64	88.13
Our Approach	94.76	95.14	94.95

4 Conclusions

An effective approach for signature recognition and verification based on text recognition has been presented. The proposed algorithm improves the character-based directional feature extraction algorithm. Experimental results show that the proposed approach has an improved of true positive rate of 94.95%. Further investigation of the effect of the proposed approach with other different signature databases is envisioned.

Acknowledgments

Portions of the research in this paper use the *Grupo de Procesado Digital de Sennales* GPDS signature database [10] collected by the Universidad de Las Palmas de Gran Canaria, Spain.

References

1. Blumenstein, M., Liu, X.Y., Verma, B.: A Modified Direction Feature for Cursive Character Recognition. IEEE Proceedings, Neural Networks 4, 2983–2987 (2004)
2. Armand, S., et al.: Off-Line Signature Verification Using an Enhanced Modified Direction Feature with Single and Multi-classifier Approaches. IEEE Computational Intelligence Magazine 2, 18–25 (2007)
3. Özgündüz, E., entürk, T., Karslýgil, M.E.: Off-Line Signature Verification and Recognition by Support Vector Machine. In: Eusipco Proceedings (2005)
4. Nguyen, V.: Off-Line Signature Verification Using Enhanced Modified Direction Features in Conjuction with Neural Classifiers and Support Vector Machines. IEEE Proceedings, Document Analysis and Recognition 2, 734–738 (2007)
5. Blumenstein, M., Verma, B., Basli, H.: A Novel Feature Extraction Technique for the Recognition of Segmented Handwritten Characters. In: Proceedings of the IEEE Conference on Document Analysis and Recognition, vol. 1, pp. 137–141 (2003)
6. Justino, E.J.R., Bortolozzi, F., Sabourin, R.: Off-Line Signature Verification using HMM for Random, Simple and Skilled Forgeries. In: International Conference on Document Analysis an Recognition, vol. 1, pp. 169–181 (2001)
7. Zhang, B., Fu, M., Yan, H.: Handwritten Signature Verification based on Neural 'Gas' Based Vector Quantization. In: IEEE International Joint Conference on Neural Networks, vol. 2, pp. 1862–1864 (1998)
8. Martinez, L.E., et al.: Parameterization of a Forgery Handwritten Signature Verification System using SVM. In: IEEE Proceedings: Security Technology, pp. 193–196 (2004)
9. Plamondon, R., Srihari, N.: On-Line and Off-Line Handwriting Recognition: A Comprehensive Survey. IEEE Transactions on Pattern Analysis and Machine Intelligence 22(1), 63–83 (2000)
10. GPDS Signature Database, http://www.gpds.ulpgc.es/download/index.htm
11. Armand, S., et al.: Off-Line Signature Verification Using the Enhanced Modified Direction Feature and Neural-based Classification. IEEE Proceedings, Neural Networks, 684–691 (2006)
12. Lin, M.W., et al.: A Texture-based Method for Document Segmentation and Classification. South African Computer Journal 36, 49–56 (2006)

Effect of Image Linearization on Normalized Compression Distance

Jonathan Mortensen[1], Jia Jie Wu[2], Jacob Furst[3],
John Rogers[3], and Daniela Raicu[3]

[1] Case Western Reserve University
[2] University California, San Diego
[3] DePaul University
Jonathan.Mortensen@case.edu,
jjw017@ucsd.edu,
jfurst@depaul.edu,
draicu@depaul.edu
http://facweb.cti.depaul.edu/research/vc/

Abstract. Normalized Information Distance, based on Kolmogorov complexity, is an emerging metric for image similarity. It is approximated by the Normalized Compression Distance (NCD) which generates the relative distance between two strings by using standard compression algorithms to compare linear strings of information. This relative distance quantifies the degree of similarity between the two objects. NCD has been shown to measure similarity effectively on information which is already a string: genomic string comparisons have created accurate phylogeny trees and NCD has also been used to classify music. Currently, to find a similarity measure using NCD for images, the images must first be linearized into a string, and then compared. To understand how linearization of a 2D image affects the similarity measure, we perform four types of linearization on a subset of the Corel image database and compare each for a variety of image transformations. Our experiment shows that different linearization techniques produce statistically significant differences in NCD for identical spatial transformations.

Keywords: Kolmogorov Complexity, Image Similarity, Linearization, Normalized Compression Distance, Normalized Information Distance.

1   Introduction

First proposed by M. Li et al., Normalized Compression Distance (NCD), a derivation of Kolmogorov complexity $(K(x))$, quantifies object similarity by calculating the compressed sizes of two strings and their concatenation [10]. NCD has been applied as a measure for image similarity because it is a universal distance measure. A successful model for image similarity will have applications in fields such as medical image retrieval, security, and intellectual property. This paper analyzes some of NCD's underlying assumptions and their effects on its resulting similarity measure between images.

D. Ślęzak et al. (Eds.): SIP 2009, CCIS 61, pp. 106–116, 2009.

In [10], [11], [4], normalized compression distance (NCD) has been shown to yield promising results in constructing phylogeny trees, detecting plagiarism, clustering music, and performing handwriting recognition. These results demonstrate the generality and robustness of NCD when comparing one dimensional data. The simplicity of NCD presents an automatic way of grouping related objects without the complicated task of feature selection and segmentation.

In [7] the compressed size of the concatenation of x and y is used to estimate $K(xy)$ as well as $K(yx)$. Compressors search for sequences shared between x and y in order to reduce the redundancy in the concatenated sequences. Therefore, if the result of this compression is much smaller than the compression of x and y separately, it implies that much of the information contained in x can be also used to describe y.

To approximate the NCD between two images, image linearization, accompanied with a variety of compression algorithms, has been applied[7]. Kolmogorov complexity is not computable but it can be approximated by using compression[10]. Image compression is the notion that given a group of neighboring pixels with the same properties, it is more efficient to use a single description for the entire region. Thus, if image x is more complex than image y, the Kolmogorov complexity of x, which is approximated by the size of compressed x, will be larger than that of y. With lossy compression, a group of pixels can be replaced with an average color to create a more compact description of the region, at the expense of slight distortions[7]. To make the best approximation of Kolmogorov complexity, the ŞbestŤ possible compression must be used. The approximation of Kolmogorov complexity is limited by the fact that it is not possible to design the ŞbestŤ system of image compression: whichever compression we use, there is always a possibility for improvement.

NCD determines image similarity based on the theory that the visual content of an image is encoded into its raw data. Theoretically, this makes NCD a suitable metric for Content-Based Image Retrieval (CBIR) which attempts to find similar images based on a query image's content. The application of NCD for CBIR in [7] has shown to produce statistically significant dissimilarity measures when tested against a null hypothesis of random retrieval. The NCD between images was used as a metric to search the visual content encoded in the raw data directly, thus bypassing feature selection and weighting. This approach performed well even when compared against several feature based methods. Although the approach in [7] uses a variety of real-world data sets, different compressors were used for each experiment and the method of linearization was unclear. In particular, the ability of a string to represent a 2D image is not clear. This paper expects to determine the effects of different methods of linearization on NCD.

Similarly, [13] investigates the parameters of visual similarity and verifies that NCD can be used as a predictor for image similarity as experienced by the human visual system. While NCD performs well as a model for similarities involving addition or subtraction of parts, it fails to determine similarity among objects that involve distortions involving form, material, and structure. Results also imply that when two images are very similar, this similarity may be better approximated by other similarity measures, such as the pixel-by-pixel difference of two images. Although [13] determined that the use of different compression algorithms and transformations has varying effects on NCD, only one photograph and one type of linearization (Row-Major) were used for NCD calculations. Therefore, it is unclear whether the results are universally applicable to images of different content. We present an experiment to illustrate the effects of different methods of linearization and transformations on a database of 100 images.

Due to the compression approximation of Kolmogorov complexity, all of the above examples create a linear string from a 2D image and then find its relative similarity to other strings and therefore other images. Each string was created with a linearization technique; however, an effective formal analysis was not conducted to evaluate the impact of the linearization method. Each linearization produces a distinctly different string and it is important to understand how the 2D signal (image) is converted to a 1D string. Thus, a question is still left standing: Can a string effectively represent a 2D image? This paper explores four different linearization techniques and their impact on the similarity measure of spatially transformed images, and tests the null hypothesis that all linearizations result in the same NCD across several transformations. Section 2 describes the basis of Kolmogorov Complexity, presents the methodology behind the four linearization techniques: Row-Major, Column-Major, Hilbert-Peano, and Self-Describing Context Based Pixel Ordering (SCPO) and presents the methodology of producing the dataset; Section 3 shows results of measuring image similarity distance between an image and a spatially transformation version of an image; and Section 4 contains discussion along with comments on future work.

2   Methodology

2.1   Kolmogorov Complexity and Derivations

To understand the necessity of linearization, a short derivation of Kolmogorov complexity is helpful. Kolmogorov complexity describes the smallest number of bits used to represent an object x [12]. $K(x)$ is the length of the shortest program, denoted string x*, to produce x. Therefore, x* is the smallest representation of x. Furthermore, conditional Kolmogorov complexity, $K(x|y)$, is the length of the shortest program to compute x given y. The conditional complexity of two objects begins the notion of similarity. Building upon conditional complexity, information distance, $E(x, y)$, is the $max\{K(x|y), K(y|x)\}$ and describes the absolute bit change needed to convert one object into another. $E(x, y)$ is not normalized. Normalization provides the Normalized Information Distance (NID),

$$NID(x, y) = \frac{\max\{K(y|x*), K(x|y*)\}}{\max\{K(y), K(x)\}} .$$ (1)

NID describes the theoretical conception of object similarity. It is shown by [10] that this measure is universal, in that it captures all other semi-computable normalized distance measures. However, because this measure is also upper semi-computable, the complexity of an object, $K(x)$, is approximated using modern compression algorithms, denoted by $C(x)$. This is referred to as the Normalized Compression Distance,

$$NCD(x, y) = \frac{C(xy) - \min\{C(x), C(y)\}}{\max\{C(x), C(y)\}}. \tag{2}$$

This is the standard formula applied to most applications of Kolmogorov complexity similarity measures and also the basic algorithm used in the CompLearn toolkit[2]. Because this approximation uses compression, and compression techniques currently depend on string inputs, the images must be transformed into linear strings [10]. We present four common linearization methods.

2.2   Linearization Methods

The scan-line is a standard scanning method that traverses an image line by line. There are two main scan-line methods: Row-Major and Column-Major. Row-Major concatenates pixel intensities to a string, row by row, starting with the upper left pixel and then continuing across the row, before proceeding to the next row. Column-Major follows much the same, but begins in lower left pixel and continues up the first column, before moving toward the right, proceeding column by column. There are variations to this, but all are in a linear fashion.

The Hilbert-Peano curve traverses all pixels in a quadrant of an image before it linearizes the next quadrant and as a result, this method of linearization has an inherently strong locality property [9]. The Hilbert-Peano space-filing curve guides the exploration of a two dimensional array and linearizes it into a continuous one dimensional string that contains information from all pixels in the plane while respecting neighborhood properties. As described by [9], a Hilbert-Peano curve subdivides a unit square into four sub-parts and after a finite number of well adapted iterations, every pixel is captured by the Hilbert-Peano curve. The Hilbert-Peano Curve can be approximated by self similar polygons and grows recursively following the same rotation and reflection at each vertex of the curve. The self similarity of polygons allows efficient computation of discrete versions of curves while preserving locality. Therefore, a search along a space-filling linearization will result in points that are also close in the original space. This type of linearization is simple and requires no contextual knowledge of the data set. In [5], the fixed space-filling curve approach has been shown to use much less computational resources compared to clustering and other context-based image linearizations. The entropy of a pixel-sequence obtained by Hilbert-Peano curves converges asymptotically to the two-dimensional entropy of the image, resulting in a compression scheme that is optimal with encoders that take advantage of redundant data [6].

[6] also proposes the use of a context-based space filling curve to exploit the inherent coherence in an image. Ideally, an adaptive context-based curve would find and merge multiple Hamiltonian circuits to generate a context-based Hamiltonian path that would traverse every pixel. [8] proposes a self-describing context-based pixel ordering for images that uniquely adapts to each image at hand and is inherently reversible. SCPO uses pixel values to guide the exploration

of the two-dimensional image space, in contrast to universal scans where the traversal is based solely on the pixel position [8]. SCPO incrementally explores the region by exploring sections with pixel intensities most similar to a current pixel and by maintaining a frontier of pixels that has already been explored. Neighboring pixels around the current pixel are added to the frontier and are concatenated to a string. Then, the pixel in the frontier with the closest intensity to the starting point is chosen as the next point about which to explore and is also removed from the frontier. The outcome is a one-dimensional representation of an image with enhanced autocorrelation. Empirical results in [8] show that this method of linearization, on average, improves the compression rate by 11.56% and 5.23% compared to raster-scans and Hilbert-Peano space-filling curve scans, respectively.

Fig. 1. Linearization of four images. Beginning in upper left, Row-Major, Column-Major, Hilbert, SCPO.

2.3   Null Hypothesis Test

To gain an understanding of the effects of linearization on an image, a similarity test is chosen in which theoretically, linearization should not affect results. Although different methods of linearizations produce distinct strings, identical spatial transformations to the image would theoretically result in the same NCD. In this experiment, a battery of spatial transformations is applied to a large standard image library. The resulting image transformations are then compared to the original and the relative similarity distance is calculated. This process is done for each of the described linearization techniques from Section 2.2. Our null hypothesis, Eq. 3, states that each linearization technique produces the same relative distance for each transformation. This would demonstrate that linearization can effectively represent a 2D image.

Our null hypothesis H_o states that different types of linearizations will produce the same NCD values across identical spatial transformations of images. Our alternative hypothesis states that at least one of the linearizations will produce a different NCD across identical spatial transformations. Formally, the null hypothesis and alternative hypothesis are presented as follows:

$$H_o : \forall s, t \; NCD_s(x, y) = NCD_t(x, y) \; s \neq t \tag{3}$$

$$H_a : \exists s, t \; NCD_s(x, y) \neq NCD_t(x, y) \; s \neq t \tag{4}$$

where s,t denote any of the 4 different linearizations, x denotes the original image and y denotes one of the 7 spatial transformation comparisons.

2.4 Dataset and Transformations

In this experiment, we selected 100 images from the Corel image database[14] and converted the color images to grayscale by calculating a weighted sum of R, G, and B components using the following equation:

$$I = 0.2989 \times R + 0.5870 \times G + 0.1140 \times B \tag{5}$$

To measure the effect of linearizations on different spatial transformations, we employed 7 different types of transformations. The original bitmap image has 8 different versions: original, left shifted, down shifted, 90°, 180°, 270° rotation clockwise, and reflections across the x and y axis. All transformations were automated for 100 original images in Photoshop 7.0, thus creating an experimental test set of 100 originals x 8 versions = 800 images. Shifted versions were translated 35 pixels. It should be noted that the empty space created by these shifts were replaced by white (255, 255, 255) pixels in Photoshop 7.0.

Using Matlab, we linearized the 800 bitmap images into text files by extracting gray level intensities from each pixel in four different fashions. Row-Major (from left to right) and Column-Major (from bottom to the top) linearizations were performed by concatenating pixel intensities to a string. Column-Major linearization, the image is flipped 90° clockwise and then linearized using the Row-Major methodology. The image is also linearized using a Hilbert-Peano space-filling curve which grows incrementally, but is always bounded by a square with finite area. The mathmatical deconstruction of the Hilbert-Peano curve can be found in [9]. For each image, a position p(j) is computed along each step in a Hilbert-Peano traversal of space. Then, our algorithm incrementally reads gray-level values from an image at these coordinates. Thus, in order for a Hilbert-Peano space-filling curve to traverse every pixel within the image, the image is resized to 128x128 pixels, potentially distorting the image. Similarly, SCPO images are reduced to 35% of its size, preserving the aspect ratio, in order to expedite the computationally-heavy SCPO process. All distorted and reduced images were compared to images with the same distortion, removing any comparison problems. In total, 800 images x 4 linearizations = 3200 text files were compared within their respective linearization groups.

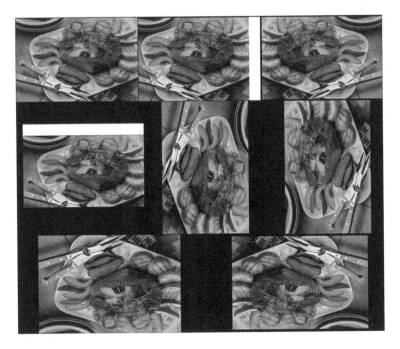

Fig. 2. Spatial Transforms of an image. From left to right: Original, Left-Shift, Ref-Y, Down-Shift, 90°, 270°, Ref-X,180°.

2.5   CompLearn

For this experiment, we used NCD(x,y) and a block sorting compression algorithm (bzip2) built into Complearn, a data analysis toolkit that provides relative compression ratios. Bzip2 exploits frequently recurring character sequences to compress data and detects patterns within a 900 kilobyte window [1]. Since this compression algorithm takes advantage of redundancy in an image to shrink its representation, it approximates the semicomputable Kolmogorov complexity of a string. The NCD(x,y) between two images is calculated using Complearn. Complearn takes several text files and compresses them, noting the bit length each and applies Eq. 2 to find the NCD between two images. Complearn ultimately creates a data set that reflects normalized compression distances between all permutations of two files. The smaller the difference or the closer the NCD is to 0, the more similar the text files. Likewise, the less redundancy between text files, the closer the NCD is to 1. Although a majority of the distances collected were between 0 and 1, some distances were slightly above 1. In [10], [4], and [3] this is not uncommon; when comparing files that share very little information, sometimes NCD can reach values greater than 1. Theoretically, this can be explained by the region in which $C(xy) - min\{c(x), c(y)\}$ is greater than $max\{c(x), c(y)\}$.

The NCD between JPEG compressed images was also calculated to provide a reference for the bzip2 compression algorithm. To measure the NCD between

JPEGs, the Complearn toolkit was not used as JPEG compression is not integrated. Instead, Eq. 2 was determined with bc (Linux). First, two images were concatenated side by side in Matlab. Next, the concatenated image and the two originals were compressed to JPEG losslessly in Matlab with 100% quality. To find $NCD(x, y)$, bc (Linux) was used with the file bit lengths of the resulting files as inputs to Eq. 2. Comparisons between 90° and 270° rotations could not be made because two images of differing dimensions could not be concatenated.

2.6 Statistical Analysis

The NCD between each of the transformations and the original image were averaged for 100 images to find the overall effect of each spatial transformation on NCD. To determine if different linearizations produced statistically different NCD values, an analysis of variance (ANOVA) test was performed. ANOVA is a well-known statistical test that compares mean square differences between groups and within groups to determine whether groups of data are statistically different from each other.

3   Results

The results are presented in Table 1, which shows the average NCD for each linearization from the original image to each transformation. This was done for 100 different images chosen from the food subset located in the Corel Image Database[14]. There are several interesting averages to note. Although Row and Column-Major linearization could not easily recognize the transformed image when it is rotated, these methods of linearization produced significantly lower NCDs when the original image was compared to the down shifted, left shifted image. In addition, Hilbert-Peano and SCPO linearizations produced NCDs consistently over 0.96 when comparing the original image to its shifted copies. With respect to transformations across the y and x axis, for Row-Major linearization, when the original was compared to itself reflected across the x axis, the mean NCD was 0.383669. For Column-Major linearization, when the original was compared to its itself reflected across the y axis, the mean NCD was 0.382232 . For 90° and 270° rotations, Hilbert-Peano linearizations produced the lowest average NCDs at 0.949334 and 0.935854 respectively, while SCPO produced average NCDs around 0.96 consistently across all transformations. For JPEG compression, nearly all NCD values are over 1, which indicated that JPEG compression found little or no redundancies between the concatenation of two images.

The ANOVA test results are shown in Table 2. The standard p-value threshold of 0.05 was used to test if the type of linearization significantly affected the NCD measure across transformations. ANOVA results show that different types of linearizations produce statistically significant differences in NCD for identical spatial transformations. Therefore, we can reject the null hypothesis that different types of linearizations will produce the same NCD values for spatial

Table 1. Mean Normalized Compression Distance (NCD)

Transform	Row	Column	Hilbert	SCPO	JPEG
Original	0	0	0	0	1.0011
Down-Shift	0.4661	0.4811	0.9697	0.9629	1.0072
Left-Shift	0.4676	0.4594	0.9683	0.9607	1.0050
90°	0.9726	0.9725	0.9493	0.9640	N/A
180°	0.9634	0.9698	0.9664	0.9658	1.0011
270°	0.9727	0.9726	0.9358	0.9630	N/A
Ref-X	0.3836	0.9697	0.9638	0.9640	1.0013
Ref-Y	0.963	0.3822	0.9353	0.9625	0.9976

Table 2. ANOVA Normalized Compression Distance (NCD)

Transform	Grouping	Sum of Sq.	df	Mean Square	F	Sig.
	Between Groups	24.29344	3	8.09781	13063.1	0*
Down-Shift	Within Groups	0.24547	396	0.0006198		
	Total	24.5389	399			
	Between Groups	25.1020	3	8.36735	14643.3	0*
Left-Shift	Within Groups	0.226278	396	0.0005714		
	Total	25.32834	399			
	Between Groups	0.036090	3	0.01203	49.1997	4.78458E-27
90°	Within Groups	0.096828	396	0.0002445		
	Total	0.132918	399			
	Between Groups	0.002070	3	0.0006	7.51982	6.62437E-05
180°	Within Groups	0.036336	396	9.17588E-05		
	Total	0.038406	399			
	Between Groups	0.090856	3	0.03028	128.202	4.90716E-58
270°	Within Groups	0.093547	396	0.0002362		
	Total	0.184404	399			
	Between Groups	25.42660	3	8.475535	55040.8	0*
Ref-X	Within Groups	0.060978	396	0.0001539		
	Total	25.48758	399			
	Between Groups	24.561154	3	8.187051	27445.8	0*
Ref-Y	Within Groups	0.118125	396	0.0002982		
	Total	24.67928	399			

*Approximates to 0

transformations. This suggests that images may not be fully expressible as a string, at least using current compression algorithms. It certainly indicates that the method of linearization does matter.

4   Discussion and Conclusion

It is shown that linearization techniques affect the measured relative distance between two images. Nearby pixels in one linearization may not be near to the same pixel in another. Thus, when applying Kolmogorov complexity as a similarity metric, careful consideration of linearization technique must used, a topic which has not been explained or explored. Also, our data suggest that the use of multiple linearizations for one comparison pair may be effective. Using Row-Major and Column-Major linearization to calculate NCD may be better at capturing similar information between shifted copies. Our results are consistent with [13], which show that Row-Major calculations of NCD can easily capitalize on shared information among shifted copies. Additionally, Row-Major linearization may best recognize images that share characteristics across the x axis while Column-Major linearization may best capture likeness among images similar across the y axis.

In addition, Hilbert-Peano and SCPO linearizations produced NCDs consistently over 0.96 when comparing the original to its shifted copies, showing that these types of linearizations may not easily capture similarity among images that are strongly shifted. Statistically significant values of NCD created by Hilbert-Peano linearization show that Hilbert-Peano linearization may capture similarity among rotated copies better than other forms of linearizations in this experiment. Although SCPO has been shown to enhance the autocorrelation and compression ratio within a single image[8], in this study it does not seem to effectively capture similarities between two images. Other conversion methods may need to be sought out to find types of linearizations that are robust to spatial transformations.

Also, JPEG compression, a standard image compression algorithm, produces large values of NCD, indicating dissimilarity, and is shown in this study to be ineffective for finding image similarity. Image compression algorithms may not yet approximate Kolmogorov complexity and future work for image compression algorithms may need to be done to bypass the potentially flawed linearization process. Nonetheless, because linearization affects the NCD, clearly a string does not fully or totally represent a 2D image. More consideration on this topic must be taken.

4.1   Future Work

Further investigation will include linearization's effect on similarity distance with regard to intensity transformations, such as intensity shifts, introduction of noise, and watermarking. The degree of a transformational change and its correlation to the degree of NCD change will also be measured. Additionally, compression algorithms need to be more rigorously compared, and as [13] demonstrated, there is a qualitative difference between the performance of different compression algorithms. [2] also mentions file size limitations to compression algorithms that may limit the sizes of files compared. We expect that efficiently linearizing an image would lead to greater compression ratio, which in turn would lead to more meaningful values of NCD if similarity exists between two images. We also expect that certain linearizations will generate NCDs that are more consistent with similarity measured by the human visual system and thus be more robust to spatial transformations. Furthermore, linearizing an image into a one dimensional string may not be the best method to represent an image. To accurately approximate the Normalized Information Distance between two images, other forms of image compression will need to be investigated.

Acknowledgments. The research in this paper was supported by NSF award IIS-0755407.

References

1. Cilibrasi, R.: Statistical Inference Through Data Compression. PhD thesis, Universiteit van Amsterdam (2007)
2. Cilibrasi, R., Cruz, A.L., de Rooij, S., Keijzer, M.: CompLearn home, http://www.complearn.org/
3. Cilibrasi, R., Vitanyi, P., de Wolf, R.: Algorithmic clustering of music. Arxiv preprint cs.SD/0303025 (2003)
4. Cilibrasi, R., Vitanyi, P.M.B.: Clustering by compression. IEEE Transactions on Information theory 51(4), 1523–1545 (2005)
5. Yeo, B.L., Yeung, M., Craver, S.: Multi-Linearization data structure for image browsing. In: Storage and Retrieval for Image and Video Databases VII, San Jose, California, January 26-29, p. 155 (1998)
6. Dafner, R., Cohen-Or, D., Matias, Y.: Context-based space filling curves. In: Computer Graphics Forum, vol. 19, pp. 209–218. Blackwell Publishers Ltd., Malden (2000)
7. Gondra, I., Heisterkamp, D.R.: Content-based image retrieval with the normalized information distance. Computer Vision and Image Understanding 111(2), 219–228 (2008)
8. Itani, A., Manohar, D.: Self-Describing Context-Based pixel ordering. LNCS, pp. 124–134 (2002)
9. Lamarque, C.H., Robert, F.: Image analysis using space-filling curves and 1D wavelet bases. Pattern Recognition 29(8), 1309–1322 (1996)
10. Li, M., Chen, X., Li, X., Ma, B., Vitanyi, P.M.B.: The similarity metric. IEEE Transactions on Information Theory 50, 12 (2004)
11. Li, M., Sleep, R.: Melody classification using a similarity metric based on kolmogorov complexity. In: Sound and Music Computing (2004)
12. Li, M., Vitanyi, P.: An Introduction to Kolmogorov Complexity and its Applications, 1st edn. Springer, New York (1993)
13. Tran, N.: The normalized compression distance and image distinguishability. In: Proceedings of SPIE, vol. 6492, p. 64921D (2007)
14. Pennsylvania State University. Corel Image Database, http://wang.ist.psu.edu/docs/related/

A Comparative Study of Blind Speech Separation Using Subspace Methods and Higher Order Statistics

Yasmina Benabderrahmane[1], Sid Ahmed Selouani[2],
Douglas O'Shaughnessy[1], and Habib Hamam[2]

[1] Institut National de la Recherche Scientifique-Energie, Matériaux,
Télécommunications, University of Quebec
800 de la Gauchetière O, H5A 1K6, Montréal, Qc, Canada
{benab,dougo}@emt.inrs.ca
[2] Université de Moncton, campus de Shippagan,
E8S 1P6 NB, Canada
selouani@umcs.ca, habib.hamam@umoncton.ca

Abstract. In this paper we report the results of a comparative study on blind speech signal separation approaches. Three algorithms, Oriented Principal Component Analysis (OPCA), High Order Statistics (HOS), and Fast Independent Component Analysis (Fast-ICA), are objectively compared in terms of signal-to-interference ratio criteria. The results of experiments carried out using the TIMIT and AURORA speech databases show that OPCA outperforms the other techniques. It turns out that OPCA can be used for blindly separating temporal signals from their linear mixtures without need for a pre-whitening step.

Keywords: Blind source separation, speech signals, second-order statistics, Oriented Principal Component Analysis.

1 Introduction

During recent decades, much attention has been given to the separation of mixed sources, in particular for the *blind* case where both the sources and the mixing process are unknown and only recordings of the mixtures are available. In several situations, it is desirable to recover all sources from the recorded mixtures, or at least to segregate a particular source. Furthermore, it may be useful to identify the mixing process itself to reveal information about the physical mixing system.

The objective of Blind Source Separation (BSS) is to extract the original source signals from their mixtures and possibly to estimate the unknown mixing channel using only the information of the observed signal with no, or very limited, knowledge about the source signals and the mixing channel. BSS techniques can be typically divided into methods using second-order [1] or higher-order statistics [2], the Kullback-Liebler distance [3], PCA methods, non-linear PCA [4] and ICA methods [5]. Most approaches to BSS assume the sources are statistically independent and thus often seek solutions with separation criteria using higher order statistical information

D. Ślęzak et al. (Eds.): SIP 2009, CCIS 61, pp. 117–124, 2009.

[2] or using only second-order statistical information in the case where the sources have temporal coherency, are non-stationary [3], or eventually are cyclo-stationary. It is worth noting that second-order methods do not actually replace higher-order ones since each approach is based on different assumptions. For example, second-order methods assume that the sources are temporally coloured, whereas higher-order methods assume white sources. Another difference is that higher-order methods do not apply to Gaussian signals but second-order methods do not have any such constraint.

This paper is organized as follows: in Section 2, we briefly present the Independent Component Analysis (ICA) model. In Section 3, we present the Higher Order Statistics (HOS) approach we have developed for blind speech signal separation. In Section 4, we describe the implementation of the OPCA method that we propose for the separation of mixed speech signals. Section 5 presents the experimental results and discusses them. Finally, Section 6 concludes and gives a perspective of our work.

2 ICA Model

We simultaneously record N conversations with N microphones. Each recording is a superposition of all conversations. The problem lies in isolating each speech signal to understand what was said. One can solve this problem merely by considering that different people produce independent speech signals. It does not take into account the semantics of these signals or even acoustics, but considers them as random signals, statistically independent. ICA is one tool for this, whose goal is to reveal factors that are independent in the full sense and not only in the sense of decorrelation. The model is usually stated as follows.

Let X denote the random vector whose elements are the mixtures $x_1,...,x_n$, and likewise by S the random vector with elements $s_1,..., s_n$. Let us denote by A a matrix with elements a_{ij}. Using vector-matrix notation, the above mixing model becomes:

$$X = AS , \tag{1}$$

where A is a square invertible matrix and S is a latent random variable whose components are mutually independent. One tries to estimate A to obtain $S=\{s_1,.....,s_n\}$, given a set $\{x_1,...,x_n\}$, randomly distributed according to the law of X, which consists in the fact that x_i is independent of x_j for $j \neq i$.

The problem of BSS is now to estimate both the source signals $s_i(t)$ and the mixing matrix A, based on observations of the $x_i(t)$ alone. In most methods, the source signals are assumed statistically independent. First, we assume all the components (except perhaps one) have non-Gaussian distributions. Second, if the components are non-stationary, smoothly changing variances and the model can be estimated as well. Therefore, ICA only considers linear superposition of independent signals, resulting from the mixing with A. This is often a legitimate restriction. For instance, transmission systems are linear media where signals act as if they were present independently of each other. For simplicity, we are assuming that the unknown mixing matrix is square. Then, after estimating the matrix A, we can compute its inverse, say W, and obtain the independent component merely by:

$$S = WX .\tag{2}$$

A very effective ICA algorithm, called Fast-ICA, was developed in 1997 by Oja and Hyvarinen [5]. It is also the most commonly used ICA algorithm for the separation of instantaneous mixtures of sources. In this paper, Fast-ICA is used as a reference to perform objective and qualitative assessments.

3   Blind Speech Separation Using High-Order Statistics

We have previously proposed this approach in the biomedical field [6]. The goal was to separate real Electrocardiological signals from power interference and other artefacts. Then, we implemented it for separation of two speech signals [7].

Consider an instantaneous mixture of a two-speaker-two-sensor system with the mixing matrix $A = [a_{11}\ a_{12}; a_{21}\ a_{22}]$. ICA works for estimating the desired de-mixing Matrix $W = [w_{11}\ w_{12}; w_{21}\ w_{22}]$, such that the separated outputs $\hat{s}_i$ represent two separated signals that are as independent from each other as possible.

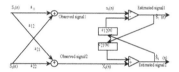

Fig. 1. Block diagram of two-source separation

In this case, the equations of the mixture at the moment n are:

$$x_1(n) = a_{11}s_1(n) + a_{12}s_2(n) ,\tag{3}$$

$$x_2(n) = a_{21}s_1(n) + a_{22}s_2(n) ,\tag{4}$$

and the output equations are:

$$\hat{s}_1(n) = x_1(n) - c_{12}\hat{s}_2(n) ,\tag{5}$$

$$\hat{s}_{12}(n) = x_2(n) - c_{21}\hat{s}_1(n) .\tag{6}$$

However, both sources s_i and mixture coefficients are unknown; from relations of $\hat{s}_1$ and $\hat{s}_2$, we cannot find the solution. These coefficients c_{ij} must therefore be adjusted by an adaptive algorithm. The coefficients c_{ij}, at each time index n, are adjusted by a rule having the following form:

$$c_{ij}(n+1) = c_{ij}(n) + \mu f(s_i(n))g(s_j(n)) ,\tag{7}$$

where the two functions f and g are different, non-linear and odd, and μ is the gain of adaptation. In our method, we choose: $f = s_i^3(n)$ and $g = s_j(n)$, $\mu = 0.9$, $c_{ij}(0) = 0$ and we get the rule adaptation as follows:

$$c_{ij}(n+1) = c_{ij}(n) + \mu . s_i^{3}(n) . s_j(n) . \tag{8}$$

4 Separation of Mixed Speech Signals by OPCA

This algorithm was previously proposed by K. I. Diamantaras and al. [8], specifically for separating four multilevel PAM (Pulse Amplitude Modulation) signals filtered by an ARMA (Auto-Regressive Moving Average) coloring filter. As a generalization of PCA, OPCA corresponds to the generalized eigenvalue decomposition of a pair of covariance matrices in the same way that PCA corresponds to the eigenvalue decomposition of a single covariance matrix. Oriented PCA describes an extension of PCA involving two signals $u(k)$ and $v(k)$. The aim is to identify the so-called oriented principal directions $e_1., ...,.e_n$, that maximize the signal-to-signal power ratio $E(e_i^T u)^2/E(e_i^T v)^2$ under the orthogonality constraint: $e_i^T R_u e_j=0$, $i \neq j$. OPCA is a second-order statistics method, which is reduced to standard PCA if the second signal is spatially white $(R_v = I)$. The solution of OPCA is shown to be the Generalized Eigenvalue Decomposition of the matrix pencil $[R_u, R_v]$. Subsequently, we shall relate the instantaneous BSS problem with the OPCA of the observed signal x and almost any filtered version of it. Some assumptions must be done as sources are pair-wise uncorrelated, with zero mean and unit variance and there exist positive time lags $l_1,....,\ l_M$ such that:

$$R_S(l_M) = diagonal \neq 0, \tag{9}$$

where R_s represents the covariance matrix. Note that the 0-lag covariance matrix of $x(k)$ is:

$$R_x(0) = AR_s(0)A^T = AA^T . \tag{10}$$

Now, consider a scalar, linear temporal filter $h = [h_0,.....,h_M]$ operating on $x(k)$:

$$y(k) = \sum_{m=0}^{M} h_m x(k - l_m) . \tag{11}$$

The 0-lag covariance matrix of y is expressed as:

$$R_y(0) = E\left\{y(k)y(k)^T\right\} = \sum_{p,q}^{M} h_p h_q Rx(l_p - l_q) . \tag{12}$$

From Eq. (1) and Eq. (5) it results:

$$D = \sum_{p,q=0}^{M} h_p h_q R_s(l_p - l_q) . \tag{13}$$

Note that according to the previous assumptions, D is diagonal. Considering A square and invertible, we find:

$$R_x(0)A^{-T} = R_x(0)A^{-T}D. \tag{14}$$

Eq. (14) expresses a Generalized Eigenvalue Decomposition problem for the matrix pencil $[R_y(0), R_x(0)]$. This is equivalent to the OPCA problem for the pair of signals $[y(k), x(k)]$. The generalized eigenvalues for this problem are the diagonal elements of D. The columns of the matrix A^{-T} are the generalized eigenvectors. The eigenvectors are unique up to permutation and scale. Therefore, the eigenvalues are distinct. In this case, for any generalized eigenmatrix W, we have $W = A^{-T}P$ with P being a scaled permutation matrix. Then, the sources can be estimated as:

$$\hat{s}(k) = W^T x(k), \tag{15}$$

which can be written as:

$$\hat{s}(k) = P^T A^{-1} A s(k) = P^T s(k). \tag{16}$$

It yields that the estimated sources are equal to the true ones except for the (unobservable) arbitrary order and scale. Note that a standard PCA can be used instead of OPCA but a spatial pre-whitening (sphering) module should be added. The OPCA method has not been applied previously for separating speech signals.

5 Experiments and Results

The source and observation data dimension is $n=2$, and the mixing matrix is chosen randomly to be:

$$A = \begin{bmatrix} 0.6787 & 0.7431 \\ 0.7577 & 0.3922 \end{bmatrix}$$

In the following experiments the TIMIT database was used. The TIMIT corpus contains broadband recordings of a total of 6300 sentences, 10 sentences spoken by each of 630 speakers from 8 major dialect regions of the United States, each reading 10 phonetically rich sentences [9]. Some sentences of the TIMIT database were chosen to evaluate our BSS methods. We tested OPCA using a filter of order 3, as mentioned in [8]. The use of more correlation matrices increases the information input in the estimation process and then it will improve the separation quality. We mixed instantly two signals: $s_1(n)$ and $s_2(n)$ that are respectively uttered by a man and a woman.

The principle of the performance measure is to decompose a given estimate $\hat{s}(n)$ of a source $s_i(n)$ as the following sum:

$$\hat{s}(n) = s_{target}(n) + e_{int\,erf}(n), \tag{17}$$

where $s_{target}(n)$ is an allowed deformation of the target source $s_i(n)$ and $e_{interf}(n)$ is an allowed deformation of the sources that accounts for the interference of the unwanted sources. Given such decomposition, one can compute the performance criteria provided by the Signal-to-Interference-Ratio (SIR):

$$SIR = 10\log\frac{\left\|S_{t\arg et}\right\|^2}{\left\|e_{int\,erf}\right\|^2}. \tag{18}$$

To evaluate this approach in the instantaneous case, we compared it with the well-known Fast-ICA algorithm and HOS method.

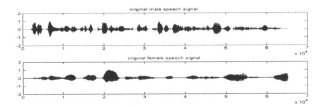

Fig. 2. Original signals, female sentence: "She had your dark suit in greasy wash water all year". Male sentence: "This brings us to the question of accreditation of art schools in general".

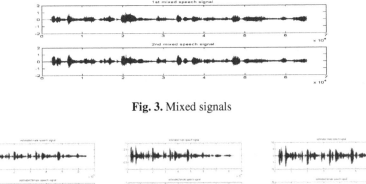

Fig. 3. Mixed signals

a)   b)   c)

Fig. 4. Estimated signals by: a) OPCA method, b) by HOS method, c) by Fast-ICA method

In comparison with HOS and Fast-ICA, the OPCA method is effective, as can be seen in the time domain, where we note that the original signals (Fig. 2) and estimated signals (Fig. 4) are very close. In Table 1, SIR_{in} and SIR_{out} represent respectively input Signal-to-Interference-Ratio before separation and output Signal-to-Interference-Ratio after separation, when S_{target} in eq. (18) is the original speech signal and the interference e_{interf}, the difference between the original and estimated speech signals.

As shown in Table 1, the ratio SIR_{out} of the OPCA method is larger than the SIR_{out} of the Fast-ICA and HOS algorithms. The improvement in the SIR_{out} ratio for male speech for example, was 6 dB for Fast-ICA and 12.45 dB for HOS and by 23 dB against the OPCA approach.

Note that Fast-ICA makes a good separation with a fast calculation, but the signals are corrupted by noise when the utterances are listened in the contexts of subjective evaluation only. In the OPCA method, that is fast too, no noise is perceptible. HOS was achieved with no fast calculation. For objective evaluation, HOS was better than Fast-ICA but not as good as OPCA, which we can see in Table 1.

Table 1. Comparison of SIR for the Fast-ICA, HOS and OPCA methods

SIR	SIR_{in}	SIR_{out} male speech	SIR_{out} female
Fast-ICA	-2.9 dB	3 dB	-3 dB
HOS	-2.9 dB	11 dB	8 dB
OPCA	-2.9 dB	20 dB	17.2 dB

A subjective method was carried out too by listening to the original signals, mixed and separated signals. A common subjective benchmark for quantifying the performance of the speech is the Mean Opinion Score (MOS) because voice quality in general is subjective to listeners. MOS is a subjective quality score that ranges from 1 (worst) to 5 (best). In our experiments, MOS tests were conducted with a group of seven listeners. The results for the three methods are close and they are summarised in Table 2.

Table 2. Comparison of MOS for the Fast-ICA, HOS and OPCA methods

Methods	Fast-ICA	HOS	OPCA
MOS	4.6	4.25	4.1

6 Conclusion

This paper reported the results of a comparative study of three blind speech separation methods of instantaneous mixtures, namely the ICA, HOS and OPCA methods. OPCA has the advantage that no pre-processing step is required. In addition OPCA is very effective to blindly estimate sources, as shown by the signal plots in both time and frequency domains. Subjective evaluation is performed through listening to estimated signals before and after mixing and after separation was used. The results are very satisfactory. OPCA has never been used before to separate speech signals from their linear mixture. We are continuing our research efforts by implementing OPCA in a mobile communication framework.

References

1. Belouchrani, S.A., Abed-Meraim, K., Cardoso, J.-F., Moulines, E.: A Blind Source Separation Technique Using Second-Order Statistics. IEEE Trans. Signal Processing 45(2), 434–444 (1997)

2. Cardoso, J.-F.: Source separation using higher order moments. In: Proc. IEEE ICASSP, Glasgow, U.K., vol. 4, pp. 2109–2112 (1989)
3. Pham, D.T., Cardoso, J.F.: Blind separation of instantaneous mixtures of non stationary sources. IEEE Transactions on Signal Processing 49(9), 1837–1848 (2001)
4. Karhunen, J., Joutsensalo, J.: Representation and separation of signals using nonlinear PCA type learning. Neural Networks 7, 113–127 (1994)
5. Hyvarinen, A., Oja, E.: A Fast Fixed-point Algorithm for Independent Component Analysis. Neural Computation 9(6), 1483–1492 (1997)
6. Benabderrahmane, Y., Novakov, E., Kosaï, R.: Méthodes de Traitement de Signaux Multidimensionnels Appliquées à l'Extraction de Micro-potentiels Electrocardiologiques, PhD thesis, Université Joseph Fourier de Grenoble I, France (1997)
7. Benabderrahmane, Y., Ben Salem, A., Selouani, S.-A., O'Shaughnessy, D.: Blind Speech Separation using High Order Statistics. In: Canadian Conference on Electrical and Computer Engineering, IEEE CCECE, St. John's, Newfoundland, Canada, May 3-6, pp. 670–673 (2009)
8. Diamantaras, K.I., Papadimitriou, T.: Oriented PCA and Blind Signal Separation. In: 4th International Symposium on Independent Component Analysis and Blind Signal Separation (ICA 2003), Nara, Japan, April 2003, pp. 609–613 (2003)
9. Fisher, W., Dodington, G., Goudie-Marshall, K.: The TIMIT-DARPA speech recognition research database: Specification and status. In: DARPA Workshop on Speech Recognition (1986)

Automatic Fatigue Detection of Drivers through Yawning Analysis

Tayyaba Azim[1], M. Arfan Jaffar[2], M. Ramzan[2], and Anwar M. Mirza[2]

[1] Institute of Management Sciences,
1-A, Sector E / 5, Phase – VII, Hayatabad, Peshawar, Pakistan
tayyaba.azim@imsciences.edu.pk
[2] FAST National University of Computer and Emerging Sciences,
A.K. Brohi Road, H11/4, Islamabad, Pakistan
{arfan.jaffar,muhammad.ramzan,anwar.m.mirza}@nu.edu.pk

Abstract. This paper presents a non-intrusive fatigue detection system based on the video analysis of drivers. The focus of the paper is on how to detect yawning which is an important cue for determining driver's fatigue. Initially, the face is located through Viola-Jones face detection method in a video frame. Then, a mouth window is extracted from the face region, in which lips are searched through spatial fuzzy c-means (s-FCM) clustering. The degree of mouth openness is extracted on the basis of mouth features, to determine driver's yawning state. If the yawning state of the driver persists for several consecutive frames, the system concludes that the driver is non-vigilant due to fatigue and is thus warned through an alarm. The system reinitializes when occlusion or misdetection occurs. Experiments were carried out using real data, recorded in day and night lighting conditions, and with users belonging to different race and gender.

Keywords: Driver Fatigue, Yawning State, Lip Detection, Mouth Openness.

1 Introduction

The increasing number of road accidents due to fatigue or low vigilance level has become a serious problem for the society. Statistics reveal that from 10% to 20% of all the traffic accidents are due to drivers with a diminished vigilance level [1]. This problem has increased the need of developing active safety systems that can prevent road accidents by warning the drivers of their poor driving condition. Drivers with a diminished vigilance level and fatigue suffer from a marked decline in their abilities of perception, recognition, vehicle control, and therefore pose serious danger to their own life and to the lives of other people. Fatigue of the driver is defined as a mental or physical exhaustion due to which he/she either feels drowsy or cannot respond to the situation. The prevention of such accidents is a major focus of effort in the field of active safety research.

Most of the previous research focuses on the collection of ocular parameters to detect the fatigue level of driver [1], [2], [3]. In most of them, the system initializes

D. Ślęzak et al. (Eds.): SIP 2009, CCIS 61, pp. 125–132, 2009.

with the detection of eyes to gather the ocular parameters like blinking frequency, PERCLOS, eye state etc. These systems may fail to predict the drivers true state if the eyes are not detected due to varying lighting conditions or vibrations experienced in real driving scenario. It happens because of their reliance on merely ocular parameters. Hence, it cannot be denied that an approach depending on different visual facial features will work more effectively than the one depending on just one feature. Yawning is another feature that helps in determining the fatigue level of driver. Yawning detection is intractable because of inter-person difference of appearance, variant illumination, and especially complex expression and widely varying pattern of mouth. Due to the space and time constraint, only the yawning part is described comprehensively in this paper.

This paper describes the idea of a non-intrusive system for monitoring fatigue of a driver based on yawning analysis. The yawning is measured by determining the mouth state of drivers. The computational overhead is reduced by decreasing the search area for the mouth and pupils. The mouth region is also detected by an improved Fuzzy C-Means clustering technique that deploys spatial as well as spectral information along with an optimal number of cluster statistics to segment the lips accurately. Other constraints imposed further increase the reliability of mouth region detection for yawning analysis.

The paper is arranged as follows: In Section 2, a review of previous studies in this line is presented. Section 3 describes the overall system architecture, explaining its main parts. The experimental results are shown in Section 4 and finally we present the conclusion and future studies in Section 5.

2 Literature Review

Different authors have adopted different procedures and combination of parameters to solve the vigilance and fatigue detection problem. The most accurate techniques are based on physiological measures like brain waves, heart rate, pulse rate, respiration, etc.[1].But these techniques are intrusive in nature as they require the drivers to wear sensors on their body. In comparison, the methods based on directly observable visual behaviors are less intrusive, but they face the challenge of choosing the right number and combination of parameters for accurate detection [1]. Notable work has been done in this stream by combining visual parameters like eye closure duration, blinking frequency, nodding frequency, face position, fixed gaze and yawning to monitor the fatigue and vigilance level [1], [2], [3], [4], [5]. The systems using right number and combination of parameters along with early detection power show more success. There also exists another hybrid approach for detecting fatigue which combines driver's state and vehicle's performance to predict the driver's fatigue [6]. Vehicle performance is judged through vehicle's lateral position, steering wheel movements and time-to-line crossings. Although these techniques are not intrusive, they are subject to several limitations such as vehicle type, driver experience, geometric characteristics, condition of the road, etc.

3 Proposed Approach

We present an algorithm that detects driver fatigue by analyzing changes in the mouth state. The method proposed can be divided into these phases: Image Acquisition, Face Detection and Tracking, Mouth Detection and Pupil Detection, Yawning Analysis/Mouth State Analysis and Driver State Analysis.

The outline of the proposed approach is shown in figure.1. In the following sections, each of the phases will be discussed in detail.

3.1 Image Acquisition

The purpose of this stage is to acquire videos of driver's face so that directly observable visual cues could be calculated for fatigue determination. The image acquisition system consists of a low-cost charged coupled device (CCD) micro camera, sensitive to near-infrared (NIR). The acquired images should be relatively invariant to changing light conditions and should facilitate pupil and mouth detection. This purpose is served by using near-IR illuminator with CCD camera [1].

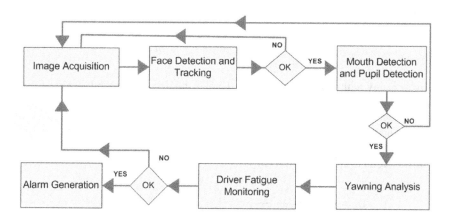

Fig. 1. Block diagram of fatigue detection system

Many of the eye trackers make use of bright pupil and dark pupil effect to calculate the difference image in which pupils are prominent [1], [2], [3]. We do not make use of the difference images; rather the data is acquired using NIR just to reduce the ambient light effects and to make the system working at night time too. The video data is collected in real driving scenarios and is obtained from [1].

3.2 Face Detection and Tracking

In order to collect the yawning information of driver, mouth should be located first. However, keeping in view the time complexity of mouth locating task in the whole frame; the area of mouth search is reduced to lower part of the face region only. This helps us in reducing the computational cost of the system, making it more suitable for use in real time. The face region is extracted using Viola and Jones real time face detection scheme based on AdaBoost algorithm [7]. AdaBoost face detection algorithm detects faces in a rapid and robust manner with a high detection rate of 15 frames per second. The face detector can detect faces that are tilted up to about ±15 degrees in plane and about ±45 degrees out of plane (toward a profile view). However, it becomes unreliable with rotation more than this. The system could be modified in future such that the camera gets rotated to get a frontal view of the driver if rotation of more than 15 degrees in plane and 45 degrees out of plane takes place.

3.3 Mouth Detection and Pupil Detection

The mouth region is extracted from the lower most part of the face bounding box to boost up the speed of fatigue detection system. After the mouth region is extracted as a preprocessing step, it is passed to the FCM module for lips segmentation. The block diagram of the mouth detection phase is shown in figure.3, whereas the functionality of FCM module is illustrated in figure.4.

Fuzzy c-means (FCM) is a method of clustering which allows one piece of data to belong to two or more clusters. It is based on minimization of the following objective function:

$$J_m = \sum_{i=1}^{N} \sum_{j=1}^{C} \mu_{ij}^m \parallel x_i - c_j \parallel^2 , 1 \le m > \infty \tag{1}$$

where m is any real number greater than 1, u_{ij} is the degree of membership of x_i in the cluster j, x_i is the ith data point of d-dimensional measured data, c_j is the d-dimension center of the cluster, and $\parallel * \parallel$ is any norm expressing the similarity between any measured data and the center. Fuzzy partitioning is carried out through an iterative optimization of the objective function shown above, with the update of membership u_{ij} and the cluster centers c_j by equation (2):

$$\mu_{ij} = \frac{1}{\sum_{k=1}^{C} \left(\frac{\parallel x_i - c_j \parallel}{\parallel x_i - c_k \parallel} \right)^{\frac{2}{m-1}}}, \quad c_j = \frac{\sum_{i=1}^{N} \mu_{ij}^m x_i}{\sum_{i=1}^{N} \mu_{ij}^m} \tag{2}$$

This iteration will stop when equation (3) is satisfied

$$\max_{ij} \{ | \mu_{ij}^{(k+1)} \mu_{ij}^{(k)} | \} < \xi \tag{3}$$

where ξ is a termination criterion between 0 and 1 and k is the iteration steps. Thus, the FCM algorithm creates a fuzzy partition that properly describes the image, using only the pixel intensity feature. This phase involves the calculation of cluster centers and membership function in the spectral domain only.

In order to keep the system totally autonomous, an automatic way to determine the right number of clusters, is needed. This was done by iterating the FCM algorithm for

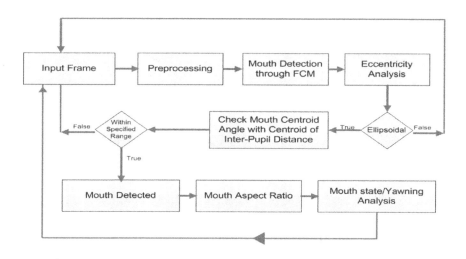

Fig. 3. Block diagram of mouth detection and yawning analysis phase

a range of hypothesized number of clusters and choosing the best option, based on a cluster validity measure. After finding out the optimal number of clusters, we pass this information to spatial FCM. Spatial FCM also consider the neighboring information of each pixel along with its spectral information, and returns fuzzy membership matrix. Spatial information incorporated by FCM helps us to remove noise in the image. A 3x3 window centered at pixel x_j was used while implementing this technique. The spatial function of a pixel x_j for a cluster j is large if the bulk of its neighborhood belongs to the same cluster. In a homogenous region, the spatial function simply fortifies the original membership, and the clustering result remains unchanged. However, for a noisy pixel, this formula reduces the weighting of a noisy cluster by the labels of its neighboring pixels. As a result, misclassified pixels from noisy regions or spurious blobs can easily be corrected. The FCM iteration proceeds with the new membership that is incorporated with the spatial function.

The iteration is stopped when the maximum difference between two cluster centers, at two successive iterations, is less than an optimal threshold. The optimal threshold is calculated through fuzzy entropy that dynamically checks the clusters found out by fuzzy c-mean. The point where maximum entropy is observed is the *optimal threshold* [8]. The fuzzy entropy based error function used is known as *logarithmic entropy*.

After the convergence, defuzzification is applied to assign each pixel to a specific cluster for which the membership is maximal. Thus, the output of FCM module is a segmented lip image in binary form [8] shown in figure.5.

The image further goes under two tests to verify that mouth has been detected successfully. The first test is based on eccentricity analysis. Since, the lips in mouth region look more like an ellipsoid making its eccentricity closer to 1, hence a check is made on the eccentricity of detected lip region; if it is closer to 1, it is checked further for the second test, other wise the next consecutive video frame is grabbed to segment

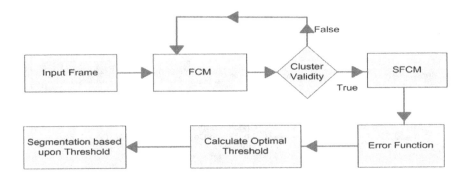

Fig. 4. Block diagram showing lip segmentation through s-FCM

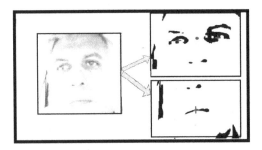

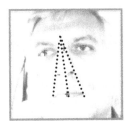

Fig. 5. Simultaneous detection of pupils and mouth in a video frame. The plus sign indicates the mouth and eye centroids, whereas the center of inter-pupil distance is indicated by red a dot.

Fig. 6. Mouth centroid angle range with inter-pupil centroid

the true lips region. The second test checks if the centroid of the detected lip region is within a specific angle range with the inter-pupil centroid. In a frontal face pose, the centroid of the lip region should be perpendicular to the point which is the centre of inter-pupil distance, as shown in figure.6. But if the whole lip region is not detected successfully, then the centroid might be misplaced; in that case the flexibility provided by angle range ensures the true presence of lip region. The pupils' locations and their centroids are calculated on the basis of radii, inter-pupil distance and the angle between the two bright blobs in upper part of the face region [1]. The formula for calculating the angle ϑ between centroids is given in equation (4) as:

$$\vartheta = \tan^{-1} \frac{|y_2 - y_1|}{|x_2 - x_1|} \qquad (4) \qquad DoO = \frac{w}{h * \cos \theta}, \qquad (5)$$

The centroid point passing the two tests is a true representative of lips. We then calculate the height and width of the lip region to determine the degree of mouth openness in different frames. To avoid scaling of mouth in different frames and the variations between subjects, we use the aspect ratio of mouth bounding rectangle to represent degree of mouth openness. The degree of mouth openness is defined as *DoO* in equation (5), where *w* is the width of mouth bounding rectangle, which is the distance between the two lip corners, and *h* is the height of mouth bounding rectangle, which is measured by the distance between two lines running through the upper and lower lip boundaries.

3.4 Yawning Analysis and Driver Fatigue Monitoring

The degree of mouth openness is constantly monitored in every fifth frame of the video sequence. Unless, the driver yawns, every fifth frame of the video is grabbed and is analyzed for the mouth state. However, when a yawning state is observed at any instance, consecutive frames in the sequence are checked instead of every fifth to see for how long the state persists.

The fatigue monitoring module looks for the result of yawning analysis at successive frames to conclude the state of the driver. If the yawning condition of the driver persists for several consecutive frames, it is concluded that the driver is fatigued. Consequently, an alarm is generated to warn the driver of his poor driving condition; otherwise the mouth state is constantly monitored at every fifth frame.

4 Experimental Results

The proposed system was implemented on Pentium 4, 1.7 GHz, using the Matlab environment. The video data was collected in real driving scenarios with different day and night lighting conditions and with drivers belonging to different ethnicity and race. The input video format is RGB with 400x320 size images. The frame rate of the video acquired from camera is 15 frames per second.

The system was tested on six different acquired videos and the accuracy found was 92% on average. The system fails to detect fatigue when the driver rotates his head to an extent that mouth or pupils occlusion takes place. The system reinitializes when the driver moves his head to the front. The mouth detection results are not affected by the exposure of teeth due to the optimal number of cluster check in spatial fuzzy c-means clustering. Moreover, the selection of dynamic threshold segments the lip region successfully and is not effected by the lighting conditions unless the brightness is so high that no contrast in the image is present. In that case, the frame under consideration is skipped and next consecutive frame is checked for yawning analysis. However, if the bad lighting condition persists, the system fails to detect the driver's fatigue.

5 Conclusion and Future Work

This paper discusses a real time fatigue detection system that makes use of yawning analysis to measure the level of inattentiveness of driver. The system was tested using different sequences recorded in real driving scenarios, with different users. The system works robustly at night time because of the use of IR illuminator. However, its performance decreases during bright time of the day. A possible solution to this problem could be the incorporation of IR filters in the car's glasses.

There are many future directions for driver's fatigue detection system. At the moment, the system does not accurately detect fatigue if the driver puts hand on his mouth while yawning (a typical yawning gesture).The mouth cannot be segmented in that frame and when the driver puts off his hand, the yawning gesture is over too. Hence, some technique must be incorporated to eliminate this shortcoming. We shall also be looking for more visual parameters to robustly detect fatigue as just a single parameter, mouth state, will require more time to raise an alarm. A combination of ocular parameters with mouth state analysis will give more accurate results as the system will not generate false positives if the driver is constantly talking.

References

1. Bergasa, L.M., Nuevo, J., Soteli, M.A., Barea, R., Lopez, M.E.: Real time system for monitoring driver Vigilance. IEEE Transactions on Intelligent Transportation Systems 7(1) (March 2006)
2. Wang, Q., Yang, J., Ren, M., Zheng, Y.: Driver Fatigue Detection: A Survey. In: Proceedings of the 6th World Congress on Intelligent Control and Automation, Dalian, China, June 2006, pp. 21–23 (2006)
3. Ji, Q., Yang, X.: Real-Time Eye, Gaze, and Face Pose Tracking for Monitoring Driver vigilance. Real Time Imaging 8(5), 357–377 (2002)
4. Park, I., Ahn, J., Byun, H.: Efficient Measurement of the Eye Blinking by Using Decision Function for Intelligent Vehicles. In: Shi, Y., van Albada, G.D., Dongarra, J., Sloot, P.M.A. (eds.) ICCS 2007, Part IV. LNCS, vol. 4490, pp. 546–549. Springer, Heidelberg (2007)
5. Saradadev, M., Bajaj, P.: Driver Fatigue Detection Using Mouth and Yawning Analysis. IJCSNS International Journal of Computer Science and Network Security 8(6), 183–188 (2008)
6. Doering, R., Manstetten, D., Altmueller, T., Lasdstaetter, U., Mahler, M.: Monitoring driver drowsiness and stress in a driving simulator. In: Proceedings of the Int. Driving Symp. Human Factors in Driver Assessment, Training and Vehicle Design (2001)
7. Viola, P., Jones, M.: Rapid object detection using a boosted cascade of simple features. In: Proceedings of IEEE Conf. Computer Vision and Pattern Recognition (2001)
8. Jaffar, M., Hussain, A., Mirza, A., Chaudary, A.: Fuzzy Entropy and Morphology based fully automated Segmentation of Lungs from CT Scan Images. International Journal of Innovative Computing, Information and Control (IJICIC) 5(12) (December 2009)
9. Fan, X., Yin, B.C., Sun, Y.F.: Yawning detection for monitoring driver fatigue. In: Proceedings of Sixth International Conference on Machine Learning and Cybernetics, Hong Kong (August 2007)

GA-SVM Based Lungs Nodule Detection and Classification

M. Arfan Jaffar[1], Ayyaz Hussain[1], Fauzia Jabeen[2] M. Nazir[1], and Anwar M. Mirza[1]

[1] FAST National University of Computer and Emerging Sciences,
A.K. Barohi Road, H11/4 Islamabad, Pakistan
arfan.jaffar@nu.edu.pk, ayyaz.hussain@nu.edu.pk,
muhammad.nazir@nu.edu.pk, anwar.m.mirza@nu.edu.pk
[2] ERP Soft, Allama Iqbal Open University,
F-7, Jinnah Super Market, Islamabad, Pakistan
fauzia.jabeen2007@gmail.com

Abstract. In this paper we have proposed a method for lungs nodule detection from computed tomography (CT) scanned images by using Genetic Algorithms (GA) and morphological techniques. First of all, GA has been used for automated segmentation of lungs. Region of interests (ROIs) have been extracted by using 8 directional searches slice by slice and then features extraction have been performed. Finally SVM have been used to classify ROI that contain nodule. The proposed system is capable to perform fully automatic segmentation and nodule detection from CT Scan Lungs images. The technique was tested against the 50 datasets of different patients received from Aga Khan Medical University, Pakistan and Lung Image Database Consortium (LIDC) dataset.

Keywords: computer aided diagnosis, mathematical morphology, segmentation, and thresholding.

1 Introduction

Lung cancer is the most important and leading cause of deaths in Western Countries. In 2004, there were 341.800 lung cancer deaths in Europe [1]. In computer vision, segmentation refers to the process of partitioning a digital image into multiple regions or sets of pixels. Each of the pixels in a region is similar with respect to some characteristic or computed property, such as color, intensity, or texture. Adjacent regions are significantly different with respect to the same characteristics [2].

In the early and untimely detection of lung abnormalities, Computer Aided Diagnosis (CAD) of lung CT image has proved to be an important and innovative development. It has been instrumental in supporting the radiologists in their final decisions [3]. The accurateness and higher decision confidence value of any lung abnormality identification system relies strongly on an efficient lung segmentation technique. Therefore it is very important for effective performance of these systems to provide them with entire and complete lung part of the image.

D. Ślęzak et al. (Eds.): SIP 2009, CCIS 61, pp. 133–140, 2009.
© Springer-Verlag Berlin Heidelberg 2009

In this paper, we propose a system capable to perform segmentation of lung images in an automatic way. We have integrated GA based optimal and dynamic threshold procedure that segment the lung part from the original image. Edge detection is performed by using morphology and thinning is performed by using Susan thinning algorithm. For nodule detection, we have used 8 directional searches on all slices and find out ROIs. After then we have pruned ROIs by using location change management of ROIs. Then we have constructed 3D ROIs image and convolve with 3D template. Main contributions of the proposed technique includes

- It is completely unsupervised and fully automatic method.
- Proposed technique finds out optimal and dynamic threshold by using fuzzy entropy.
- Eight directional searches have been used to extract ROIs and FCM have been used to classify ROI that contain nodule.

This proposed method ensures that no nodule information is lost after segmentation. This will help us for better classification of nodules.

The paper is organized as follows. Section II contains a survey on previous research that is most closely related to the present work and finds out problems. This is followed by a detailed description of the proposed system follows (Section III). Implementation and relevant results are presented in (Section IV). Finally, Section V ends the paper with several conclusions drawn from the design and the work with the proposed system.

2 Related Work

Shiying et al. [4] have developed an automatic method for identifying lungs by using gray-level thresholding for the extraction of lung region from CT-Scan image. Ayman El-Baz et al [5] have applied optimal gray-level thresholding to extract thorax area. Binsheng et al. [6] have used histogram for threshold selection and this threshold is then used to separate the lung from the other anatomical structures on the CT images. Michela Antonelli, et al [7] has used iterative gray level thresholding for lungs segmentation. Samuel et al. [8] have used gray level thresh-holding and Ball-Algorithm for the segmentation of lungs.

Some techniques have used iterative gray level thresholding for the extraction thorax and lungs from the CT scan images. These techniques did not perform well, when there is overlapping of intensities in lung parenchyma and surrounding chest wall. This is very serious shortcoming of these techniques which may lead to inaccurate diagnosis of the disease [9]. Some techniques have used morphology and the main limitation of these techniques is the size of the ball selected for morphological operation. It was observed that ball size selected for morphological closing did not work for the entire database of a single patient. Some times a ball of specified size worked for one patient while it was not sufficient for another patient. Thus we have to vary the ball size patient by patient which makes the algorithm semi-automatic. It can also be observed that the ball algorithm includes the unnecessary areas as the lung region, (not actually the part of the lung) [3], [9]. Thus if such a situation arises, these approaches are most likely to yield poor results. Thus we can say that these methodologies only work partially.

3 Proposed Method

The proposed system consists of combination of image processing and genetic algorithm that performs automatic and optimal threshold of the image. A block diagram of the proposed algorithm is depicted in Figure 4.

3.1 Threshold Block

In an ideal case, for images having two classes, the histogram has a sharp and deep valley between two peaks representing two different objects. Thus the threshold can be chosen at the bottom of this valley [10]. CT scanned lung images have also two objects. Background and lungs is considered as a single object because both have similar grey levels while other object is the middle portion of images. When we draw histogram and normalize it, then there are two clearly peaks and a single valley can be seen as shown in Fig. To find out these two peaks and valley, we have applied GA on the normalized histogram.

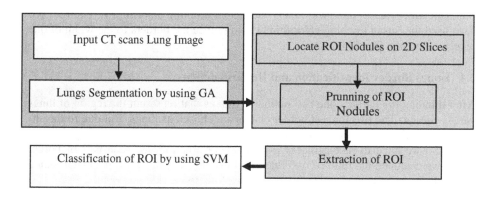

Fig. 1. Block diagram of our proposed method

Genetic Algorithms are used for getting the global solution. But in this problem we need global as well as local solutions. To implement GA, on image segmentation first we create chromosomes using input image. For this problem 8 bit length chromosomes are created due to color range (0-255). Draw histogram on the bases of frequency of color intensity on image. From Fig. 2(a) we clearly show that histogram has only two peaks. To find two peaks we implement Genetic Algorithm. To maintain two classes peak we use the crowding method in GA. After finding peaks, we select specific range to find out valley between two peaks. Then again apply Genetic Algorithm to find out valley. After finding valley, this value is well defined threshold to separate out background and object. Details of the algorithm can be found in [10].

3.2 Background Removal

By simply applying threshold to the image, we cannot get whole lungs part from background because there are high similarity between the gray levels of the lungs and the image background as shown in the Fig. 2(c). So there is need a mechanism to remove background [9]. For this purpose, we have started an operator that starts from the four corners of the image moves along the four directions identifying those pixels as background pixels with a gray level within a range and remove that pixels till pixels go outside the range or rows or columns ends. We also traverse the image from middle of the top and bottom so that any remaining background pixels remove. The image resulting from the application of this operator consists of just the chest and the lungs, as shown in Fig. 2(c).

3.3 Edge Detection and Thinning

In this step we have done three operations. First, to filter noise an opening-closing operation is used as a morphological preprocessing. Second to smooth the image, we have used closing followed by dilation. Third, to detect edges, we have found the difference between the processed image by above process and the image before dilation. But edges due to this method are thick. So a thinning algorithm is required. We have used the Susan thinning algorithm [11] to reduce the borders to the width of one pixel as shown in Fig. 2(d).

3.4 Lungs Border Identification and Reconstruction

After thinning the borders, the two pulmonary lobes that represent the region of lungs part is chosen. The two longest border chains are chosen as lungs. But due to thresholding and edge detection, it is possible that border of lungs part is wrongly eliminated. So we have to reconstruct it [9]. For reconstruction, for each pair of border pixels we calculate the ratio of two Euclidean distances. If this ratio is greater than a threshold then it is considered for reconstruction otherwise skip that pixel.

3.5 Filling and Lungs Part Extraction

To identify the set of pixels belonging to the internal part of the two pulmonary Lobes and to restore their initial gray level, we use a region filling technique [10]. Then we superimposed the two thinned chains that represent the lung lobes on the original image to show the accuracy of our method in identifying the pulmonary regions and extract by using that boundary. Extracted part is only lung image that contains all information. There is no loss of nodules within the lungs part. The region to examine for nodule detection has now been highly reduced.

3.6 Histogram Based Threshold and Cleaning

After extracting lungs part, we calculate histogram of lungs part only and calculate threshold. We segment the lungs part by using threshold. Nodules are of limited in size (3.5 to 7.3 mm) and most of the areas smaller than 3 pixels are false areas. Thus there is need a cleaning process that eliminate objects with area of 1 or 2 pixels. Therefore we have used an erosion filter of 3x3.

0	1	0
1	1	1
0	1	0

In a typical image, the number of objects after the erosion decreases from 40/130 to 10/65, and most of the deleted objects are 1 or 2 pixels in area. The filter was set 3 by 3 because heavier erosion by a larger filter would have deleted too many pixels. The diamond shape proved to delete less objects than the square one, so it was chosen for precaution.

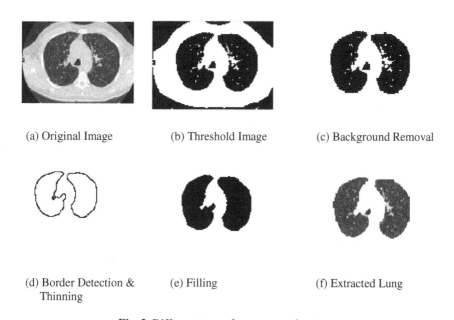

(a) Original Image   (b) Threshold Image   (c) Background Removal

(d) Border Detection &   (e) Filling   (f) Extracted Lung
    Thinning

Fig. 2. Different steps of our proposed system

3.7 Extraction of Region of Interests (ROIs)

To reduce the complexity, we have to extract ROIs. Instead of scanning the whole lung image with the template pixel by pixel, only ROIs were considered in the scan. As a result the computation time and the detection time were reduced. To extract ROIs, it was considered that pixels which form a ROI must be members of a set of adjacent neighbor pixels with suitable intensities. We have used 8 directional searches. It has been observed that diameters of nodules are between the upper and lower boundaries. Therefore, to understand whether a pixel was in the center region of the ROI, diameter of the ROI was considered initially. In this stage, we introduce two more thresholds which form the boundaries, one is the "minimum distance threshold minDist" representing the lower boundary and the other is the "maximum distance threshold maxDist" representing the upper boundary. These threshold values dealt with the resolution of the CT scan image and used to avoid very big or very small objects that are not part of lung nodule [12]. We have to count adjacent neighbors of a pixel adjCount.

> If adjCount < minDist OR adjCount > maxDist then
>> that pixel not considered for ROI
> else
>> considered for ROI

3.8 Pruning of ROIs

On serial images, vessels maintain a similar cross-sectional size and their in-plane circular appearance changes its location. But the true lung nodules remain at the same location from slice to slice. To check whether a ROI is on same location on consecutive slices, we calculate Euclidian distance of current ROI to the all ROIs in the upper slice and lower slice. From these distances, we find out the minimum distance from upper slice and minimum distance from the lower slice. If these minimum distances are less then a threshold then that ROI is considered as a nodule otherwise deleted from that image. As a result, we got an image with less number of ROIs in the image. Results have been shown in Fig. 3(d).

3.9 Features Extraction and Classification

After pruning ROIs, we have extracted features to classify remaining ROI. ROI is classified whether ROI is nodule or not. So it is a binary classification. For binary classification, we have used Support Vector Machine (SVM) as a binary classifier. We have used libsvm with rbf kernel [13].

Following features have been extracted of all ROIs after prunning. Perimeter, area, major axis length, minor axis length, equivalent diameter, circularity, compactness, mean Grey Level and standard deviation of Grey Level are the features calculated for each ROI nodule.

4 Experimental Results and Discussion

We have implemented the proposed system by using the MATLAB environment on PENTIUM IV, processor speed 1.7 GHz. We obtained datasets from Aga Khan Medical University, Pakistan and Lung Image Database Consortium (LIDC) dataset. In this work, we have studied the performance of different segmentation techniques that are used in Computer Aided Diagnosis (CAD) systems using thorax CT Scans.

Figure 2(f) gives the result of our proposed technique that segments the lungs part on the test image. It is evident through observation that the proposed system produces much smoother results. Results of our proposed method that are shown in fig 3(d) demonstrate the lungs nodule detection. The figure shows that using our method, we are able to segment and extract nodules from that segment part is very good and promising. There is also no loss of lung nodules in our proposed method. The essence of our segmentation method lies in its ability to fully automatically segment the lungs part from whole CT scan image and detect nodules and classify these. We have tested on 50 patient's data set from Aga Khan Medical University. Each patient has 100 different slices. We have made a features data set of all these patients. We have used 9 fold cross validation. Then classify it by using libsvm. We have found 97% ROIs contain nodules.

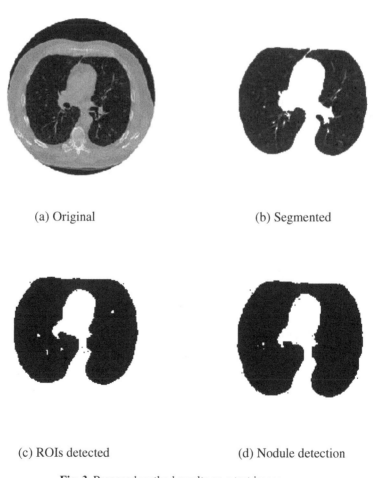

(a) Original   (b) Segmented

(c) ROIs detected   (d) Nodule detection

Fig. 3. Proposed method results on a test image

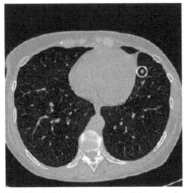 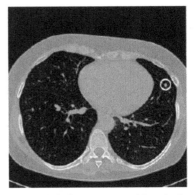

(a) Nodule is represented by circle   (b) Nodule is represented by circle

Fig. 4. Nodule detection by Proposed Method

References

1. Boyle, P., Ferlay, J.: Cancer incidence and mortality in Europe, 2004. Annals of oncology 16(3), 481–488 (2005)
2. American Association for Cancer Research, http://www.aacr.org
3. Memon, N.A., Mirza, A.M., Gilani, S.A.M.: Segmentation of Lungs from CT Scan Imges for Early Diagnosis of Lung Cancer. In: Proceedings of World Academy of Science, Engineering and Technology, vol. 14 (August 2006)
4. Hu, S., Huffman, E.A., Reinhardt, J.M.: Automatic Lung Segementation for Accurate Quantitiation of Volumetric X-Ray CT images. IEEE Transactions on Medical Imaging 20(6) (June 2001)
5. El-Baz, A., Farag, A.A., Falk, R., Rocca, R.L.: A Unified Approach for Detection, Visualization and Identification of Lung Abnormalities in Chest Spiral CT Scan. In: Proceedings of Computer Assisted Radiology and Surgery, London (2003)
6. Zhao, B., Gamsu, G., Ginsberg, M.S.: Automatic detection of small lung nodules on CT utilizing a local density maximum algorithm. Journal of Applied Clinical Medical Physics 4(3) (Summer 2003)
7. Antonelli, M., Lazzerini, B., Marcelloni, F.: Segmentation and reconstruction of the lung volume in CT images. In: ACM Symposium on Applied Computing (2005)
8. Armato III, S.G., Giger, M.L., Moran, C.J.: Computerized Detection of Pulmonary Nodules on CT Scans. RadioGraphics 19, 1303–1311 (1999)
9. Arfan Jaffar, M., Hussain, A., Mirza, A.M., Asmat ullah, C.: Fuzzy Entropy based Optimization of Clusters for the Segmentation of Lungs in CT Scanned Images. In: Knowledge and Information Systems
10. Arfan Jaffar, M., Hussain, A., Nazir, M., Mirza, A.M.: GA and Morphology based fully automated Segmentation of Lungs from CT scan Images. In: International Conference on Computational Intelligence for Modeling, Control and Automation, Vienna, Austria, December 10-12 (2008)
11. Smith, S.M., Brady, J.M.: SUSAN - a new approach to low level image processing. Int. Journal of Computer Vision 23(1), 45–78 (1997)
12. Ozekes, S., Osman, O., Ucan, O.N.: Nodule Detection in a Lung Region that's Segmented with Using Genetic Cellular Neural Networks and 3D Template Matching with Fuzzy Rule Based Thresholding. Korean J. Radiol. 9(1) (February 2008)
13. http://www.csie.ntu.edu.tw/~cjlin/libsvm/

Link Shifting Based Pyramid Segmentation for Elongated Regions

Milos Stojmenovic, Andres Solis Montero, and Amiya Nayak

SITE, University of Ottawa, Ottawa, Ontario, Canada K1N 6N5
{mstoj075,amon,anayak}@site.uottawa.ca

Abstract. The goal of image segmentation is to partition an image into regions that are internally homogeneous and heterogeneous with respect to other neighbouring regions. An improvement to the regular pyramid image segmentation scheme is presented here which results in the correct segmentation of elongated and large regions. The improvement is in the way child nodes choose parents. Each child node considers the neighbours of its candidate parent, and the candidate parents of its neighbouring nodes in the same level alongside the standard candidate nodes in the layer above. We also modified a tie-breaking rule for selecting the parent node. It concentrates around single parent nodes when alternatives do not differ significantly. Images were traversed top to bottom, left to right in odd iterations, and bottom to top, right to left in even iterations, which improved the speed at which accurate segmentation was achieved. The new algorithm is tested on a set of images.

Keywords: image segmentation, pattern recognition.

1 Introduction

Image segmentation is an important step in many computer vision applications. It decomposes an image into homogeneous regions, according to some criteria [6]. It is rarely achieved comprehensively for any single application, and algorithms that do perform well in one application are not suited for others.

Pyramid segmentation was proposed in [7, 10], and further elaborated on in [3]. Pyramids are hierarchical structures, which offer segmentation algorithms with multiple representations with decreasing resolution. Each representation or level is built by computing a set of local operations over the level below (with the original image being at base level or level 0 in the hierarchy). Level L consists of a mesh of vertices. Each vertex at level L is linked to at most one parent node at level $L+1$. The value of the node is derived from the values of all its children nodes at the level below. When this is applied to children nodes transitively down to the base level, the value of each node at a given level is decided by the set of its descendent pixels at the base level (its receptive field). Each node at level L also represents an image component of level L segmentation, consisting of pixels belonging to its receptive field (if nonempty). Thus each level has its predefined maximum number of components. However its minimum number of components is left open and depends on a concrete algorithm.

Pyramids were roughly classified into regular and irregular. Regular pyramids have a well-defined neighbourhood intra-level structure, where only natural neighbours in

D. Ślęzak et al. (Eds.): SIP 2009, CCIS 61, pp. 141–152, 2009.
© Springer-Verlag Berlin Heidelberg 2009

the mesh that defines a level of a pyramid are considered. Inter-level edges are the only relationships that can be changed to adapt the pyramid to the image layout. Literature also places a condition of a constant reduction factor for regular pyramids [6]. Regular pyramids can suffer several problems [1, 5, 6]: non-connectivity of the obtained receptive field, shift variance, or incapability to segment elongated objects.

To address these problems, irregular pyramids were proposed which vary in structure, increase the complexity and run time of the algorithms, and/or are dependant on prior knowledge of the image domain to be designed successfully. In the irregular pyramid framework, the spatial relationships and reduction factors are non-constant. Existing irregular pyramid based segmentation algorithms were surveyed in [6, 5]. The described methods all appear very complex, have higher time complexities compared to regular pyramids and all consider the connectivity of the receptive field (base layer) as a design goal.

In this research, we first observe that the connectivity of the receptive field (region in segmentation) is not always a desirable characteristic. For instance, in forestry applications, the same type of vegetation could be present in several parts of the image, and treating them as single segment may in fact be preferred for further processing. In other applications an object that is partially obscured and 'divided' by other objects may also be more naturally identified as a single object instead of processing it in multiple pieces. Our proposed segmentation algorithm allows for non-connectivity of receptive fields.

Our algorithm is inspired by the regular pyramid linked approach (PLA) originally proposed by Burt et al. [3]. Their approach differs from previous pyramid structures in that nodes from one level may alter their selection of parent at each iteration of the segmentation algorithm. Starting from level 0, links between level L and level $L+1$ are decided. Each vertex at level L has four fixed candidate parents, and chooses the one which is the most similar from the higher level. The local image property value of each parent is recalculated by averaging the local image property of its current sons. The process continues until the son-parent edges do not vary, or a certain number T of iterations is reached. The process then continues at the next higher level. The main advantage of this method is that it does not need any threshold to compute similarity between nodes. In the original method [3], each node must be linked to one parent node. Antonisse [1] removed the need by introducing 'unforced linking' which allows the exclusion of some vertices from the linking scheme, and the presence of small components in the segmented image. This method has two parameters, one for the degree of similarity for decision linking, and one for the minimal size of a component (that is, from which level unforced linking is permitted). Antonisse [1] also proposed path fixing (some arbitrary path from the image level to the top level is fixed). This was indirectly applied here by the initial selection of random numbers for each pyramid vertex and the tie breaking rule is applied on them. Antonisse [1] also proposed randomized tie-breaking: when potential parent nodes have the same value, the choice of the parent is randomized. We have changed this rule in this paper.

Burt et al [3] however limit the choice of parent to 4 fixed nodes directly above the child. This approach has contributed to more successful segmentation, but due to the limited selection of parents, elongated and generally large segments that cover a significant portion of the image are not considered 'joined'. Figure 1 below demonstrates this weakness in their algorithm, even at the top level of the segmentation process.

Since each node can only chose among the parents in its immediate neighbourhood, nodes at opposite ends of an image, (or an elongated segment) can never point to the same parent node.

Fig. 1. Elongated regions pose challenges for regular pyramid based segmentation

We strive to design an algorithm that would properly handle elongated objects, while not enforcing connectivity of the receptive field and preserving shift invariance (the stability when minor shifts occur), having favourable time complexity (receiving an answer within seconds, depending on the image size), and overall simplicity, so that the algorithm can be easily understood, implemented, and used in practice.

To achieve our goals, we have made some simple changes in the way parent nodes are selected in the regular pyramid framework, which resulted in major improvements in their performance, including reduced shift variability and handling elongated objects. Instead of always comparing and selecting among the same four candidate parent nodes, each vertex at the current level will compare and select the best among the current parent, its neighbours, and current parents of its neighbouring vertices at the same level. Either of 4- or 8-connectivity neighbours can be used.

Experiments were conducted on both simple shapes, to validate the proposed methods, and on several real images. Evaluating segmentation quality in imagery is a subjective affair, and not easily done. "The ill-defined nature of the segmentation problem" [6] makes subjective judgment the generally adopted method (existing quantitative measurements are based on subjective formulas). In general, it is not clear what a 'good' segmentation is [6]. For this reason, no quantitative evaluation measure is applied to verify the results. The quality of the obtained segmentations appears visually satisfying for at least one level in each image.

2   Literature Review

2.1   Overview

There exists a vast literature on the segmentation problem, and hundreds of solutions were proposed. A list of references and descriptions of basic approaches (thresholding, edge-based, region-based, matching, mean-shift, active contour, fuzzy, 3D graph-based, graph cut and surface segmentation) are given in [8]. Marfil et al. [6] present a concise overview of existing pyramid segmentation algorithms. Our discussion is concentrated to such schemes which, compared to others, offer segmentations with low or no parameters, and flexibility with respect to the output (one answer is given at each level of pyramid).

We found one existing publicly available image segmentation software solution by [2], which is used here for comparing our results. It provides segmentation at various levels and appears to generally work well. The algorithm itself is sophisticated and based on linear algebra and statistics tools. In our opinion it is currently the most complete non supervised image segmentation system available.

2.2 Segmentation Algorithms

A plethora of segmentation algorithms exists [8]. However, their practicality is often not clear. For example, Shi and Malik [9] propose a 'normalized cut' that calculates both similarities and differences between regions. The normalized cut in an image revolves around calculating eigenvalues of matrices of the size n x n, where n is the number of pixels in the image. Their results appear to work relatively well only for very small images, and when segment connectivity is enforced, but are impractical for processing images larger than roughly 100 x 100 pixels, which is the size of their test set.

Pyramid based segmentation algorithms are described in [6], together with the data structures and decimation procedures which encode and manage the information in the pyramid. Our new algorithm is based on the one originally proposed by Burt et al [3]. They further the idea that the original image should be down sampled in successive steps and that pixels from one level should point to parents in the next. Segmentation can be done on the whole pyramid, and a given segmentation result is achieved on each level. Details are given in the next section, together with the changes made by our new algorithm.

2.3 Region Similarity and Unforced Linking

Antonisse [1] introduced the 'unforced linking' concept of holes in the segmentation process. This means that a pixel in one level does not necessarily need to select any parent in the above level if all of the proposed parents are sufficiently different from it. Antonisse uses a 2 step segmentation process, one of which is a clustering step on each layer which we do not employ in our work. Let include(u) be m times the standard deviation of a 3x3 neighbourhood around the node u (m is a parameter, =2 for 95% confidence and =3 for 99% confidence assuming normal distribution of pixel intensities). Find the most similar parent q. If it has the value which differs by at most include(u) of from u's value, u links to q. Otherwise, u fails to link to q and becomes the root of a new sub pyramid [1]. The similarity between two regions u and q can be measured also directly using their pixel distributions. For example, [4] propose a statistical test that analyzes two regions for homogeneity. Their test involved comparing the grey level distribution of pixel intensities of the two regions and evaluating the level of overlap of the 2 distributions. The test measures the similarity of variances of image intensities of two regions. Thus even two regions with the same mean values can be declared as dissimilar only due to different variances in their regions. In our implementation, we use a simple threshold for both region similarity and unforced linking.

3 Pyramid Image Segmentation Algorithm

Here we describe the proposed pyramid segmentation algorithm. It is nearly identical to that of [3], except in the way that children choose their parent nodes. This change resulted in major improvements in its performance, while preserving its simplicity. Another enhancement includes a tie breaking rule for parent selection.

The input to the algorithm is a greyscale, single channel image of dimensions $2^N \times 2^N$ pixels, for $N \geq 2$. Each pixel (at level 0) u has integer value $I(u)$ in interval [0,255]. The output is the original image overlaid with the resultant segmentations at each level. In the pseudo code and discussion below, L is the level of the pyramid, $L = 0, 1..., N$. The bottom level ($L = 0$) is a matrix $2^N \times 2^N$ pixels, representing the original image. Level L is matrix with 2^{N-L} rows and columns, $i, j = 0, 1, 2 ... 2^{N-L}-1$. Top level $L=N$ has one element.

3.1 Creating the Initial Image Pyramid

We describe and use the overlapping image pyramid structure from [3]. Initially, the children of node $[i, j, L]$ are: $[i',j',L-1]=[2i+e, 2j+f, L-1]$, for $e,f \in \{-1,0,1,2\}$. There are a maximum of 16 children. That is, each $\{2i-1, 2i, 2i+1, 2i+2\}$ can be paired with each $\{2j-1, 2j, 2j+1, 2j+2\}$. For $i=j=0$ there are 9 children, and for $i=0$ and $j>0$ there are 12 children. There are also maximum index values $2^{N-L+1}-1$ for any child at level $L-1$ which also restricts the number of children close to the maximum row and column values. For $L=N$ there are four children of single node $[0, 0, N]$ on the top: $[0, 0, N-1]$, $[0, 1, N-1]$, $[1, 0, N-1]$, $[1, 1, N-1]$ since minimum index is 0 and maximum is 1 at level $N-1$. Two neighbouring parents have overlapping initial children allocations. Conversely, each child $[i, j, L]$ for $L<N$ has 4 candidate parent nodes (if they exist) $[i'', j'', L+1]=[(i+e)/2, (j+f)/2, L+1]$, for $e,f \in \{-1,1\}$, where integer division is used. Pixels at the edges of the image would have fewer parents to choose from. The average intensity of all possible children is set as the initial intensity value $I(v)$ of the parent v, for nodes at levels $L>0$. This calculation is relatively slow since almost every parent has 16 table lookups to perform; therefore, an improvement was built into the system. Integral images were built for every layer in the pyramid, and for each set of points that needed to be summed, the number of table lookups was reduced from 16 to 4. This resulted in a 4x improvement in the construction of the image pyramid. The integral image technique first calculates cumulative sums of intensity values of all pixels with a first index $\leq i$ and second index $\leq j$. The summary value inside a rectangle is then found easily from these cumulative sums at the four rectangle corners.

3.2 Testing Similarity of Two Regions, Unforced Linking and Tie-Breaking Rule

Our algorithm makes use of a procedure for comparing the similarity of two regions, namely receptive fields of a vertex u and its candidate parent v. These two vertices are similar (*similar(u,v)=true*) if their intensities are roughly the same. In the current implementation of the algorithm, we use a simple threshold based (the threshold value used in algorithm is $S=15$) algorithm for testing similarity. We also use a test for dissimilarity of two regions, to decide whether of not the best parent is an acceptable link, for the unforced linking option. We use simple threshold based comparison against threshold value $D=100$.

Each pixel chooses the parent with the closest intensity to itself out of all candidate parents for its initial choice. The tie-breaking rule in [BHM] is as follows. When two or more 'distances' are equally minimal, one is chosen at random (randomization is done when needed). However, if one of them was selected parent in the previous iteration, it is not then changed. We observed a slow merging process when this rule is employed. For example, two neighbouring pixels would not change and merge two neighbouring parents when all of them have the same intensities. We have modified this rule as follows. Each vertex v is initially assigned a random number $r(v)$ in $[0,1]$ which was never changed later on. Also, each vertex v at level L has a counter $c(v)$ which contains the number of children that selected that parent in the last iteration. Before the first iteration, $c(v)=0$ for each vertex. Let w be a child node that compares parent candidates u and v, and let better(w,u,v) be one of u or v according to the comparison. The function is as follows.

If $|I(w)-I(u)| < |I(w)-I(v)|$ **then** *better=u* **else** *better=v*;
If similar(w,u) and similar(w,v) **then** {
If $c(u)>c(v)$ **then** *better=u*; **If** $c(u)<c(v)$ **then** *better=v*;
     If $c(u)=c(v)$ **then** {IF $r(u)>r(v)$ **then** *better=u* **else** *better=v*}}.

3.3 Candidate Parents

Among candidate parents, each vertex selects the one which the closest to it, using a tie-breaking rule to decide among them. For the first iteration, each vertex has up to four candidate parents, as per initial setup [3] described above. This fixed set of candidate parents has been changed (for further iterations) in our algorithm by a dynamic flexible set of candidate nodes that revolves around the current parent selection and the parent selection of neighbouring vertices at the same level. Suppose node $w = [i, j, L]$ is currently linked to parent $u = [i", j", L+1]$ in iteration t, which we will denote simply by $p(w)=u$. The full notation would lead to $p[i, j, L][t]=[i", j", L+1][t]$, and is convenient for easy listing of candidate parents. One set of candidate parents consists of the current parent and its 8 neighbours at the same level. Thus, in our notation, the candidate parents for the next iteration are: $[i"+e, j"+f, L+1]$, where $e, f \in \{-1,0,1\}$. This produces a maximum of 9 candidate parents, which is a 3x3 grid centered at currently linked parent. We also added four additional parent candidates, by considering the current selection (from the previous iteration) of neighbouring vertices at the same level. For $w=[i, j, L]$ and iteration $t+1$, we also consider $p[i+1, j, L][t]$, $p[i-1, j, L][t]$, $p[i, j+1, L][t]$, and $p[i, j-1, L][t]$ as parent candidates, if they exist. Both sets allow us to shift the parent further away in the next iteration, and possibly link the current child to a remote parent after the iterative process stabilizes (with no more changes in selected parents). This change of parent selection nodes is directly responsible for the ability of our new algorithm to handle elongated objects. Note that, alternatively, we could consider 5 candidate parents, or 8 parents of the neighbouring nodes. The current selection is an initial option, as a compromise between time complexity, and the quality of segmentation.

3.4 Pyramid Segmentation

Once the image pyramid and pointer pyramid structures have been initialized, the segmentation procedure may begin. The decisions are attained at a given level (starting at layer 0, and working toward the top of the pyramid) after several iterations, before the process advances to the next level, where each pixel modifies its selection of parent in the next layer until equilibrium is reached. At layer L, each pixel points to the parent which has the closest intensity to itself in layer $L+1$, among the 9+4 candidate parents. The intensities of the parents in layer L are recalculated based on the average intensity of the pixels in its current receptive field at the end of each iteration. Since these averages are calculated from the averages of its children, they must be appropriately weighted (by the number of pixels in the receptive fields of the children). This cycle of choosing parents, recalculating intensities, and reassigning parents continues for $T=5$ iterations per pair of layers, or until there is no change in parent selection between layers. Each node u also has an associated random number $r(u)$, counter $c(u)$ of the number of children that selected it as unique parent, and $n(u)$, number of pixels in its receptive field. Note that $n(u)$ is used in calculating $I(u)$ from intensity values of children, e.g. if w_1, w_2 and w_3 are children of u then $I(u) = (I(w_1)n(w_1)+I(w_2)n(w_2)+ I(w_3)n(w_3))/n(u)$, $n(u) = n(w_1) + n(w_2) + {}_n(w_3)$. The algorithm can be, at the top level, described as follows.

Create initial image pyramid (section 3.1)

For levels $L = 1$ **to** N **Do**
$t = 0$ (iteration counter)
    Repeat $t = t+1$
each vertex u at level L selects parent $p(u)$ among the 9+4 candidate ones
apply unforced linking (however 'broken' $p(u)$ is still needed in the next iteration)
each vertex v at level $L+1$ recalculates $I(v)$ based on its current receptive field
    Until $t=T$ or none of links changed in iteration t.
Endfor; Display segmentation for each level in pyramid.

We also mention a simple modification in the general parent selection algorithm that has secured declaration of elongated shapes as single component at a very early stage. Initially, the traversal of pixels was always top to bottom, left to right. Elongated shapes along the NW-SE direction were immediately discovered as single components; however shapes in the NE-SW direction remained multi-componential for several levels. To avoid this problem, pixels were traversed top to bottom, left to right in odd iterations, and bottom to top, right to left in even iterations. The elongated components were then immediately recognized, at the first level of the process, as illustrated in Figure 2. Without this correction in traversal, all of the levels seen below (b-h) would have been obtained. However, with this change in traversal pattern, the whole shape, as seen in Fig. 2 -h was obtained in the first level of the segmentation.

This greatly improved the accuracy of the segmentation algorithm for more complex shapes as well.

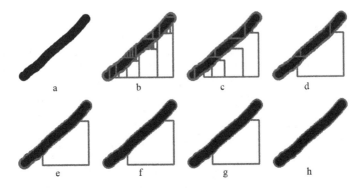

Fig. 2. [SNH] segmentation results per layer

4 Experimental Results

The algorithm presented here was designed to solve the problem of correctly segmenting elongated shapes in the framework of regular pyramid segmentation. As such, we demonstrate the results of the algorithm on a set of images, containing simple black shapes on a white background, of size 256x256 pixels. The processing time per image is 16 seconds on a single core of a Pentium 1.66 GHz dual core machine, implemented in C# on the Windows XP operating system. Our algorithm was also applied to several realistic images of size 256x256 pixels and were reduced to single channel greyscale representation as input. Although it is relatively easy for us to judge whether or not a simple shape laid against a high contrast background is segmented properly, it is far more difficult to evaluate the precision of a segmentation of real imagery. The judgement of the quality of the segmentation is subjective.

4.1 Segmentation of Basic Shapes

The shapes below are listed in 3 columns. The first column shows the original image, the second column shows the results of our algorithm proposed here, with segmentation result outlined in red, and the third column shows the results of the original pyramid segmentation algorithm proposed by [3], outlined in green. Since the shapes below have few segments, only the highest layer in the resultant segmentation pyramids is displayed, and chosen to represent the segmentation of each image.

We notice that our algorithm segmented the shapes correctly all of the time when considering basic shapes. The [3] algorithm does not work well with elongated shapes in images. The shapes listed above all fit into this category, so it is not surprising that this algorithm performs poorly on them. The [3] algorithm also has difficulties even with some non-elongated shapes. For example, it did not detect a hole in example 8, while the other hole, just a bit larger, is divided into several segments.

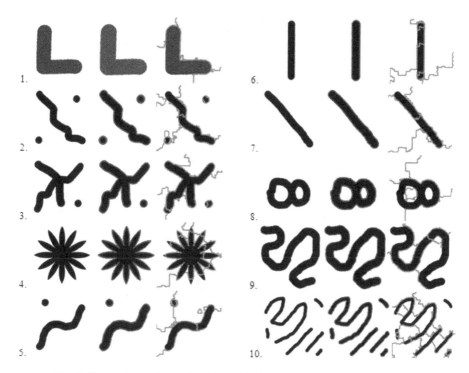

Fig. 3. Comparison of new algorithm with [3] algorithm on simple examples

4.2 Segmentation of Other Imagery

The elongated segment finding procedure was applied to some examples of everyday imagery. We compared our results with those of a [2], who also employ a type of hierarchical segmentation structure, but take into considerations texture as well as colour. Both approaches are relatively parameterless, and form a good comparison basis. The images in Figure 4 are all 256x256 pixels, their segmentation pyramids contain 8 levels in our representation and 13 in that of [2].

We show only the best level of segmentation of each pyramid since they best reflect the desired segmentation results for these images. We see that both methods segment relatively well. They segment out the wolf, high grass in the foreground and the fence posts. Our algorithm had more success at finding the sheep, the middle fencepost, and the sky in the top right corner. Both the teddy bear and the cat are equally well isolated by the algorithms, yet [2] does not segment out the cat's tail, and adds 3 other unnecessary segments around the bed sheets. This is due to superior texture matching property of [2], which in this example hinders more than helps its segmentation ability. The small dog was equally well found by both algorithms, yet they equally failed to segment the yellow container the dog sits in. This is because the container is round, and as such produces a gradient colour effect: from light yellow to dark yellow. Our algorithm fails to pick up on this change in tones, and cuts the segment into two pieces, while the [2] algorithm fails because there is no texture there to

Fig. 4. Sample images and segmentation of [2] and our method

match, and primarily relies on its colour matching ability, which is no better than that of our algorithm. The last example was nearly identically segmented by the algorithms except for the large tree in the bottom left corner. This tree is found by the [2] algorithm, but not by the one we propose. This once again illustrates the superior texture matching ability of their solution. However, based on this test set, the algorithms are fairly competitive.

5 Future Work

The algorithm shows a promising way forward in the quest for successful image segmentation. Since this method relies on colour intensities, it has a tendency of over-segmenting covered regions that appear homogenous to human observers, yet distinguishably heterogeneous to the program, at (the lower layers of the segmentation pyramid). It also has the opposite tendency at the higher levels. The selection of the proper level appears possible, but is not the same for every picture, and eventually a human observer may be needed to select the best outcome for each picture, according to its further processing needs.

To improve the outcome of this segmentation algorithm, one would have to have at least some prior knowledge of the scene that is to be segmented. Such knowledge includes the minimum possible segment size, and possibly a range of pixel intensities within a region that could be considered homogenous. Other solutions may include considering more than just greyscale intensities of input data. In the current implementation, just the RGB layers are considered, and they are combined into just a single layer greyscale representation of the original image. By considering the Euclidean distance between two 3D points in an RGB space instead of simply considering greyscale differences, more accurate parent selection could be achieved at the expense of increased computation time. Another point to consider is to include some way of grouping similar textures together along with colours, such as the algorithm proposed by [2]. We are currently considering such an addition of texture to color in region similarity tests.

We are planning to enhance the similarity and dissimilarity of two regions (to avoid using an arbitrary threshold) by using a statistical testing, which is more intuitively acceptable than the one described in [4]. It was not done for this version because this new algorithm will slow down the execution of the program, which at this testing stage is also an important consideration.

We have used and experimented with the overlapping image pyramid structure as originally proposed in [3]. This refers to the fact that parent vertices at level 1 have overlapping receptive fields. Antonisse [1] already argued that perhaps a non-overlapping structure could perform better. We left this modification for further study, so that we can first investigate the impact of single major change proposed here, the use of flexible parent links. We may further experiment with the versions with different candidate parent sets, such as 5 candidate parents instead of 9 around current parent, or 8 parents of neighbouring nodes instead of 4.

References

1. Antonisse, H.: Image Segmentation in Pyramids. Computer Graphics & Image Processing 19, 367–383 (1982)
2. Alpert, S., Galun, M., Basri, R., Brandt, A.: Image segmentation by probabilistic bottom-up aggregation and cue integration. In: IEEE Conf. on Computer Vision and Pattern Recognition, CVPR 2007 (2007)

3. Burt, P., Hong, T., Rosenfeld, A.: Segmentation and estimation of image region properties through cooperative hierarchical computation. IEEE Trans. Systems, Man, and Cybernetics 11(12) (1981)

4. Chen, S., Lin, W., Chen, C.: Split-and-Merge Image Segmentation Based on localized feature analysis and statistical tests. Graphical models and Image Processing 53(5), 457–475 (1991)

5. Fu, K., Mui, J.: A survey on image segmentation. Pattern Recognition 13, 3–16 (1980)

6. Marfil, R., Molina-Tanco, L., Bandera, A., Rodriguez, J., Sandoval, F.: Pyramid segmentation algorithms revisited. Pattern Recognition 39, 1430–1451 (2006)

7. Riseman, E., Hanson, A.: Design of a semantically-directed vision processor, Tech. Rep. 74C-1, Department of Computer Information Science, University of Massachusetts, Amherst, MA (1974)

8. Sonka, M., Hlavac, V., Boyle, R.: Image Processing, Analysis, and Machine Vision. Thompson (2008)

9. Shi, J., Malik, J.: Normalized Cuts and Image Segmentation. IEEE Transactions on Pattern Analysis and Machine Intelligence 22(8) (2000)

10. Tanimoto, S., Pavlidis, T.: A hierarchical data structure for picture processing. Computer Graphics and Image Processing 4, 104–119 (1975)

A Robust Algorithm for Fruit-Sorting under Variable Illumination

Janakiraman L. Kausik[1] and Gopal Aravamudhan[2]

[1] Department of Electrical & Electronics Engineering, Birla Institute of Technology & Science, Pilani 333 031, Rajasthan, India
kaushik_l@yahoo.com
[2] Central Electronics Engineering Research Institute, CSIR Madras Complex, Taramani, Chennai 600 013, Tamil Nadu, India

Abstract. Computer vision techniques are essential for defect segmentation in any automatized fruit-sorting system. Conventional sorting methods employ algorithms that are specific to standard illumination conditions and may produce undesirable results if ideal conditions are not maintained. This paper outlines a scheme that employs adaptive filters for pre-processing to negate the effect of varying illumination followed by defect segmentation using a localized adaptive threshold in an apple sorting experimental system based on a reference image. This technique has been compared with other methods and the results indicate an improved sorting performance. This can also be applied to other fruits with curved contours.

Keywords: Computer Vision, On-line fruit sorting, Surface defect, Adaptive thresholding.

1 Introduction

Fruit inspection and grading is an indispensable horticultural procedure. Uniformity in size, shape and colour are few of the many parameters that are vital in determining consumer acceptance. While the task at hand is to develop a machine vision system that identifies defective fruits based on odd shapes and surface defects, and to categorize them depending on consumer acceptability, the objective has to be accomplished with certain constraints [1]. Such a system has to be operable at high speeds suitable for real-time processing yielding a high throughput, must inspect the entire fruit surface, must be adaptable to varying fruit size, shape etc., and be applicable under various physical conditions like brightness, luminance etc.

Over the past decade, various techniques have been proposed for defect segmentation and grading. Reference [2] uses flooding algorithm to identify and characterize different types of defects based on perimeter, area etc. The snake algorithm discussed in [3] can be used to localize the defect and reduce false positives. Reference [4] employs a raw approach based on colour frequency distribution to associate pixels to a specific class while [5] accomplishes the same using 'Linear discriminant analysis'.

D. Ślęzak et al. (Eds.): SIP 2009, CCIS 61, pp. 153–160, 2009.
© Springer-Verlag Berlin Heidelberg 2009

Hyper-spectral and multispectral imaging systems have also been proposed for sorting various food commodities as discussed in [6]-[7].

An inherent limitation in most of the existing techniques is their sensitivity to changing illumination conditions. Any flash of external stimulus can result in bright patches in the captured image which could result in misclassification. Practical considerations dictate that any technique should be immune to occasional changes in external conditions and deliver acceptable performance. This paper incorporates the use of adaptive filters based on the conventional LMS algorithm as a pre-processing step prior to segmenting defects using an adaptive threshold. This paper has been organized as follows. Section 2 explains the components of the practical set-up used to capture images of the fruit to be sorted. Section 3 discusses the proposed methodology for pre-processing and defect segmentation. Results of the experiment have been tabulated and discussed towards the end.

2   Experimental Set-Up

The prototype system[8] used for testing apples comprises of individually designed and integrated sub-systems which include a system for fruit separation, a conveyor system, illumination and imaging system, a processor for computing and a sorting system.

The fruits are dumped for sorting in an open container which forms the head of the feeding system, where a baffle mechanism is employed to guide the fruit, one at a time. A singulating drum feeds each fruit to the orienting section. The conveyor chain consists of a series of axles bearing aluminium rollers rotating freely about their axes. This roller mechanism ensures that apples are aligned to the stem-calyx position for ease of processing. The imaging system is placed in the illumination chamber towards one end of the conveyor.

The illumination unit consists of incandescent lamps and compact fluorescent lamps to provide uniform lighting. However, illumination may vary with time due to voltage fluctuations, ambient temperature etc. The camera and lens system captures 6 orientations of each fruit 60° apart. A frame grabber controlled by a series of rectangular pulses synchronised with the fruit movement, captures one frame per trigger. A standard personal computer is used to run the algorithm for processing the captured image. This involves a series of steps summarized in the next section. In the final model, this unit may be replaced by a dedicated Digital signal processor equipped to handle specialized image processing functions efficiently. A series of cups aligned and molded with solenoid actuated strips controlled by the computing system is used to sort fruits into different grades. Each fruit is categorized and the solenoid associated with the corresponding grade gets activated and enables the collection of the fruit into the cup using a tilting mechanism.

3   Methodology

The captured image goes through a pre-processing routine before defect segmentation and grading. These steps are discussed below:

3.1 Pre-processing

As discussed, variation in intensity of illumination may result in varied levels of brightness and contrast. Under these conditions, the techniques for background removal and defect segmentation, which have been developed for standard illumination, may behave strangely. Hence, practicable solution is to employ adaptive normalization. To accomplish this, two images of a reference apple are used. One of the images is captured under standard conditions to eternally serve as the standard reference 'S'. An image of this reference fruit is also captured alongside the fruit to be graded, so that the reference fruit and the test fruit ('T') to be graded are both subjected to similar conditions. Let this image be labelled 'R'. It is essential to understand that this image of the reference fruit is not merely a scaled version of the one taken under ideal conditions as the intensity distribution is non-uniform. Hence, soft scaling and normalization techniques do not guarantee reliable results. On the other hand, illumination-normalization techniques discussed in [9]-[10] are computationally intensive and adds to the complexity of the existing system. A technique better suited for this purpose uses two-dimensional LMS adaptive filters with reduced complexity [11].

In this technique, a window 'w' is applied on 'R' and the filter coefficients are adaptively modified using the ideal reference image 'S' to obtain the predicted image, ($\hat{S}$) after a few iterations. The method is described below:

- A filter window 'w' of size N x N satisfying the linear-phase constraint (1) and (2) with initial weight (3) is chosen.

$$w(x, y) = w(N - 1 - x, N - 1 - y) . \tag{1}$$

where, $(x, y) \in$

$$\left[\left[\left(0, \frac{N-1}{2}\right), \left(0, \frac{N-1}{2}\right)\right] \cup \left[\left(\frac{N+1}{2}, N - 1\right), \left(0, \frac{N-3}{2}\right)\right]\right] . \tag{2}$$

- The filter is applied on R(i,j) to obtain the predicted output $\hat{S}$(i,j). This convolution operation is summarized in (4).
- The output is compared with the desired response S(i,j) and the error e(i,j) is used to successively update the coefficient w(x,y) in accordance with (6).
- After the required number iterations [11], the predicted output ($\hat{S}$) is perceivably similar to the desired (S) as shown in Fig. 1.

$$w_0(x, y) = \frac{1}{N^2} \; \forall \, (x, y) . \tag{3}$$

$$\hat{S}(i, j) = \sum_{x=0}^{N-1} \sum_{y=0}^{N-1} w(x, y) R(i - x, j - y) . \tag{4}$$

$$e(i,j) = S(i,j) - \hat{S}(i,j).\tag{5}$$

- The sequence of updated filter coefficients thus obtained, can now be applied on the image of the test fruit to be graded (T) to obtain a normalized version (T_n) suitable for further processing which includes background removal and adaptive thresholding as discussed below.

$$w_{k+1}(x,y) = w_k(x,y) + e(i,j) * [\hat{S}(i\text{-}x,j\text{-}y) + \hat{S}(i\text{+}x\text{-}N\text{+}1,j\text{+}y\text{-}N\text{+}1)].\tag{6}$$

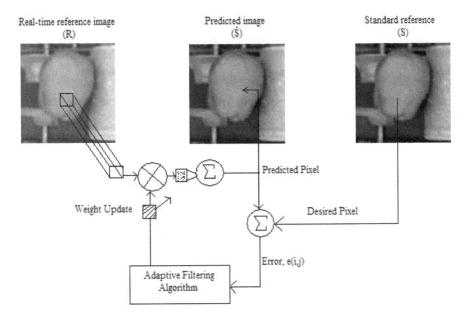

Fig. 1. A schematic of the 2D LMS Adaptive filter used for pre-processing

3.2 Defect Segmentation

The normalized image T_n obtained after pre-processing is subjected to background removal followed by comparison with a reference image. Finally, the defect is segmented by using a local adaptive threshold based on Sauvola's Technique. The steps involved in this process are outlined below:

Background Removal. The technique used for background removal is specific to the conveyor used to capture the frames. If the conveyor is of a uniform background, then a simple global threshold would do the task of segmenting the fruit provided it is normalized to rid the effect of varying illumination. This is performed on both the standard reference(S) and the normalized image (T_n) to obtain (S_{br}) and (T_{nbr}) respectively.

Subtraction. Once the background has been removed, the fruit to be inspected is subtracted from the reference and the residual periphery is removed to obtain the difference (D). This process is carried out on each of the 3 R, G and B components separately rather than on the gray-scale image. Although this results in a slight increase in computational complexity, preserving the colour information helps characterize the type of defect.

$$D(i,j) = S_{br}(i,j) - T_{nbr}(i,j).$$ (7)

Local Adaptive Thresholding. Defect segmentation is accomplished by applying Sauvola's binarization technique [12]. This is preferred over Otsu's method [1, 13] for retaining local characteristics as discussed in [14].

- A window of size N x N is chosen where N depends on the area of the fruit and the extent of localization sought. While choosing a smaller N guarantees better localization, the complexity introduce into the system would affect its throughput.
- The window is then slid across different points in the image D(i,j). Now local mean 'm(i,j)' and standard deviation 'σ(i,j)' are obtained for each pixel.

$$t(i,j) = m(i,j)\left[1 + k\left(\frac{\sigma(i,j)}{R} - 1\right)\right].$$ (8)

where, R=128 for an 8-bit representation and $k \in [0.2, 0.5]$.

$$B(i,j) = \begin{cases} 0, \text{if } D(i,j) < t(i,j) \\ 255, \qquad \text{otherwise} \end{cases}.$$ (9)

- Now, local threshold 't' is calculated for each window by applying (8) for all three colour components and the final image B containing the defect patches is obtained using (9).

4 Results

A Database of 160 colour images of apples, 110 of which had mechanical defects of some sort was used to test the algorithm offline. The images had a resolution of 120 x 120. A 2.4 GHz Pentium 4 machine supporting Windows Platform was used to run the tests. A standard reference was used to test all fruits. A window size of 11 x 11 was used for the 2D LMS adaptive filter in the pre-processing stage. The value of N and k for Sauvola's technique were taken as 14 and 0.34 respectively for optimum performance [15]. The results were compared with Otsu's adaptive threshold algorithm, with and without the proposed pre-processing routine. Based on the segmented defect patch, fruits were classified as either 'acceptable' or 'defective'. As shown in Table 1, superior performance in terms of accuracy was recorded for the proposed

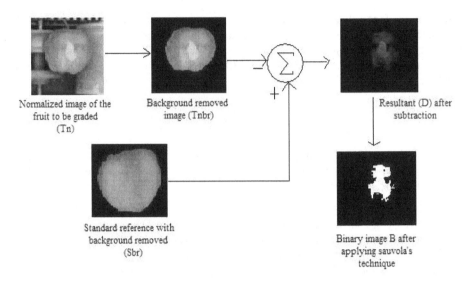

Fig. 2. An Illustration depicting the steps involved in defect segmentation

Table 1. Comparison of results of various algorithms

Test Fruit Quality	Output Grade	Algorithm used			
		Otsu's adaptive threshold without pre-processing (A)	Otsu's adaptive threshold with pre-processing (B)	Sauvola's adaptive threshold without pre-processing (C)	Sauvola's adaptive threshold with pre-processing (Proposed)
Healthy (50 nos.)	Acceptable	29	39	31	42
	Defective	21	11	19	8
Defective (110 nos.)	Acceptable	26	19	20	17
	Defective	84	91	90	93
Percentage Accuracy (%)		70.625	81.25	75.625	84.375

method. Moreover, this technique was found to eliminate false segmentations. This is reflected in the number of instances of healthy fruits being misclassified as defective, which was found to reduce drastically. An analysis of the results indicates that this improvement can be attributed to the proposed pre-processing technique which accomplishes 'active' normalization based on 2D LMS filters as against the conventional 'soft' or 'passive' normalization techniques.

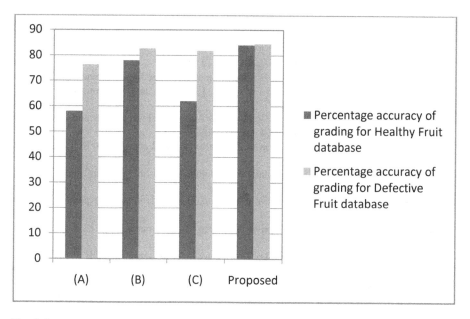

Fig. 3. Bar graph illustrating the relative accuracy of different algorithms for healthy and defective fruits. (A) - Otsu's Adaptive Thresholding without pre-processing, (B) – Pre-processing followed by Otsu's Threshold method, (C) – Sauvola's Threshold method without pre-processing.

5 Conclusion

A reliable method for defect segmentation has been introduced. An effective pre-processing technique equipped to handle variable illumination has also been suggested. These two techniques can be used independent of each other and the choice of using either one or both is left to the user depending on the practical set-up in place. Although the reported success rate is lower than the industry standard, the proposed technique offers scope for improving upon the existing accuracy rates. The complexity issues have been left unaddressed and will be treated in future.

References

1. Li, Q., Wang, M., Gu, W.: Computer vision based system for apple surface defect detection. Computers and Electronics in Agriculture 36(2-3), 215–223 (2002)
2. Yang, Q.: Approach to apple surface feature detection by machine vision. Computers and Electronics in Agriculture 11, 249–264 (1994)
3. Yang, Q., Marchant, J.A.: Accurate blemish detection with active contour models. Computers and Electronics in Agriculture 14(1), 77–89 (1996)
4. Leemans, V., Magein, H., Destain, M.F.: Defect segmentation on 'Jonagold' apples using colour vision and a Bayesian classification method. Computers and Electronics in Agriculture 23(1), 43–53 (1999)

5. Molto, E.: Multispectral inspection of citrus in real-time using machine vision. Computers and Electronics in Agriculture 33 (2002)
6. Chen, Y.-R., Chao, K., Kim, M.S.: Machine vision technology for agricultural applications. Computers and Electronics in Agriculture 12 (2002)
7. Ariana, D., Guyer, D.E., Shrestha, B.: Integrating multispectral reflectance and fluorescence imaging for defect detection on apples. Computers and Electronics in Agriculture 50, 148–161 (2006)
8. Iqbal, M., Sudhakar Rao, P.: Mechanical systems for on-line Fruit Sorting and Grading using Machine Vision Technology. ISOI Journal 34(3)
9. Huang, C.: Region based illumination-normalization method and system. ROC patent and United States Patent #7263241
10. Huang, C.: Novel illumination-normalization method based on region information. SPIE proceedings series (2005)
11. Pei, S.C., Lin, C.Y., Tseng, C.C.: Two-dimensional LMS adaptive linear phase filters. In: ISCAS proceedings (1993)
12. Sauvola, J., Pietikäinen, M.: Adaptive document image binarization. Pattern Recognition 33(2), 225–236 (2000)
13. Otsu, N.: A threshold selection method from gray-level histograms. IEEE SMCG, 62–66 (1979)
14. Shafait, F., Keysers, D., Breuel, T.M.: Efficient implementation of local adaptive thresholding techniques using integral images. In: SPIE proceedings, vol. 6815 (2008)
15. Badekas, E., Papamarkos, N.: Automatic evaluation and document binarization results. In: CGIM 2008. 10th Iberoamerican congress on pattern recognition, pp. 1005–1014 (2005)

Selection of Accurate and Robust Classification Model for Binary Classification Problems

Muhammad A. Khan, Zahoor Jan, M. Ishtiaq, M. Asif Khan,
and Anwar M. Mirza

Department of Computer Science,
FAST-National University of Computer & Emerging Science
A.K Brohi road, H-11/4, Islamabad, 44000, Pakistan
{aamir.khan,zahoor.jan,m.ishtiaq,muhammadasif.khan,
anwar.m.mirza}@nu.edu.pk

Abstract. In this paper we aim to investigate the trade off in selection of an accurate, robust and cost-effective classification model for binary classification problem. With empirical observation we present the evaluation of one-class and two-class classification model. We have experimented with four two-class and one-class classifier models on five UCI datasets. We have evaluated the classification models with Receiver Operating Curve (ROC), Cross validation Error and pair-wise measure Q statistics. Our finding is that in the presence of large amount of relevant training data the two-class classifiers perform better than one-class classifiers for binary classification problem. It is due to the ability of the two class classifier to use negative data samples in its decision. In scenarios when sufficient training data is not available the one-class classification model performs better.

Keywords: One Class classifier, Two-Class Classifier, Binary classification, classification model, ROC, Evaluation, Q- Statistics.

1 Introduction

In Pattern recognition literature and practice binary classification is a well known classification problem. Over the years researcher have proposed solutions for this classification problem. In binary classification problem we have objects belonging to two categories or groups and a corresponding category or group for a new previously unseen pattern has to be determined. The search for classification model that is robust and accurate for binary classification is eminent because of its application in the machine learning field. To asses, analyze and compare classification models from the literature and know their merits and demerits are very necessary. In many classification problems the application of a model is not a matter of choice but a result of detail study of the model and the pertinence of the model to be applied. [9]

As far our study goes no classification model in the literature can be said as the best classification model available for all problems. In reality for classification

D. Ślęzak et al. (Eds.): SIP 2009, CCIS 61, pp. 161–168, 2009.

problems we have to limit ourselves to a tradeoff. For example, achieving 100% accuracy and avoiding the chance of over fitting and over training of the classification model; is a scenario where we compromise on one of the parameter. The research question that is tried to answer in this paper is 1. In which cases either of the classification models (multi-class or one-class) is better? 2. What is the plausible explanation of the better performance of the multi-class classifier on the dataset? 3. Evaluating the models on different metrics to have a fair and objective comparison [15].

In this paper we have thouroghly investigated and studied the available classifier model. Mainly there are two classification models namely (1) multi-class classifier model and (2) one-class classifier models [14]. In many cases, a one-class classifier is used in preference to a multi-class classifier because it may be difficult to use non-target data in training or only data of a single category is available. Some examples of the use of one-class classification are password hardening [3], typist recognition [1] and authorship verification [4]. One class classification is referred to as outlier detection because it classifies the given with respect to its training data.

In section 2 the literature review of one-class and two class classification model is presented. Experimental setup is described in section 3. Results and discussion of the empirical observation are given in section 4. The paper is concluded in section 5.

2 Literature Review

In this section, a brief description is presented on one-class and two class classifiers.

2.1 One Class Classifier

In one class classification problem the description of a particular class called as target class is learnt. New patterns are classified according to that learnt description of the target class and assigned the label of the target class if belong to the target class and the patterns that don't belong to the target class as labeled as outliers [12] [13]. In the training set patterns from the target class are presented only.

Mathematically a one class classifier is expressed as

$$h(x) = \begin{cases} target & if \ f(x) \leq \theta \\ outlier & if \ f(x) > \theta \end{cases} \tag{1}$$

Where θ is a threshold calculated according to the maximum allowable error on target class defined.

For example, the gender classification problem is a typical binary classification problem, assume for training the female subjects data is available, a one class classifier x will be trained on the female subjects data and the female class will be the target class of this classifier. The new incoming patterns in the testing phase will be classified on the basis of the description learned from target class. The new patterns will be assigned the label of the female class (target class) or they will be classified as outlier (any pattern not belonging to the target class will be assigned the label outlier).

2.2 Two Class Classifier

In two class classification problem the classifier learns from the training patterns of both the classes. The boundary of the two the classes is constructed from the description extracted from the training patterns of the two classes. In contrast the one class classifier constructs the boundary of one class because it has no input from the outlier class.

3 Experiments

3.1 Experimental Setup

We have used four dataset from the UCI data set repository and one from the Stanford University Medical Student Dataset. The two-class classifiers are used from the PRTools [6] and the one-class classifiers are used from the ddTools [7] software packages. All the experiments are coded in MATLAB 7.3 on Intel core 2 duo machine with 1 GB of RAM.

3.2 Data Sets

The dataset were used in a ratio of 65% for Training and the remaining 35% for testing. In the WPBC four records contains missing values for certain attributes. All the four records were eliminated from the dataset. The Data sets used in the experiments are listed in Table no.1.

Table 1. This table list the datasets used in the experiment

Dataset	Classes	Instances	Dimension
SUMS	2	400	25
WDBC	2	569	30
WPBC	2	198	32
Ionosphere	2	351	34
German Credit Data	2	1000	34

3.3 Multi Class Classifier

We have used four different multi class classifiers listed in table no. 2. Classifiers are of diverse nature and suited for binary classification. All the classifiers are briefly described below.

KNN is a well known classifier it takes into account the distance of the neighboring objects. Decision tree is a well known two class classifier well suited for binary classification problem. Linear perceptron is a neural network with proved classification performance. Support Vector Machine (SVM) is a maximum margin classifier and regarded as one of the best binary classifier in literature. SVM with RBF kernel is used in our experiments.

Table 2. List of Multi-class Classifier used in the experimetns

Classifier Name	Description
K- Nearest Neighbors	With 3 nearest neighbors
QDC	Quadratic bayes normal classifier
Linear perceptron	A Neural Network
Support Vector Machine	With RBF Kernel

3.4 One Class Classifier

The one-class classifiers used in the experiment are described one by one.

The Gaussian Data descriptor classifier models the target class as a Gaussian distribution based on the mahalanbolis distance [7]. Equation no.2 represents a Gaussian data descriptor

$$f(x) = (x - \mu)^T \sum {}^{-1}(x - \mu) \tag{2}$$

Where μ is mean and $\sum$ is the covariance matrix.

Kmeans classifies the data into K clusters by means of standard K-means clustering procedure [5] [7]. The average distance of data objects $\mathbf{x}$ to the cluster center $\mathbf{c}$ is minimized. The target class is defined as

$$f(x) = \min(\mathbf{x} - \mathbf{c_i})^2 \tag{3}$$

Table 3. List of One Class Classifier

Classifier Name	Description
Gaussian Data Description	Based on Gaussian distribution
K means	Standard K means clustering
SVDD	Support vector data description
Knndd	One class knn classifier

Support vector data descriptor (SVDD) [7, 8 and 14] fits a hyper sphere around the target class. The hyper sphere is optimized using variety of kernels. In knndd the distance to kth nearest neighbor is calculated and the minimum is selected.

4 Results and Discussion

The fundamental difference between one class and two class classifier is the utilization of the negative data instances in training by the two class classifier function. The one class classifier approach has the advantage of using a smaller training set, less space and lesser training time. In some problems there exist a large amount of known data and it is not desirable to use all the data in training or we may not even know the relevant data in such problems only the data of the class to discover is used. Following are the evaluation results of the classifiers presented with respect to the measure used.

4.1 Receiver Operating Curve (ROC)

The Receiver Operating Curve (ROC) performance of one class and two class classifier is presented on all of the five data sets. ROC is a preferred technique of comparing two classifiers. The ROC with the larger area under the convex hull the better the classifier is.

Detailed and through observation reveals that the multi class classifier exhibits better ROC on all the datasets as compare to the one class classifier. Two-class classifier takes into account the 'negative data' or the data about the *'non target'* class which make it able to classify the objects with a broader knowledge about both the classes [13]. The ROC of some of the classifier on the six dataset is presented in Fig no. 1.

4.3 Q Statistics (Pair Wise Measure)

For two classifiers D_i and D_k the Yule's Q-Statistics [2] is expressed as

$$Q_{i,k} = \frac{ad - bc}{ad + bc} \tag{5}$$

Where a is the probability of both classification models being correct, d is the probabilty of both the classifier being incorrect, b is the probability that first classifier is correct and secong is incorecnt, c is the probability second classifier is correct and first is incorrect. Value of Q varies between -1 and 1. Classifier that tends to classifiy the same patterns correcly will have a value of Q > o and the classifiers that gives erroneous results on different patterns will have balue of Q <0. The average value of Q is calculated as shown in [2]. For a pair of classifiers the average Q values is calcualted as follows

$$Q_{av} = \frac{2}{L(L-1)} \sum_{i=1}^{L-1} \sum_{k=i+1}^{L} Q_{i,k} \tag{6}$$

Table 4. List of Q-Statistics value for the two-class classifiers

Classifier Model	WPBC	WDBC	German Credit	SUMS	IONOSPHERE
KNN	0.5042	0.9824	0.5810	0.8271	0.9863
QDC	0.6979	0.9909	0.5589	0.8692	0.9933
Perceptron	0.7636	0.9903	0.6406	0.8066	0.9944
SVM	1	1	0.6157	0.9180	0.9792

Table 5. List of Q-Statics value for the one-class classifier

Classifier Model	WPBC	WDBC	German Credit	SUMS	IONOSPHERE
Gauss_dd	0.8255	0.9857	0.8887	0.9670	1
kmeans	0.8689	0.9874	0.9175	0.9619	0.9986
svdd	1	1	1	0.9786	0.9986
knnd	0.9565	0.9983	0.9175	0.9694	1

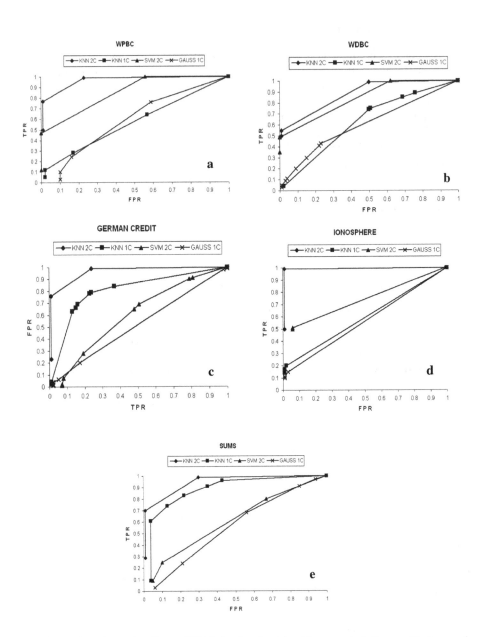

Fig. 1. Receiver Operating Curve (ROC) for the six dataset is presented. Two multi-class and two one-class classifiers are compared. (a) Wisconsin Prognosis Breast Cancer (WPBC) dataset (b) Wisconsin Diagnosis Breast Cancer (WDBC) dataset (c) German Credit dataset (d) Ionosphere (e) Stanford Medical University Students Images Data (SUMS) dataset.

4.4   Cross Validation Error

Cross validation error are reported for both the classification models in table 6 and table 7.

Table 6. Cross validation Error for the two-class classifiers

Classifier Model	WPBC	WDBC	German Credit	SUMS	IONOSPHERE
KNN	0.2319	0.0738	0.303	0.4475	0.135
QDC	0.2319	0.0456	0.231	0.4350	0.132
Perceptron	0.2938	0.0281	0.293	0.4675	0.144
SVM	0.2268	0.0492	0.231	0.4750	0.125

Table 7. Cross validation Error for the one-class classifiers

Classifier Model	WPBC	WDBC	German Credit	SUMS	IONOSPHERE
Gauss_dd	0.1090	0.5014	0.2400	0.3750	0.7607
kmeans	0.6539	0.4114	0.2500	0.4750	0.7607
svdd	0.1186	0.2839	0.4200	0.5000	0.7009
knnd	0.7084	0.4836	0.2100	0.4250	0.6709

5   Conclusion

In this paper we have presented the lessons learned from the application of two-class and one-class classification model to binary classification problem. One-class classification model is good at outlier detection and in scenarios when only the training data of the target class is available. Two-class classification model are more versatile and they construct the class boundaries using the information from the training data of both the classes. Training data presented to the two class model has instances from both the classes. Overall the performance of the two class classification model was better than one class model on the six datasets. The plausible explanation is that due to the knowledge of both the classes the two-class model achieves better performance but in case of unavailability of sufficient data the one class model is preferred.

References

1. Nisenson, M., Yariv, I., El-Yaniv, R., Meir, R.: Towards behaviometric security systems: Learning to identify a typist. In: Lavrač, N., Gamberger, D., Todorovski, L., Blockeel, H. (eds.) PKDD 2003. LNCS (LNAI), vol. 2838, pp. 363–374. Springer, Heidelberg (2003)
2. Kuncheva, L.I., Whitaker, C.J.: Measures of diversity in classifier ensembles and their relationship with the ensemble accuracy. Machine Learning 51, 181–207 (2003)
3. Yu, E., Cho, S.: Novelty detection approach for keystroke dynamics identity verification. In: Proceedings of the 4th International Conference on Intelligent Data Engineering and Automated Learning, Berlin, Germany, pp. 1016–1023. Springer, Heidelberg (2003)

4. Koppel, M., Schler, J.: Authorship verification as a one-class classification problem. In: Proceedings of the 21st International Conference on Machine Learning, pp. 489–495. ACM Press, New York (2004)
5. Bishop, C.: Neural Networks for Pattern Recognition. Oxford University Press, Oxford (1995)
6. Pattern Recognition Toolbox (PRTOOLS) for Matlab implemented by R.W.P.Duin
7. One Class classification toolbox (ddtools) for Matlab implemented by D.M.J Tax
8. Tax, D.M.J., Duin, R.W.P.: Support Vector Domain Descriptor. Pattern Recognition Letters 20(11-12), 1191–1199 (1999)
9. Duda, R.O., Hart, P.E., Stork, D.G.: Pattern classification, 2nd edn. John Wiley and Sons, New York (2001)
10. Webb, Statistical Pattern Recognition. John Wiley & Sons, New York (2002)
11. Liu, C., Wechsler, H.: Robust Coding Schemes for Indexing and Retrieval from Large Face Databases. IEEE Transactions on Image Processing 9(1), 132–136 (2000)
12. Tax, D.M.J., Duin, R.P.W.: Combining One-Class Classifiers. In: Kittler, J., Roli, F. (eds.) MCS 2001. LNCS, vol. 2096, p. 299. Springer, Heidelberg (2001)
13. Tax, D.M.J.: One-Class Classification, Concept Learning in the Absence of Counter Examples. Ph.D. Thesis, Delft University of Technology, Delft, Netherland (2001)
14. Tax, D.M.J., Duin, R.P.W.: Support Vector Data Description. Machine Learning 54(1), 45–66 (2004)
15. Kuncheva, L.I.: Combining Pattern Classifiers: Methods and Algorithms. Wiley, Chichester (2004)

Robust Edge-Enhanced Fragment Based Normalized Correlation Tracking in Cluttered and Occluded Imagery

Muhammad Imran Khan[1], Javed Ahmed[1], Ahmad Ali[2], and Asif Masood[1]

[1] Department of Computer Science, NUST Military College of Signals, Rawalpindi, Pakistan
{mimran,javed,asifmasood}@mcs.edu.pk
[2] Pakistan Institute of Engineering and Applied Sciences, Islamabad, Pakistan
ahmadali1655 @hotmail.com

Abstract. Correlation trackers are in use for the past four decades. Edge based correlation tracking algorithms have proved their strength for long term tracking, but these algorithms suffer from two major problems: clutter and slow occlusion. Thus, there is a requirement to improve the confidence measure regarding target and non-target object. In order to solve these problems, we present an "Edge Enhanced Fragment Based Normalized Correlation (EEFNC)" algorithm, in which we: (1) divide the target template into nine non-overlapping fragments after edge-enhancement, (2) correlate each fragment with the corresponding fragment of the template-size section in the search region, and (3) achieve the final similarity measure by averaging the correlation values obtained for every fragment. A fragment level template updating method is also proposed to make the template adaptive to the variation in the shape and appearance of the object in motion. We provide the experimental results which show that the proposed technique outperforms the recent Edge-Enhanced Normalized Correlation (EENC) tracking algorithm in occlusion and clutter.

Keywords: fragment, correlation, template, occlusion, clutter, Kalman filter.

1 Introduction

A visual tracking system automatically finds the location of target in the consecutive frames of a video. This task becomes difficult when the target is changing its orientation, shape and size. The presence of clutter (i.e. other objects near the target) and occlusion (i.e. other objects in front of the target) in tracking environment makes the problem even more difficult. Computational cost of the algorithm is also important for real time tracking applications.

While performing tracking in an environment where abrupt changes in the background are not expected, modeling of background is normally a preferred approach. For this purpose, Gaussians Mixture Model [4] technique is very successful but this technique has limited capabilities when used alone. Tracking of objects, when the background is not static and changing dynamically, becomes more challenging. In this situation, we can not develop model of the background as new objects quickly become part of the background and then disappear. Furthermore, the moving target

D. Ślęzak et al. (Eds.): SIP 2009, CCIS 61, pp. 169–176, 2009.
© Springer-Verlag Berlin Heidelberg 2009

can change its orientation which is an additional problem. Lucas Kanade tracking algorithm [2] uses different feature points of object to be tracked in the next frame, but dynamic selection and then tracking of these feature points in real-world scenarios is very difficult, especially in case of illumination variation.

The histogram matching based trackers [10, 11, 12] can also work without requiring background model, but they suffer from the inherent problem with the histogram that two different images can have similar histograms because the histogram does not preserve the pixel location. To some extent, this problem has been addressed in [7] by dividing the object template into multiple non-overlapping fragments and using the histograms of those fragments in the matching process, but the same problem with the histograms can occur in the fragment level.

Since the correlation [1, 3] process does not lose the spatial information of the pixels, they are more robust to clutter than the histogram based trackers. Edge Enhanced Normalized Correlation (EENC) tracker [5] has significantly solved the real-world problems of orientation, illumination, obscuration, intermittent occlusion, complex object motion, object fading, and noise. In EENC, the template and the search image are edge-enhanced before performing normalized correlation between them. The best match of the template in the search image is found at the location corresponding to the location of the peak value in the correlation surface (matrix). The best match region and the current template are then linearly combined to prepare the new template to be searched for in the next frame. This technique of template updating plays a vital role in long term tracking as it caters for the variation in the object shape, appearance, and orientation. The EENC tracker also handles the intermittent fast occurring occlusion using Kalman filter [6], but it fails in case of slow occurring occlusion and strong clutter. In order to overcome these issues, we propose Edge-Enhanced Fragment Based Normalized Correlation (EEFNC), in which we divide the template into nine non-overlapping fragments. Then, we correlate every fragment of template with the corresponding fragment of the template-size section in the search region of the video frame. Furthermore, in order to address the varying appearance, shape, and orientation of the object and reject the effect of clutter and occlusion further, we update every fragment of the template independently.

The next section discusses in detail the proposed EEFNC tracking framework. Section 3 presents the experimental results. Finally, the conclusion is drawn in Section 4.

2   EEFNC Tracking Framework

The proposed Edge-Enhanced Fragment Based Normalized Correlation (EEFNC) tracking framework consists of edge-enhancement, fragment based normalized correlation, fragment level template updating, and Kalman predictor.

Edge Enhancement. The most commonly used similarity measure is normalized correlation coefficient (NCC), when the images to be correlated are gray-level images. However, it has been reported in [5] that normalized correlation (NC) is more robust than NCC, when the images to be correlated are edge-enhanced images. Therefore, we use the latter technique. The edge-enhancement procedure comprises: (1) RGB to gray level conversion as in [9] for reducing computational cost without significantly affecting the tracking performance, (2) Gaussian smoothing with adaptive standard deviation parameter as in [5] for attenuating noise without introducing noticeable blur, (3) gradient magnitude computation using Sobel edge detector masks in x and y directions as in [5], and (4) normalization to stretch the pixel values in the gradient images in the whole range of 0 to 255 as in [5], so the object may stand out in low contrast imagery.

Fragment Based Normalized Correlation. As a result of conventional normalized correlation, a correlation surface is developed which provides matching values between template t, and search image s, when the template is placed at every pixel of search image as [5]:

$$C(m,n) = \frac{\sum_{i=0}^{K-1}\sum_{j=0}^{L-1}s(m+i,n+j)\,t(i,j)}{\sqrt{\sum_{i=0}^{K-1}\sum_{j=0}^{L-1}s^2(m+i,n+j)}\sqrt{\sum_{i=0}^{K-1}\sum_{j=0}^{L-1}t^2(i,j)}}, \tag{1}$$

where $C(m,n)$ is an element of correlation surface (matrix) at row m and column n, where $m = 0, 1, 2,...M\text{-}K$, $n = 0, 1, 2,...N\text{-}L$, and K and L are the height and width of the template, respectively.

In order to make the edge-enhanced correlation tracker more robust to strong clutter and slow occlusion, we propose to divide the edge-enhanced template and template-size patch in the search region into nine non-overlapping fragments, $F_{(a,b)}$, where $a = 0, 1, 2$ and $b = 0, 1, 2$, as depicted in Fig. 1.

Then, we propose to correlate every fragment of the edge-enhanced template with the corresponding fragment of the current template-size patch in the edge-enhanced search image, and compute the average value of all nine correlation results to get the final correlation value at the position (m, n) in the search image. Mathematically, the fragment based normalized correlation can be formulated as in (2):

$F_{(0,0)}$	$F_{(0,1)}$	$F_{(0,2)}$
$F_{(1,0)}$	$F_{(1,1)}$	$F_{(1,2)}$
$F_{(2,0)}$	$F_{(2,1)}$	$F_{(2,2)}$

Fig. 1. Nine non overlapping fragments, $F_{(a,b)}$ with 2D indexing

$$C(m,n) = \frac{\sum_{a=0}^{2}\sum_{b=0}^{2}C_{a,b}(m,n)}{9}, \tag{2}$$

where $a = 0, 1, 2$, $b = 0, 1, 2$, and $C_{a,b}$ is the correlation value corresponding to fragment $F_{(a,b)}$ computed as in (3), where h_a and w_b are the height and width, respectively, of the fragment at (a, b), and the sign '$\wedge$' represents logical AND operation. After obtaining the correlation surface, $C(m, n)$, we get the best-match location in the search image by finding the (m^*, n^*) position of the peak value, c_{max}, in $C(m, n)$. The basic difference between EENC and EEFNC is that of normalization technique used in the correlation. In EEFNC, the normalization effect is local to each fragment, thus producing better results than EENC in cluttered imagery. Furthermore, by dividing the template into fragments, the effects of occlusion also become local to each fragment and fragments that are affected from occlusion could be separated from non affected fragments. Therefore, instead of updating the whole template, fragment level updating (discussed in the next sub-section) supports in occlusion handling.

$$C_{a,b}(m,n) = \begin{cases} \dfrac{\sum\limits_{i=0}^{h_g-1}\sum\limits_{j=0}^{w_b-1}s(m+i,n+j)\,t(i,j)}{\sqrt{\sum\limits_{i=0}^{h_g-1}\sum\limits_{j=0}^{w_b-1}s^2(m+i,n+j)}\sqrt{\sum\limits_{i=0}^{h_g-1}\sum\limits_{j=0}^{w_b-1}t^2(i,j)}}, & if\,(a=0)\wedge(b=0) \\[4ex] \dfrac{\sum\limits_{i=0}^{h_g-1}\sum\limits_{j=0}^{w_b-1}s(m+i,n+j+bw_{b-1})\,t(i,j+bw_{b-1})}{\sqrt{\sum\limits_{i=0}^{h_g-1}\sum\limits_{j=0}^{w_b-1}s^2(m+i,n+j+bw_{b-1})}\sqrt{\sum\limits_{i=0}^{h_g-1}\sum\limits_{j=0}^{w_b-1}t^2(i,j+bw_{b-1})}}, & if\,(a=0)\wedge(b>0) \\[4ex] \dfrac{\sum\limits_{i=0}^{h_g-1}\sum\limits_{j=0}^{w_b-1}s(m+i+ah_{a-1},n+j)\,t(i+ah_{a-1},j)}{\sqrt{\sum\limits_{i=0}^{h_g-1}\sum\limits_{j=0}^{w_b-1}s^2(m+i+ah_{a-1},n+j)}\sqrt{\sum\limits_{i=0}^{h_g-1}\sum\limits_{j=0}^{w_b-1}t^2(i+ah_{a-1},j)}}, & if\,(a>0)\wedge(b=0) \\[4ex] \dfrac{\sum\limits_{i=0}^{h_g-1}\sum\limits_{j=0}^{w_b-1}s(m+i+ah_{a-1},n+j+bw_{b-1})\,t(i+ah_{a-1},j+bw_{b-1})}{\sqrt{\sum\limits_{i=0}^{h_g-1}\sum\limits_{j=0}^{w_b-1}s^2(m+i+ah_{a-1},n+j+bw_{b-1})}\sqrt{\sum\limits_{i=0}^{h_g-1}\sum\limits_{j=0}^{w_b-1}t^2(i+ah_{a-1},j+bw_{b-1})}}, & otherwise \end{cases} \quad (3)$$

Fragment Level Template Updating. In order to make the template adaptive to the variation in the object shape and appearance in the real world scenarios, we must update the template. In [5], the template is updated as:

$$t[n+1] = \begin{cases} \lambda c_{max}b[n]+(1-\lambda c_{max})t[n] & if\ c_{max}>\tau_t \\ t[n] & otherwise \end{cases}, \quad (4)$$

where $t[n]$ is the current template, $t[n+1]$ is the updated template for next iteration, $b[n]$ is the current best match section, and C_{max} is the peak value in the correlation surface. The value of λ is 0.16, and τ_t is the threshold of which value is 0.84.

We propose to update the fragments independently instead of updating the whole template, as:

$$F_{(a,b)}[n+1] = \begin{cases} \lambda f_{(a,b)max}b_{(a,b)}[n]+(1-\lambda f_{(a,b)max})F_{(a,b)}[n], & if\ (f_{(a,b)max}>\tau_t)\wedge(c_{max}>\tau_t) \\ F_{(a,b)}[n], & otherwise \end{cases}, \quad (5)$$

where $F_{(a,b)}[n]$ is fragment at (a,b) position of the current template, $b_{(a,b)}[n]$ is the fragment at (a,b) position of the current best-match, $f_{(a,b)max}$ is the correlation value between $F_{(a,b)}[n]$ and $b_{(a,b)}[n]$, and $F_{(a,b)}[n+1]$ is the fragment at (a,b) position of the updated template to be used in the next iteration.

There are two differences between (4) and (5). Firstly, in (4) the updating is performed at template level, while in (5) the updating is performed at fragment level. Secondly, in (4) the template is updated if the best-match correlation value C_{max} is greater than threshold, τ_t, but in (5) the fragment level correlation value is also considered. This way the fragment containing major portion of the target is updated with higher weight while the fragment containing short-term background clutter or occluding object is updated with lower weight (or even not updated). For better understanding, consider a situation when a slow moving object is occluding the target. In this situation, C_{max} will be higher than the threshold τ_t and the whole template will be allowed to update in case of EENC (in which case the occluded object will become part of the template, resulting in target loss later). However, in fragment level updating, the fragments that have been occluded will not be updated, because the fragment level correlation value is dropped below τ_t for the occluded fragments.

Kalman Predictor. EENC [5] has been strengthened by the use of Kalman predictor [6]. When the process of normalized correlation produces the peak value below τ_t, the target coordinates estimated by it are disregarded the target coordinates predicted by the Kalman filter in the previous iteration are utilized in the current iteration, and the process of template updating is bypassed. This technique provides support in case of occlusion. This advantage of Kalman predictor has also been exploited in the proposed EEFNC. We have used constant acceleration with random walk model with six states: position, velocity and acceleration in x and y directions. Furthermore, the position and the dimensions of the search window for the next iteration are also dynamically updated using the predicted position and its error, as in [5], to reduce the computational complexity and cater for object maneuvering with variable velocity.

3 Experimental Results

In this section, we compare EENC with EEFNC using different publicly available standard image sequences. We will further analyze the behavior of both the algorithms in the presence of clutter and occlusion using post regression analysis technique [14], which compares the calculated target coordinates with the ground truth target coordinates and provides three parameters m (regression slope), b (regression Y-intercept), and R (regression correlation coefficient). The ideal values of these parameters (when the calculated and the ground truth coordinates match perfectly) are $m = 1.0$, $b = 0.0$, and $R = 1.0$.

Frame 92   Frame285   Frame456

Fig. 2. The first row presents the results of EENC and the second row presents the results of EEFNC. Both algorithms have visually performed well.

The first experiment is performed on a publicly available *Walking Woman* image sequence [7]. The tracking results from both the algorithms are visually almost same, as shown in Fig. 2. However, the accuracy of the target coordinates provided by the EEFNC is better than that of the target coordinates provided by the EENC, as illustrated in Table 1. In the second experiment we test the trackers on *Three Men Crossing* sequence from AV16.3 v6 dataset[13], EEFNC has survived the occlusion and clutter, and has provided much longer tracking than EENC, as shown in Figs. 3 and 4. In Fig. 3, the target (face of the person with white shirt) is being tracked by EEFNC successfully even during occlusion in Frame 451 and Frame 587. When an occlusion event is sensed automatically because the peak correlation value is dropped below the threshold, the color of the overlaying text and reticule is changed to golden from yellow for demonstration purpose. The occlusion is then handled using Kalman filter and the proposed fragment level template updating method. Once the object comes into view again, the tracking is resumed in normal mode. Moreover, in Frame 763, the target is passing through background clutter; even then the tracking is not disturbed. If the tracking is performed using EENC along with its Kalman filter and template updating method, the tracking is lost after Frame 587, when the target is partially out of view, as shown in Fig. 4. The regression analysis in Table 1 for Three Men Sequence also illustrates that the EEFNC tracker outperforms the EENC tracker also for this sequence.

Third experiment is performed on *Shop Assistant* sequence from the CAVIAR database [8]. Figure 5 (upper row) illustrates, that while tracking the person with dark shirt, the EENC did not survive the oc-

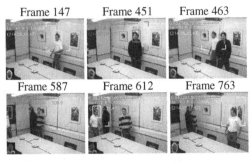

Fig. 3. Tracking results of EEFNC for Three Men Crossing sequence

Fig. 4. Tracking results of EENC for Three Men Crossing sequence

Table 1. Post regression results

Tracker	m	b	R
Walking Woman Sequence			
EENC	0.5825	75.03	0.6882
EEFNC	0.7530	41.78	0.8790
Three Men Crossing Sequence			
EENC	0.6428	43.74	0.6232
EEFNC	0.9710	7.98	0.9759
Shop Assistant Sequence			
EENC	-0.1026	148.98	-0.3732
EEFNC	0.9344	12.72	0.9309
F16 Take-off Sequence			
EENC	0.0514	95.6092	0.0830
EEFNC	0.9955	-10.349	0.9404

clusion produced by the person with white shirt. EEFNC is, however, able to track the target object successfully even during and after the occlusion, as shown in Fig. 5 (lower row). Table 1 also illustrates that the EEFNC performs much better than EENC in terms of tracking accuracy.

The fourth experiment has been performed on F-16 Take-off sequence, which has been used in [5], which proved robustness of EENC in heavily cluttered imagery. The same sequence has been used here to test EEFNC based tracking. It is observed that the robustness of EENC depends on how accurately the template is initialized. Typically, when we selected the template from initial frame and started the tracking session, the EENC tracker was disturbed by the clutter (white roof of small shed), and the track is lost, as shown in the first row in Fig. 6. However, when we selected the template of the same size from the same place in the initial frame, and started the EEFNC tracker, the airplane was tracked successfully throughout the whole image sequence, as shown in the last two rows in Fig. 6. Post regression analysis results of EEFNC are also better than those of EENC as illustrated in Table 1.

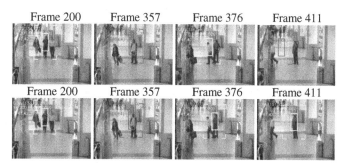

Fig. 5. Tracking result of EENC is shown in the first row and tracking result of EEFNC is shown in the second row

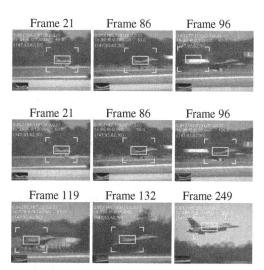

Fig. 6. Tracking result of EENC is shown in first row and tracking result of EEFNC is shown in second row and third row

4 Conclusion

We presented Edge Enhanced Fragment Based Normalized Correlation (EEFNC) algorithm and fragment level template updating method accompanied with Kalman filter to address the problems of strong clutter and slow occlusion that the recent Edge Enhanced Normalized Correlation (EENC) method could not reliably handle. As far as the computational speed is concerned, the EENC works at the speed of about 75 fps [5] when the template size is typically 25×25 pixels. However, the proposed EEFNC for the same size template is about 25 fps, which was achieved when the search was performed using pyramid search technique up to two course levels.

Although 25 fps is enough for the standard PAL cameras, the speed can be further increased using optimization techniques, if the higher frame rate cameras are used.

The concept of BMRA (Best Match Rectangle Adjustment) is presented in [15]. This technique adjusts the template size while minimizing background from the template and improves the tracking performance significantly. Use of BMRA with the proposed EEFNC algorithm can further enhance its performance. Furthermore, the exploitation of color components instead of single gray scale component can further make the EEFNC algorithm more robust to complex situations in which the clutter object look exactly same as the target object even when their color is different.

References

1. Lipton, A.J., Fujiyoshi, H., Patil, R.S.: Moving Target Classification and Tracking from Real-time Video. In: IEEE Workshop on Applications of Computer Vision (1998)
2. Lucas, B., Kanade, T.: An Iterative Image Registration Technique with an Application to Stereo Vision. In: 7th International Joint Conference on Artificial Intelligence (IJCAI), pp. 674–679 (1981)
3. Wong, S.: Advanced Correlation Tracking of Objects in Cluttered Imagery. In: Proceedings of SPIE, vol. 5810 (2005)
4. Stauffer, C., Grimson, W.: Learning Patterns of Activity Using Real Time Tracking. IEEE Transactions on Pattern Analysis and Machine Intelligence 22(8), 747–767 (2002)
5. Ahmed, J., Jafri, M.N., Shah, M., Akbar, M.: Real-Time Edge-Enhanced Dynamic Correlation and Predic-tive Open-Loop Car Following Control for Robust Tracking. Machine Vision and Applications Journal 19(1), 1–25 (2008)
6. Kalman, R.E., Bucy, R.S.: New Results in Linear Filtering and Prediction Theory. Transactions of the ASME - Journal of Basic Engineering 83 (1961)
7. Adam, A., Rivlin, E., Shimshoni, I.: Robust Fragments-based Tracking using the Integral Histogram. In: IEEE Conference on Computer Vision and Pattern Recognition, June 17-22 (2006)
8. Caviar datasets,
   `http://groups.inf.ed.ac.uk/vision/caviar/caviardata1/`
9. Ahmed, J., Jafri, M.N., Ahmad, J., Khan, M.I.: Design and Implementation of a Neural Network for Real-Time Object Tracking. In: International Conference on Machine Vision and Pattern Recognition in Conjunction with 4th World Enformatika Conference, Istanbul (2005)
10. Porikli, F.: Integral histogram a fast way to extract histograms in cartesian spaces. In: IEEE Conference on Computer Vision and Pattern Recognition (2005)
11. Comaniciu, D., Visvanathan, R., Meer, P.: Kernel based object tracking. IEEE Trans. Pattern Anal. Mach. Intell. 25(5), 564–575 (2003)
12. Comaniciu, D., Ramesh, V., Meer, P.: Real-time tracking of non rigid objects using mean shift. In: Proceedings, IEEE Conference on Computer Vision and Pattern Recognition, Hilton Head, vol. 1, pp. 142–149 (2000)
13. Lathoud, G., Odobez, J., Gatica-Perez, D.: AV16.3: An Audio-Visual Corpus for Speaker Localization and Tracking. In: Bengio, S., Bourlard, H. (eds.) MLMI 2004. LNCS, vol. 3361, pp. 182–195. Springer, Heidelberg (2005)
14. MATLAB 7.0 On-line Help Documentation
15. Ahmed, J., Jafri, M.N.: Best-Match Rectangle Adjustment Algorithm for Persistent and Precise Correlation Tracking. In: Proc. IEEE International Conference on Machine Vision, Islamabad, Pakistan, December 28-29 (2007)

Robust and Imperceptible Watermarking of Video Streams for Low Power Devices

Muhammad Ishtiaq[*], M. Arfan Jaffar, Muhammad A. Khan,
Zahoor Jan, and Anwar M. Mirza

Department of Computer Science
National University of Computer & Emerging Sciences
A.K. Brohi Road H-11/4, Islamabad Pakistan
{m.ishtiaq,arfan.jaffar,aamir.khan,zahoor.jan,
anwar.m.mirza}@nu.edu.pk

Abstract. With the advent of internet, every aspect of life is going online. From online working to watching videos, everything is now available on the internet. With the greater business benefits, increased availability and other online business advantages, there is a major challenge of security and ownership of data. Videos downloaded from an online store can easily be shared among non-intended or unauthorized users. Invisible watermarking is used to hide copyright protection information in the videos. The existing methods of watermarking are less robust and imperceptible and also the computational complexity of these methods does not suit low power devices. In this paper, we have proposed a new method to address the problem of robustness and imperceptibility. Experiments have shown that our method has better robustness and imperceptibility as well as our method is computationally efficient than previous approaches in practice. Hence our method can easily be applied on low power devices.

Keywords: Invisible watermarking, Robust watermarking, Steganography, Data Hiding, Multimedia security, Copyright protection.

1 Introduction

Watermarking is a method used to embed information in the host medium. Medium can be any digital photograph, text, audio, software or a video [11]. The main reason to watermark the video is to protect it from illegal or non-copyright use [13],[17-19]. Other uses of watermarking are in field of defence, business, and areas where information needs to be protected at all costs from attackers. Watermarking has seven properties [3]. The main three properties are Imperceptibility, Robustness and Payload or Capacity [12]. Imperceptibility means whether our watermark make significant changes to the host medium, that can be perceived by the human visual system or not. It depends on the strength of the watermark. We should only embed watermark information in the regions of the video frame in which imperceptibility is least affected. Robustness means that our watermark will sustain any intended or unintended attack

---

[*] The work is funded by Higher Education Commission of Pakistan and National University of Computer & Emerging Sciences, Islamabad, Pakistan.

D. Ślęzak et al. (Eds.): SIP 2009, CCIS 61, pp. 177–184, 2009.

on the video that tries to destroy or remove watermark information [5],[6]. Major attacks on the videos are filtering, adding noise, compression, scaling, rotation and cropping. The watermarking method should be robust enough to tolerate these attacks. If we increase robustness, imperceptibility decreases and vice versa. Both robustness and imperceptibility are also affected if we embed the watermark information in overlapping regions of the video frame. Payload means the amount of information that can be successfully embedded and retrieved. Increase in payload affects imperceptibility negatively and if we increase imperceptibility then we can only hide few bits (lower payload). These three requirements of watermarking make a triangle. Improvement in any one of them, affects the other two negatively. We have to find the correct balance between these conflicting requirements of watermarking [4]. Our method embeds the watermark in each video frame. Frame is divided into equal sized segments. Rather than embedding watermark randomly, it is distributed throughout the frame in equal proportion. This helps us achieve better robustness and imperceptibility.

The rest of the paper is organized as in Section 2 a brief overview of other watermarking techniques is given. Our proposed methods is explained in section 3, while section 4 contains discussion on the results obtained which will follow conclusions and future work in section 5.

2 Background

In the past people had used cryptographic techniques to embed some information into the medium and send it through a transmission channel. Cryptographic techniques basically rely on the fact that the information is being carried as it is and the information can only be viewed with the help of some secure encrypted key. The information itself is not changed and anyone with the proper key can gain access. Once the key is lost, all of the information security is compromised.

Comparing this with the techniques where the host medium contains the key. The hidden information remains with the host medium despite some alterations and attacks on the host medium. Large number of methods can be found in the literature based on this approach [14-16].

In [7] the most simple and straightforward method i.e. Least Significant Bit (LSB) embedding method of watermarking is given. As it is obvious from the name, this method embeds the watermark information in the least significant bit of the cover medium. In this way a smaller object can be embedded several times to improve robustness. If a single copy of watermark survives, the method passes the robustness test. LSB method is robust to cropping but lossy compression or addition of noise will completely destroy the watermark. Another disadvantage of this method is that if the embedding algorithm is revealed to attackers, the watermark can easily be modified.

Another watermarking technique which exploits the correlation properties of Pseudo Random Patterns (PRPs) is given in [8]. PRPs are added to cover video frame according to 1.

$$VF_W(x, y) = VF(x, y) + k * W(x, y) \tag{1}$$

Here VF is the video frame to be watermarked, k is the intensity factor, W is the watermark and VF_w is the watermarked video frame. The robustness of the method can be highly increased by taking larger k values at the expense of imperceptibility. At the time of retrieval the same PRPs are generated and a correlation between patterns and the watermarked video frame is computed. If the correlation is greater than a pre-defined threshold then the embedded dada is detected. The major problem of this technique is the selection of the threshold. With higher difference degree between the successive frames of the video the method will require multiple thresholds to success-fully extract the watermark. This method can be improved if we use separate PRPs for '0' and '1'. This way we do not need to determine the threshold and the method can be improved significantly.

Watermarking methods can also be applied in the frequency domain [13]. In these methods the medium is first transformed to frequency domain. Then watermark is inserted in the transformed domain and in the end, the inverse frequency transform is performed to obtain the watermarked medium. The most popular of the frequency domain methods is Discrete Cosine Transform (DCT). DCT divides an image into different frequency bands. This help in embedding watermark in the middle frequency bands. Middle frequencies are chosen rather than low or high frequencies. If we select low frequencies, the imperceptibility affects negatively and in case of high bands the details of the edges and other information is affected. Since watermarks applied in the frequency domain will be distributed in the entire frame in spatial domain upon in-verse transformation, this method is more robust to cropping as compared to spatial domain. But this method is not better then spatial when it comes to imperceptibility. Also frequency domain method is more computational intensive then its spatial counterpart.

Another possible domain used for watermarking is that of wavelet domain. In [8],[19] they have proposed a watermarking method based on wavelets. In this method first we perform wavelet transformation of the host medium. Wavelet trans-form divides the image into 4 quadrants. LL quadrant contains the low resolution approximation image of the host medium, HL contains the horizontal details, LH con-tains the vertical details and HH has diagonal details. LL contains the low frequencies i.e. smoother regions of the image, hence we cannot embed watermark in this quad-rant. HH are the highest frequencies present in the image, if we embed the watermark in this quadrant, major details of the image will be lost. So we select HL and LH for watermarking.

By embedding the watermark into these quadrants we can obtain high impercepti-bility and better robustness as well. For embedding purpose we can use similar kind of technique which is used in DCT based methods. Equation 2 is used for embedding. Here $VF_w(x,y)$ is the watermarked video frame, VF_i is the transformed video frame, α is the scaling factor and x_i is the embedding bit. The disadvantage of wave-let based methods is that they are computationally expensive when compared with spatial and DCT based methods.

$$VF_W(x,y) = \begin{cases} VF_i(x,y) + \alpha * |VF_i| x_i, & x,y \in HL, LH \\ VF_i & x,y \in LL, HH \end{cases} \qquad (2)$$

The DCT and wavelet based methods have acceptable imperceptibility and very good robustness against cropping and noise attack, but due to their computational cost they are not suitable for low power devices.

Spatial domain methods are best when it comes to imperceptibility, if we can improve its robustness, it will better suit low power devices than other watermarking domains. Dittman [2] has given a method of watermarking in spatial domain. If we distribute the watermark in the host medium in spatial domain and add redundant bits to increase robustness, this shows better results than [2].

3 Proposed Method

The proposed technique is designed to embed information in low resolution videos that can be played on mobile devices. In [1] authors insert redundant watermark in the video to improve robustness. Our method needs data (watermark), host medium, strength of the watermark n (the amount by which luminance values of the video are modified) and a stego key from the user to embed the watermark into the medium provided. High value of n increases robustness but decreases imperceptibility and vice versa. Using stego key we generate the same pseudo random patterns at both embedding and detection. The algorithm is simple and can easily run on devices having low processing capabilities.

Proposed Embedding Algorithm

```
n = Watermark strength
BCH encode the input data
For each video frame
  For each bit of data do
    • Generate 8x8 pseudo random pattern of 0 and 1
    • Apply low pass filter on the pattern
    • Replace in pattern 0 with −n and 1with n
    • Divide the frame into equal size blocks.
    • Select a block of the frame to embed pattern
      If embed bit = 1
          Add pattern to selected block of video frame
      Else
            Subtract pattern from the selected block
  End for
End for
```

$$n = \left\lfloor \sqrt{m} \right\rfloor + 1 \tag{3}$$

Here n is the number of divisions on each axis and m is the size (in bits) of the message to embed. According to 3 the frame is divided in the equal size blocks, as shown in fig. 1(a). Then within each block a position is selected randomly to embed the watermark in it. In this way we can embed watermark in non-overlapping regions and the watermark will spread evenly in the whole frame which give more robustness and enhances the imperceptibility of the proposed method. In fig. 1(b) the distribution of watermark across different regions of the video frame is shown, the intensity of the watermark is kept high for better visibility. The flow diagram of watermark embedding process is given in fig. 2.

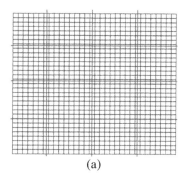
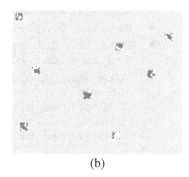

(a)   (b)

Fig. 1. (a) Frame divided into different regions (b) Watermark effect on gray frame, n=100 is used for better visibility

Watermark Extraction

For detection we select the same blocks of each frame using the same user provided stego key. Then with the help of stego key we generate the same pseudo random patterns and apply low pass filter to it. Blocks in which pseudo random watermark was embedded are selected. In the pseudo random pattern replace 0 with -1. Dot product of this pattern and the selected block is computed and all the values are added together. If the result is positive then the detected bit is 1 otherwise it is 0. Figure 3 depicts the watermark extraction process.

4   Test Results

Proposed method is tested on a standard video of the museum [2] and on polyak's [1] video. Polyak's method is based on Dittman`s [2] algorithm. Main two types of tests were conducted on the watermarked video, Imperceptibility tests and robustness tests. To measure imperceptibility, PSNR caused due to the watermarked video was calculated. The comparison with [1] shows that our technique has better imperceptibility. In fig. 4 (a) are the imperceptibility test results. It is clear from the figure that our method has better imperceptibility then [1]. The watermark was embedded with different strengths in the video.

Robustness was checked by adding Gaussian noise of zero mean and 0.2 variance, salt & pepper noise with 0.5 density and applying averaging filter of 7X7. The results are given in fig. 4 (b). The robustness obtained is acceptable for videos and the method is better suited for devices having low computational power.

Proposed Extraction Algorithm

```
For each video frame
  Divide the frame into equal size blocks
  For each block
    • Generate 8x8 pseudo random pattern of 0 and 1
    • Apply low pass filter on the pattern
    • Replace 0 by -1 in the pattern
    • Take the dot product of the pattern and se-
      lected video block.
    • Take Sum of all the values of the product.
      If sum > 0
        Detected_bit = 1
      Else
        Detected_bit = 0
  End for
  BCH decode the detected binary pattern
End for
```

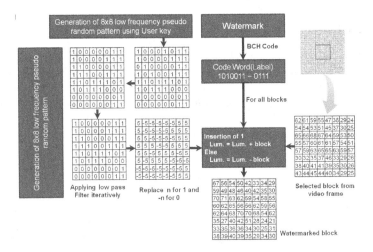

Fig. 2. Watermark embedding process

5   Conclusions and Future Work

DCT and Wavelet based watermarking techniques are good in robustness but due to their computational cost they are not suitable for low power devices. Spatial domain watermarking techniques are imperceptible and computationally efficient and better robustness can be achieved even in spatial domain using correlation and equal distribution of watermark in the host medium. From the results comparison, we can conclude that our technique is more robust and imperceptible at the simultaneously. The algorithm is not complex and hence can easily be used in systems having low computational power. In future the effect of computational intelligence techniques such as genetic programming and others can be investigated to further improve the robustness and imperceptibility.

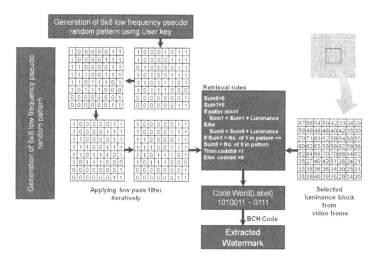

Fig. 3. Watermark extraction process

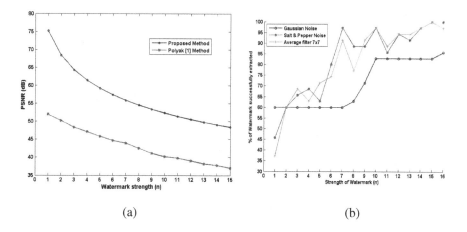

(a)   (b)

Fig. 4. Test results (a) PSNR of video watermarked with different strengths (b) % of watermark successfully extracted after adding Gaussian, salt & pepper noise and average filtering (7x7)

References

[1] Polyak, T.: Robust Watermarking of Video Streams. ACTA Polytechnica 46(4), 49–51 (2006)

[2] Dittmann, J., Stabenau, M., Steinmetz, R.: Robust MPEG video Watermarking Technologies. In: ACM Multimedia, Bristol, UK, August 1998, pp. 71–80 (1998)

[3] Cox, I.J., Miller, M.L., Bloom, J.A., Fridrich, J., Kalker, T.: Digital Watermarking and Steganography, 2nd edn. Morgan Kaufmann, San Francisco (2008)

[4] Dittmann, J., Megias, D., Lang, A., Joancomarti, J.H.: Theoretical framework for a practical evaluation and comparison of audio/video watermarking schemes in the triangle of robustness, transparency and capacity. In: Shi, Y.Q. (ed.) Transactions on Data Hiding and Multimedia Security I. LNCS, vol. 4300, pp. 1–40. Springer, Heidelberg (2006)

[5] Smirnov, M.V.: Holographic approach to embedding hidden watermarks in a photographic image. Journal of Optical Technology 72(6), 464–468 (2005)

[6] Usman, I., Khan, A., Ali, A., Choi, T.-S.: Reversible watermarking based on Intelligent coefficient selection and integer wavelet transform. International Journal of Innovative Computing, Information and Control (IJICIC) 5(12) (December 2009)

[7] Johnson, N.F., Katezenbeisser, S.C.: A survey of steganographic techniques. In: Information Techniques for Steganography and Digital Watermarking, December 1999, pp. 43–75 (1999)

[8] Langelaar, G., Setyawan, I., Lagendijk, R.L.: Watermarking digital image and video data. IEEE Signal Processing Magazine 17, 20–43 (2000)

[9] Hernandez, J.R., Amado, M., Perez-Gonzalez, F.: DCT-Domain watermarking techniques for still images: Detector Performance Analysis and a New Structure. IEEE Trans. Image Processing 9, 55–68 (2000)

[10] Sheomaker, C.: Hidden Bits: A survey of techniques for digital watermarking. In: Independent Survey (July 2002)

[11] Piva, A., Bartolini, F., Barni, M.: Managing copyright in open networks. IEEE Transactions on Internet Computing 6(3), 18–26 (2002)

[12] Cox, I.J., Miller, M.L., Bloom, J.A.: Digital Watermarking and fundamentals. Morgan Kaufmann, San Francisco (2002)

[13] Hernandez, J.R., Amado, M., Perez-Gonzelez, F.: DCT-Domain watermarking techniques for still images: Detector performance analysis and a new structure. IEEE Transactions on image processing 9(1), 55–68 (2000)

[14] Khan, A., Mirza, A.M., Majid, A.: Intelligent perceptual shaping of a digital watermark: Exploiting Characteristics of human visual system. International Journal of Knowledge based and Intelligent Engineering Systems 10(3), 213–223 (2006)

[15] Khan, A., Mirza, A.M., Majid, A.: Automatic Visual Masking of a Digital Watermark Using Genetic Programming. International Journal of Knowledge based and Intelligent Engineering Systems (KES) 9(1), 1–11 (2005)

[16] Liu, Z., Karam, L.J., Watson, A.B.: JPEG2000 Encoding With Perceptual Distortion Control. IEEE Transactions on Image Processing 15(7), 1763–1778 (2006)

[17] Khan, A., Mirza, A.M.: Genetic Perceptual Shaping: Utilizing Cover Image and Conceivable Attack Information Using Genetic Programming. Information Fusion, Elsevier Science 8(4), 354–365 (2007)

[18] Hassanien, A.E.: A Copyright Protection using Watermarking Algorithm. Informatica 17(2), 187–198 (2006)

[19] Yen, E., Tsai, K.-S.: HDWT-based grayscale watermark for copyright protection. An International Journal Source Expert Systems with Applications 35(1-2), 301–306 (2008)

A New Ensemble Scheme for Predicting Human Proteins Subcellular Locations

Abdul Majid[1,2] and Tae-Sun Choi[1]

[1] Department of Mechatronics, Gwangju Institue of Science and Technology
[2] Department of Information and Computer Sciences, Pakistan Institute of Engineering and Applied Sciences
{abdulmajiid,tschoi}@gist.ac.kr

Abstract. Predicting subcellular localizations of human proteins become crucial, when new unknown proteins sequences do not have significant homology to proteins of known subcellular locations. In this paper, we present a novel approach to develop *CE-Hum-PLoc* system. Individual classifiers are created by selecting a fixed learning algorithm from a pool of base learners and then trained by varying feature dimensions of Amphiphilic Pseudo Amino Acid Composition. The output of combined ensemble is obtained by fusing the predictions of individual classifiers. Our approach is based on the utilization of diversity in feature and decision spaces. As a demonstration, the predictive performance was evaluated for a benchmark dataset of 12 human proteins subcellular locations. The overall accuracies reach upto 80.83% and 86.69% in jackknife and independent dataset tests, respectively. Our method has given an improved prediction as compared to existing methods for this dataset. Our *CE-Hum-PLoc* system can also be a used as a useful tool for prediction of other subcellular locations.

Keywords: Subcellular location, ensemble classifier, individual classifier, Amphiphilic Pseudo Amino Acid Composition.

1 Introduction

The function of a protein is closely correlated with its specific location in the cell. It means precise protein function requires to locate properly its subcellular location, otherwise there is danger that protein may lose its function [1]. Information about subcellular location give understanding about their engagement in specific metabolic pathways [2]. The locations of proteins with known function help in understanding its biological function [3] and proteins interaction [4]. Newly synthesized proteins are localized to the appropriate subcellular spaces to perform their biological functions. Therefore, in large-scale genome analysis, demand to develop more accurate and reliable predictor is increasing [5].

In the literature, research related to accurately predict human proteins into various subcellular localizations has gained much importance. Researchers have proposed

D. Ślęzak et al. (Eds.): SIP 2009, CCIS 61, pp. 185–192, 2009.

both individual and fusion of classifier strategies. Early attempts were based on the decision of a single learner. Covariant Discriminant Classifier (CDC) was attempted using different feature extraction techniques [5], [6], [7], [8], [9], [10]. Support Vector Machines (SVM) classifier was tried with Functional Domain Composition [11] features. A SVM based prediction model was developed by constructing new Amino Acid Composition (AAC) distribution features [12]. The prediction of a single classifier is limited due to large variation in length and order of protein sequences. Therefore, researchers have also proposed fusion of classifiers strategies [4], [13], [14], [15]. In this way, fusion of diverse types of classifiers often yield better prediction than the individual ones [16]. An ensemble of CDC classifiers is developed using Amphiphilic Pseudo Amino Acid Composition (PseAAC) [13]. An ensemble of KNN classifiers was built by fusing individual KNN classifiers to develop *Hum-PLoc* system [14]. In this work, individual classifiers are trained on the hybridized features of Gene Ontology and Amphiphilic PseAAC.

The above predicator schemes do not combine classifiers that are individually trained on different feature spaces and at the same time posses different learning mechanisms. In this paper, we propose a novel ensemble approach called *CE-Hum-PLoc*. The main idea of this scheme is based on the utilization of diversity in feature and decision spaces simultaneously. In the work, comparative analysis shows improved prediction accuracy than the existing approaches.

2   Materials and Methods

In this work, four learning mechanisms are selected as base learners: 1) 1-Nearest Neighbor (NN), 2) Probabilistic Neural Network (PNN), 3) SVM and 4) CDC. All learning mechanisms, except SVM, are inherently based on proximity. SVM is a margin based binary classifier that constructs a separation boundary to classify data samples. CDC and NN classifiers are commonly used for predicting protein sequences. The prediction of CDC is found by exploiting the variation in the PseAA features of protein sequence [13]. NN is reported to perform well on classification tasks regarding protein sequences [13], [17], [18]. PNN classifier is based on the Bayes theory to estimates the likelihood of a sample being part of a learned class [19]. The detail of benchmark datasets is provided in Table 1. This dataset was developed by Chou and Elrod [7]. Later on, researchers adopted this dataset to compare the results of their proposed methods. To reduce redundancy and homology bias, this datasets was passed through window screening.

2.1   Proposed Method

Individual ensembles (IEs) are produced by exploiting diversity in feature spaces. Combined ensemble (CE) is then developed by fusing predictions of IEs classifiers. CE classifier is expected to be more effective as compared to IEs classifiers.

Suppose we have N proteins feature vectors (P_1, P_2 . . . P_N) derived from protein dataset. Each P_i belong to one of V classes with labels Q_1, Q_2,..,Q_V. A kth subcellular protein feature vector from a class v can be expressed as:

Table 1. Number of proteins sequences in each subcellular location

Sr. no.	Subcellular locations	Dataset	
		Jackknife test	Independent test
1	Chloroplast	145	112
2	Cytoplasm	571	761
3	Cytoskeletons	34	19
4	Endo. Reticulum	49	106
5	Extracell	224	95
6	Golgi Apparatus	25	4
7	Lysosome	37	31
8	Mitochondria	84	163
9	Nucleus	272	418
10	Peroxisome	27	23
11	Plasma Memb.	699	762
12	Vacuole	24	---
	Total	**2,191**	**2,494**

$$\mathbf{P}_v^k = \left[p_{v,1}^k \ p_{v,2}^k \cdots p_{v,20}^k \cdots p_{v,\Phi}^k \right]^{\mathbf{T}} \tag{1}$$

where $p_{v,1}$, $p_{v,2}$, ..., $p_{v,20}$ are the frequencies of occurrence of 20 amino acid sequences. The elements $p_{v,21}$, $p_{v,22}$, ..., $p_{v,\Phi}$ are the 1st-tier to $(\xi-1)$-tier correlation factors of an amino acid sequence in the protein chain based on two indices of hydrophobicity and hydrophilicity.

In order to develop IEs, first, *PseAA* composition with varying dimensions ranging from 20 to 62 is utilized, i.e. $\Phi=20+2(i-1)$, where $i=1,2,..., \xi$. Here, $\xi=22$, represents the number of *IE* classifiers.

The individual predictions R_i of IE classifiers can be expressed as:

$$\{R_1, R_2, R_3 \cdots, R_\xi\} \in \{Q_1, Q_2, Q_3, \cdots, Q_V\} \tag{2}$$

Now IE based voting mechanism for a protein can be formulated as:

$$Z_j^{IE} = \sum_{i=1}^{\xi} w_i \Delta(R_i, Q_j), \quad j=1,2,...V \tag{3}$$

where w_i represents weight factor. Here, for simplicity, its value is set to unity and function $\Delta(R_i, Q_j)$ is defined as: $\Delta(R_i, Q_j) = \begin{cases} 1, & \text{if } R_i \in Q_j \\ 0, & \text{otherwise} \end{cases}$.

Finally, the query protein is assigned the class γ that obtains maximum votes:

$$Z_\gamma^{IE} = Max\left\{ Z_1^{IE}, Z_2^{IE}, ... Z_V^{IE} \right\} \tag{4}$$

In the second step, the aim was to combine the diverse decision spaces generated by *IE* classifiers. In this way, the shortcoming of one classifier can be overcome by the advantage of others. Let $l=1,2,3,..., L$ represents the number of different base learners in the entire-pool voting. We compute the votes of each class for *CE* as:

$$Z_j^{CE} = \sum_{i=1}^{L*\xi} w_i \Delta(R_i, Q_j), \quad j = 1, 2, ... V \tag{5}$$

The predicted class τ by the *CE* classifier will be decided by using the *Max* function:

$$Z_\tau^{CE} = Max\left\{Z_1^{CE}, Z_2^{CE}, ... Z_V^{CE}\right\} \tag{6}$$

In jackknife test, if a tie occurs for a query protein; then decision of the highest performing IE^{SVM} classifier is taken. If the highest performing ensemble also delivers a tie, then the vote of the 2nd highest performing IE^{NN} ensemble is considered. Results reported in the Table 2 justify this action.

2.2 Evaluation Methods

In the literature of Bioinformatics, both independent and jackknife tests are used by the leading investigators to evaluate the performance of their prediction methods [4], [5], [6], [7], [8], [9], [10], [12], [13], [14], [15], [20]. The jackknife test is considered as the most rigorous and objective [21]. This test is conducted for the cross validation based performance analysis. During this test, each protein sequence is singled out as a test sample and remaining samples are used to train.

The overall percent accuracy (*acc.%*) is calculated as:

$$acc.\% = \frac{\sum_{i=1}^{V} p(i)}{N} \times 100 \tag{7}$$

where, V is the 12 class number, $p(i)$ is the number of correctly predicted sequences of location i.

The numerical value of Q-statistic indicates the independency of component classifiers [22]. For any two base classifiers C_i and C_j, the Q-statistic is defined as:

$$Q_{i,j} = \frac{ad - bc}{ad + bc} \tag{8}$$

where, a and d represent the frequency of both classifiers making correct and incorrect predictions, respectively. However, b shows the frequency when first classifier is correct and second is incorrect; c is the frequency of second classifier being correct and first incorrect. The average value of Q-statistic among all pairs of L base classifiers in CE ensemble is calculated as:

$$Q_{avg} = \frac{2}{L(L-1)} \sum_{i=1}^{L-1} \sum_{k=i+1}^{L} Q_{i,k} \tag{9}$$

The positive value of Q_{avg} shows that classifiers recognize the same objects correctly. The positive value of Q_{avg} (<1) shows the diversity level of base classifiers in the CE.

3   Results and Discussions

In this section, the overall accuracy and Q-statistic of IEs and CE are computed and results are given in the Table 2. This table indicates average Q-statistics of IEs and CE, in jackknife test, are in the range of 0.94-0.74 and 0.63, respectively. In case of independent test, Q-values of IEs and CE are in the range of 0.96-0.69 and 0.86, respectively. This highlights sufficient diversity in IEs and CE. This diversity in individual learners is accumulated that result improvement in CE. Further, in jackknife test, this table shows better prediction accuracy of IE^{SVM} among other IEs. IE^{SVM} predicts correctly 1738 out of 2191 sequences. Thus it gives an overall accuracy of 79.32%. Therefore, if SVM based learning mechanism is incorporated as a base classifier, then chance of CE improved enhances. IE^{NN} predicts correctly 1665 out of 2191 sequences and it gives overall accuracy of 76.01%. This predicted accuracy of IE^{NN} is comparable with IE^{PNN}. The prediction accuracies of IEs are also investigated on independent dataset containing 2494 protein sequences. The results show IE^{NN} correctly predicts 2115 protein sequences and gives an overall accuracy of 84.85%. In this case, the prediction accuracy of IE^{PNN} is lower than IE^{NN}. For independent dataset test, overall prediction of IE^{SVM} is not appreciable (69.52%). However, to improve SVM prediction, there is a need to find optimal kernel parameters.

Table 2. Performance comparison of IEs vs. CE

IEs/CE	Jackknife test			Independent test		
	Correct Prediction	Acc. %	Avg. Q statistics	Correct Prediction	Acc. %	Avg. Q statistics
IE^{SVM}	1738	79.32	0.94	1734	69.52	0.96
IE^{CDC}	1600	73.60	0.74	1965	78.79	0.69
IE^{NN}	1665	76.01	0.92	2115	84.85	0.93
IE^{PNN}	1686	76.99	0.85	2057	82.48	0.93
CE	1771	80.83	0.63	2162	86.65	0.86

In Table 3, we summarize the results of existing approaches and then compare with our method. The basic reason for this comparison was; these researchers have used the same human protein dataset and estimation tests to evaluate their prediction algorithm, except Shi et al. [12]. This table indicates that our CE classifier, in jackknife test, correctly classifies 1771 protein sequences out of 2191 giving an overall accuracy of 80.83%. However, for independent test, CE correctly classifies 2162 protein sequences out of 2494 to give an overall accuracy of 86.69%. This shows an improved in prediction accuracy of our approach as compared to the existing approaches proposed, except [12].

Table 3. Summary of comparative analysis

Prediction methods		Jackknife test	Independent test	Ref
Input features form	Prediction algorithm	Acc. %	Acc. %	
28 PseAAC component	Aug. CDC	1590/2191=72.6	1865/2494=74.8	[20]
PseAAC formation via three types filters	Aug. CDC	1532/2191=69.9	----------	[6]
Simple AAC	CDC	1492/2191=68.	1888/2494=75.7	[7]
PseAAC	CDC	1600/2191=73	2017/2494=80.9	[8]
PseAAC generated by DSP	Aug. CDC	1483/2191=67.68	1842/2494=73.86	[9]
Lempel-Ziv complexity	Aug.CDC	1612/2191=73.6	1990/2494=79.8	[10]
Quasi-sequence-order	Aug. CDC	1588/2191=72.5	1985/2494=79.6	[5]
Functional Domain Composition	SVM	1461/2191=66.7	2037/2494=81.7	[11]
AAC distribution	SVM	1800/2191=82.15	2132/2494=85.49	[12]
PseAAC	CE-classifier	1771/2191=80.83	2162/2494=86.69	Our method

By comparing our results with Shi et al., we obtained a comparable performance. Average accuracies of both methods come out to be nearly equal, i.e. 83.74% and 83.82%. Currently, we have utilized a simple feature extraction strategy and exhaustive single-one-out cross validation test. However, Shi et al. have developed a complex feature extraction and prediction results are estimated using 5-fold cross validation.

4   Conclusion

In this paper, we have developed a new ensemble in predicting human protein into 12 subcellular localizations. The proposed *CE-Hum-PLoc* system delivers more accurate predictions than existing approaches, except [12]. This improvement was made possible by exploiting diversity in feature and decision spaces simultaneously. Currently, we have attempted with four base classifiers and one feature extraction strategy. However, by adding more base learners, further improvement is possible.

Acknowledgement

This research was supported by the Bio Imaging Research Center at Gwangju Institute of Science and Technology (GIST), South Korea.

References

1. Bjorses, P., Halonen, M., Palvimo, J.J., Kolmer, M., Aaltonen, J., Ellonen, P., Perheentupa, J., Ulmanen, I., Peltonen, L.: Mutations in the AIRE gene: Effects on subcellular location and transactivation function of the autoimmune polyendocrinopathy-candidiasis - Ectodermal dystrophy protein. Am. J. Hum. Genet. 66, 378–392 (2000)

2. Garg, A., Bhasin, M., Raghava, G.P.S.: Support Vector Machine-based Method for Subcellular Localization of Human Proteins Using Amino Acid Compositions, Their Order, and Similarity Search. J. Biol. Chem. 280, 14427–14432 (2005)

3. Reinhardt, A., Hubbard, T.: Using neural networks for prediction of the subcellular location of proteins. Nucleic Acids Res. 26, 2230–2236 (1998)

4. Shen, Y., Burger, G.: 'Unite and conquer': enhanced prediction of protein subcellular localization by integrating multiple specialized tools. BMC Bioinformatics 8, 420 (2007)

5. Chou, K.C.: Prediction of Protein Subcellular Locations by Incorporating Quasi-Sequence-Order Effect. Biochem. Biophys. Res. Commun. 278, 477–483 (2000)

6. Gao, Y., Shao, S., Xiao, X., Ding, Y., Huang, Y., Huang, Z., Chou, K.C.: Using pseudo amino acid composition to predict protein subcellular location: Approached with Lyapunov index, Bessel function, and Chebyshev filter. Amino Acids 28, 373–376 (2005)

7. Chou, K.C., Elrod, D.W.: Protein subcellular location prediction. Protein Eng. 12, 107–118 (1999)

8. Kuo-Chen, C.: Prediction of protein cellular attributes using pseudo-amino acid composition. Proteins: Structure, Function, and Genetics 43, 246–255 (2001)

9. Pan, Y.X., Zhang, Z.Z., Guo, Z.M., Feng, G.Y., Huang, Z.D., He, L.: Application of Pseudo Amino Acid Composition for Predicting Protein Subcellular Location: Stochastic Signal Processing Approach. J. Protein Chem. 22, 395–402 (2003)

10. Xiao, X., Shao, S., Ding, Y., Huang, Z., Huang, Y., Chou, K.C.: Using complexity measure factor to predict protein subcellular location. Amino Acids 28, 57–61 (2005)

11. Chou, K.C., Cai, Y.D.: Using functional domain composition and support vector mchines for prediction of protein subcellular location. J. Biol. Chem. 277, 45765–45769 (2002)

12. Shi, J.Y., Zhang, S.W., Pan, Q., Zhou, G.P.: Using pseudo amino acid composition to predict protein subcellular location: approached with amino acid composition distribution. Amino Acids 35, 321–327 (2008)

13. Chou, K.C., Shen, H.B.: Predicting protein subcellular location by fusing multiple classifiers. J. Cell. Biochem. 99, 517–527 (2006)

14. Chou, K.C., Shen, H.B.: Hum-PLoc: A novel ensemble classifier for predicting human protein subcellular localization. Biochem. Biophys. Res. Commun. 347, 150–157 (2006)

15. Shen, H.B., Chou, K.C.: Virus-PLoc: A fusion classifier for predicting the subcellular localization of viral proteins within host and virus-infected cells. Biopolymers 85, 233–240 (2007)

16. Dzeroski, S., Zenko, B.: Is combining classifiers with stacking better than selecting the best one? Machine Learning 54, 255–273 (2004)

17. Jia, P., Qian, Z., Zeng, Z., Cai, Y., Li, Y.: Prediction of subcellular protein localization based on functional domain composition. Biochem. Biophys. Res. Commun. 357, 366–370 (2007)

18. Khan, A., Fayyaz, M., Choi, T.S.: Proximity based GPCRs prediction in transform domain. Biochem. Biophys. Res. Commun. 371, 411–415 (2008)

19. Duda, R.O., Hart, P.E., Stork, D.G.: Pattern Classification. John Wiley & Sons, Inc., New York (2001)

20. Xiao, X., Shao, S., Ding, Y., Huang, Z., Chou, K.C.: Using cellular automata images and pseudo amino acid composition to predict protein subcellular location. Amino Acids 30, 49–54 (2006)
21. Chou, K.C., Zhang, C.T.: Prediction of Protein Structural Classes. Critical Reviews in Biochemistry and Molecular Biology 30, 275–349 (1995)
22. Kuncheva, L.I., Whitaker, C.J.: Measures of diversity in classifier ensembles and their relationship with the ensemble accuracy. Mach. Learn. 51, 181–207 (2003)

Designing Linear Phase FIR Filters with Particle Swarm Optimization and Harmony Search

Abdolreza Shirvani[1], Kaveh Khezri[2], Farbod Razzazi[3], and Caro Lucas[4]

[1] Islamic Azad University of Tehran Science and Research Branch/
Computer Engineering Department Iran
abdols@gmail.com
[2] Islamic Azad University of Tehran Science and Research Branch/
Biomedical Engineering Department, Iran
kaveh.khezri@gmail.com
[3] Islamic Azad University of Tehran Science and Research Branch/
Electrical Engineering Department, Iran
farbod_razzazi@yahoo.com
[4] University of Tehran/Electrical and Computer Engineering Department,
Control and Intelligent Processing Center of Excellence, Iran
lucas@ipm.ir

Abstract. In recent years, evolutionary methods have shown great success in solving many combinatorial optimization problems such as FIR (Finite Impulse Response) filter design. An ordinary method in FIR filter design problem is Parks-McClellan, which is both difficult to implement and computationally expensive. The goal of this paper is to design a near optimal linear phase FIR filter using two recent evolutionary approaches; Particle Swarm Optimization (PSO) and Harmony Search (HS). These methods are robust, easy to implement, and they would not trap in local optima due to their stochastic behavior. In addition, they have distinguishing features such as less variance error and smaller overshoots in both stop and pass bands. To prove these benefits, two case studies are presented and obtained results are compared with previous implementations. In both cases, better and reliable results are achieved.

Keywords: Harmony search, particle swarm optimization, meta-heuristic algorithms, optimization, FIR filters, Parks-McClellan.

1 Introduction

Although FIR filters have simple structures, many important concepts of digital signal processing such as wavelet transforms [1] and multi-rate signal processing [2] refer to these primitive filters. Designing FIR filters by windowing method is an effortless and quite easy task but has its own drawbacks. The filter is often designed based on a specific order M. From the theory of Fourier series, the rectangular window has the best mean-squared error with respect to specified frequency response for a given value of M [4].

D. Ślęzak et al. (Eds.): SIP 2009, CCIS 61, pp. 193–200, 2009.

$$h[n] = \begin{cases} h_d[n], & 0 \le n \le M, \\ 0, & o.w. \end{cases} \tag{1}$$

which minimizes

$$\varepsilon^2 = \frac{1}{2\pi} \int_{-\pi}^{\pi} \left| H_d(e^{j\omega}) - H(e^{j\omega}) \right| d\omega \tag{2}$$

This estimation criterion leads to undesired behavior at discontinuities of $H_d(e^{j\omega})$ and does not permit individual control over the approximation errors in different bands [4].

In the following discussion the PM algorithm is considered as a commonly used algorithm in FIR filter design with generalized linear phase due to its efficiency and flexibility among traditional algorithmic methods. The impulse and frequency response of an FIR filter of type I are as (4), (5).

$$h_e[n] = h_e[n-1], \quad -\frac{M}{2} \le n \le \frac{M}{2} \tag{3}$$

$$A_e(e^{j\omega}) = \sum_{n=-\frac{M}{2}}^{\frac{M}{2}} h_e[n]e^{-j\omega n} = h_e(0) + 2\sum_{n=1}^{N} h_e[n]\cos(\omega n) \tag{4}$$

$$H(e^{j\omega}) = A_e(e^{j\omega})e^{-j\omega \frac{M}{2}} \tag{5}$$

According to PM with $M/2$, ω_p and ω_s fixed, the frequency selective filter design becomes a Chebyshev approximation over disjoint sets [3]. So with respect to [3], [4] and [5] the approximation error function is defined as (6).

$$E(\omega) = W(\omega)[H_d(e^{j\omega}) - A_e(e^{j\omega})] \tag{6}$$

where $W(\omega)$ is the weighting function. $E(\omega)$, $W(\omega)$ and the desired frequency response $H_d(e^{j\omega})$ are defined over closed subintervals of $0 \le \omega \le \pi$. The minimax criterion is used in the PM algorithm, so that it minimizes the maximum weighted approximation error of equation (6).

$$\min_{\left\{ h_e[n]:0 \le n \le \frac{M}{2} \right\}} \left(\max_{\omega \in F} |E(\omega)| \right) \tag{7}$$

where F is closed subset of $0 \le \omega \le \pi$ such that $0 \le \omega \le \omega_p$ or $\omega_s \le \omega \le \pi$, the set of impulse responses that minimize δ. Although only filters of type I are addressed here, the results and algorithm can be generalized to include types II to IV linear-phase filters [3], [4].

To achieve an efficient implementation using PM algorithm, one must overcome the difficulties of robust interpolation methods which are computationally expensive [3].

In contrast with traditional optimization approaches, evolutionary optimization methods are easy to implement and reliable. Two evolutionary design methods for LTI (Linear Time Invariant) FIR filters are considered; Particle Swarm Optimization (PSO) and Harmony Search (HS). PSO utilizes particles of candidate solutions to evolve an optimal or near optimal solution to a problem. The degree of optimality is measured by a user defined fitness function [6]. PSO is a powerful and computationally efficient method which has relatively few parameters. This method is resistant to being trapped in local optima. On the other hand, HS is a relatively new meta-heuristic optimization algorithm. It has been approved as an effective method in many optimization problems [7], [8] and [9]. The effort to find harmonies in the music is analogous to discover the optimality in an optimization problem. Harmonies are regarded as solution vectors, therefore the perfect harmonies are optimal solutions in which their fitness are measured by a function defined by the user [7]. The aim of this paper is to improve the results of [5] and compare them with HS. PSO algorithm is shortly discussed in third section and HS is briefly introduced in fourth section. In section five, a case study is presented including a comparison between PM, PSO and HS, and a discussion on the efficiency of the algorithms. Section six is dedicated to the conclusion.

2 Particle Swarm Optimization

PSO has its roots in social psychology, engineering and computer science [6]. Particles defined as initial solutions of an optimization problem, are flown through the problem space. Each iteration, every particle's velocity is stochastically accelerated toward its previous best position (where it had its highest fitness value) and toward its globally seen best position (where it is the best seen fitness value by all particles). The original PSO algorithm [6] initializes the position of each particle (solution) with random values and updates them using an individual term (individual learning attitude toward a behavior) and a social term (cultural transmission or subjective norm) [6].

The PSO algorithm samples the search space by modifying the velocity term. The direction of the movement is a function of the current position and velocity, which is defined as the weighted sum of the difference between the particles current position and it's locally best seen position by the particle and the difference between the particles current position and globally best seen position by all particles.

The pseudo code of the basic PSO in the space of real numbers is:

$\forall \vec{x}_i \in U[\vec{a}, \vec{b}] \wedge \forall \vec{v}_i = \vec{0}, i = $ number of particles, $\vec{a}, \vec{b} \subset R^n$ do

while not converged do

$\quad for\, i = 1$ to number of particles do

$\qquad \vec{r}_1, \vec{r}_2 \in \vec{U}[0,1]$

$$\vec{v}_i(t) = \omega \vec{v}_i(t-1) + c_1 \vec{r}_1.(\vec{p}_i - \vec{x}_i(t-1)) + c_2 \vec{r}_2.(\vec{p}_g - \vec{x}_i(t-1)).\ \vec{x}_i(t) = \vec{x}_i(t-1) + \vec{v}_i(t).$$

$\quad$ *if* $fitness(\vec{x}_i(t)) < \vec{p}_i \rightarrow \vec{p}_i = \vec{x}_i(t)$

$\quad$ *if* $fitness(\vec{x}_i(t)) < \vec{p}_g \rightarrow \vec{p}_g = \vec{x}_i(t)$

$\quad$ end

$\quad$ end

$\quad$ $\vec{p}_g$ is the solution

where $\vec{a}, \vec{b}$ are input boundaries, ω is the momentum coefficient, $\vec{p}_i$ is the local best vector, $\vec{p}_g$ is the global best vector, *fitness* is a user defined function to evaluate the optimality of the solution.

3 Harmony Search Optimization

Geem et al. formalized harmony search into three quantitative optimization processes: usage of harmony memory, pitch adjustment and randomization [7]. The usage of harmony memory (HM) is important because it ensures that good harmonies are considered as elements in the new solution vectors. In order to use this memory effectively, the HS algorithm uses the harmony memory accepting rate. Low values of this parameter causes selection of elite harmonies and it may converge slowly. If this parameter is set to high values then pitches in the memory is often used and other are not explored well. The second component is the pitch adjustment which has parameters such as pitch bandwidth and pitch adjustment rate. As pitch in music means changing the frequency, it is slightly different in HS algorithm [7], [8]. The third component is randomization, which is to increase the diversity of the solutions. Although the pitch adjustment has a similar role, it is limited to certain area and thus corresponds to a local search. The use of randomization can derive the system further to more exploration to attain global optimality. The harmony search algorithm is summarized below.

$\quad$ *begin*

$\quad$ Define objective function $f(\vec{x}), \vec{x} = (x_1, x_2, ..., x_d)^T$

$\quad$ Define harmony memory accepting rate (r_{accept}).

$\quad$ Define pitch adjustment rate (r_{pa}) and other parameters

$\quad$ Generate harmony memories with random harmonies

$\quad$ *while* ($t <$ max number of iterations)

$\quad\quad$ *while* ($i <=$ number of variables)

$\quad\quad\quad$ *if* ($rand < r_{accept}$),

$\quad\quad\quad$ Choose a value from HM for the variable i

$\quad\quad\quad$ *if* ($rand < r_{pa}$),

$\quad\quad\quad\quad$ Adjust the value by adding certain amount

$\quad\quad\quad$ *endif*

> *else* Choose a random number
> > *endif*
> *endwhile*
> Accept the new harmony (solution) if better.
> > *endwhile*
> Find the current best solution.
> *end*

4   Case Studies and Discussion

Given a linear phase FIR filter of order M with ω_p and ω_s specified, the frequency response of the corresponding filter designed by PM algorithm is given by (6). We would like to obtain the coefficients that minimize (7). The fitness functions, *fitness1* and *fitness2*, are defined as (12) and (13).

$$fitness1 = \max_{\omega \in F_p \cup F_s} \left(\left| E(\omega) \right| \right) \tag{8}$$

$$fitness2 = \left| \max_{\omega \in F_p} \left(\left| E(\omega) \right| - \delta_p \right) \right| + \left| \max_{\omega \in F_s} \left(\left| E(\omega) \right| - \delta_s \right) \right| \tag{9}$$

A linear phase FIR filter with order $M=20$, $\omega_p =0.4$, $\omega_s =0.45$ $\delta_p/\delta_s =1$ is considered. Tables 1 and 3 demonstrate the results of application of *fitness1* and *fitness2* for PSO and PM with 300 iterations and 20 particles. Tables 2 and 4 depict the outcomes of using *fitness1* and *fitness* with 400 iterations for PSO of 75 particles. Harmony memory size is 30 and 10000 iterations are considered for HS. Figures 1 to 3 shows the frequency responses of each method. Overshoots in pass and stop bands and variance error at these regions are comparable. Results are achieved with lower number of particles, less iteration numbers and smaller variance error and overshoots for case studies 1 and 2 than [5], which uses 75 particles and 500 iterations for PSO and limited stop band of 0 to 0.5. It also reveals that 20 particles (instead of 75 in this case) suffice to achieve a good result.

Table 1. The first case study results for PSO and PM

Method	*fitness1*	pbve[1]	sbve[2]	pbo[3]	sbo[4]
PSO	0.171291	0.008166	0.001453	0.339790	0.150619
PM	0.1663	0.013692	0.002620	0.332161	0.165710

---

[1] Pass band variance error.
[2] Stop band variance error.
[3] Pass band overshoot.
[4] Stop band overshoot.

Table 2. The first case study results for PSO, HS and PM

Method	*fitness1*	pbve	sbve	pbo	sbo
PSO	0.158356	0.008634	0.001518	0.311970	0.155654
HM	0.157399	0.002198	0.002351	0.200313	0.176859
PM	0.1663	0.013692	0.002620	0.332161	0.165710

Table 3. The second case study results for PSO

Method	*fitness2*	pbve	sbve	pbo	sbo
PSO	0.295253	0.006436	0.001550	0.290326	0.140993
PM	0.3326	0.013692	0.002620	0.332161	0.165710

Table 4. The second case study results for PSO, HS and PM

Method	*fitness2*	pbve	sbve	pbo	sbo
PSO	0.288781	0.004435	0.001804	0.253827	0.144400
HS	0.284451	0.004487	0.001333	0.270248	0.141455
PM	0.3326	0.013692	0.002620	0.332161	0.165710

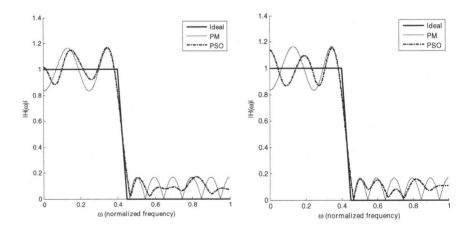

Fig. 1. Frequency response of tables 1 and 2 for PM and PSO (left and right)

A brief analysis of computational complexity also shows that HS is much simpler in implementation (additions, multiplications and inner loops) than PSO, thus in every iteration of HS, it has less run time than PSO. The error plots illustrates that the results are not premature and early convergence is avoided. Conclusions are accomplished through several repetitions of PSO and HM.

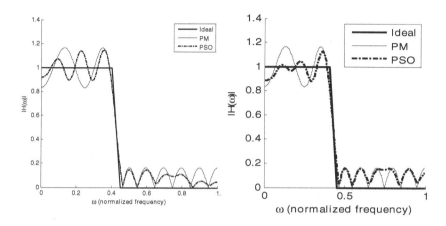

Fig. 2. Frequency response of the tables 3 and 4 for PM and PSO (right and left)

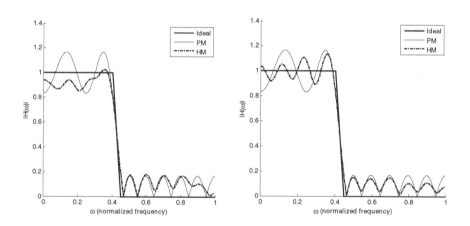

Fig. 3. Frequency response of the tables 2 and 4 for PM and HS (left and right)

5   Conclusion

This paper presented two meta-heuristic optimization methods, PSO and HS, to obtain optimal coefficients of a linear phase FIR filter. The results were compared with a prior application of PSO and a traditional PM approach. It was elucidated that the obtained results were superior in both quantitative and qualitative measures such as number of iterations, particle count, variance error and overshooting in stop-band and pass-band. An analogy between PSO and HS revealed that although PSO is fast, it has higher computational order than HS. Furthermore, the variance error acquired by HS is less than PSO. Both methods are easy to implement and have few parameters. By adjusting these parameters at each iteration, the convergence rate can be changed to make a tradeoff between exploration and exploitation. The changes in the parameters directly affect the convergence rate and as mentioned earlier, these methods would not trap in local optima theoretically because of stochastic behavior.

These approaches can be used in wide spectrum of filter design applications. Working on benchmark results, obtaining quantized coefficients, and other error analyses are the interests in further research efforts.

References

1. Vetterli, M.: Wavelets and filter banks: theory and design. IEEE Trans. Signal Processing 40 (1992)
2. Wornell, G.W.: Emerging applications of multi-rate signal processing and wavelets in digital communications. IEEE proceedings 84(4) (1996)
3. Parks, T.: Chebyshev approximation for non-recursive digital filters with linear phase. IEEE Trans. Circuit Theory 19 (1972)
4. Oppenheim, V.: Discrete-time signal processing, 2nd edn. Prentice-Hall, New Jersey (1998)
5. Ababneh, J.I.: Linear phase FIR filter design using particle swarm optimization and genetic algorithms. Elsevier Digital signal processing 18 (2007)
6. Kenndy, J.: Swarm intelligence. Morgan Kaufmann, San Francisco (2001)
7. Geem, Z.W.: A new heuristic optimization algorithm: harmony search. Simulation 76 (2001)
8. Mahdavi, M.: An improved harmony search algorithm for solving optimization problems. Elsevier Applied mathematics and computation 188 (2007)
9. Lee: A new meta-heuristic algorithm for continuous engineering optimization: harmony search theory and practice. Comput. Methods Appl. Mech. Engrg. 194 (2005)

Blind Image Steganalysis Based on Local Information and Human Visual System

Soodeh Bakhshandeh[1], Javad Ravan Jamjah[2], and Bahram Zahir Azami[2]

[1] Islamic Azad University Science and Research Branch, Tehran, Iran
[2] University of Kurdistan, Sanandaj, Iran
{s.bakhshandeh,j.ravanjamjah,zahir}@ieee.org

Abstract. In this paper, a new steganalysis method is introduced based on human visual system. Steganalysis uses the effect of steganography on the statistical characteristics to detect if such effect exists or not. Steganography methods do not have the same effect on all of the pixels of an image. We use local information to select the best area. We cannot use each individual pixel for feature extraction, so we use blocks. At first, segmentation and clustering algorithm are employed to detect the best segments for steganalysis. In the next step, the features based on wavelet are extracted. At the end, Support Vector Machine is applied as the classifier. The performance of this algorithm is verified by experimental results. The results show that the detection accuracy of our method reaches 98.67% for true positive and 90.67% for true negative when 100% capacity of the image is used with spread spectrum steganography algorithm.

Keywords: Human Visual System, Segmentation, Fuzzy Clustering, Support Vector Machine.

1 Introduction

Steganography is the art of transferring hidden data without arising any suspicion. Steganography algorithms usually use a cover media for transferring and hiding the information. While the users and applications of steganography are very diverse, unfortunately terrorist groups and malicious users show special interest in it. Therefore, law enforcement agents should find a way to detect the existence of secret communication.

Steganalysis is the art of detection of such communications. Perfect steganography doesn't have any effects on the cover [1]. Therefore, in theory it isn't detectable. However, all practical steganography algorithms have some kind of effects on the cover and hence can be detected. Based on these residual effects many algorithms have been proposed for steganalysis. Some mathematical approaches such as [2, 3] are proposed for detecting the hidden communication and information extraction. However, the mathematical approach has many assumptions about the steganography algorithm and in real world such information isn't available.

D. Ślęzak et al. (Eds.): SIP 2009, CCIS 61, pp. 201–208, 2009.

In [4], the visual attack was used for steganalysis. This type of steganalysis algorithms was limited to some steganography methods such as Least Significant Bit (LSB) steganography.

However, most successful steganalysis algorithms were proposed based on machine learning. To apply a learning algorithm, we should first extract a feature vector. After feature extraction, we need to train a classification algorithm. In [5], it was assumed that stego-image has more complexity than clean-image. The authors used Laws' image texture to measure the complexity of the image and at the end, used SVM as the classifier. In [6], the authors used the correlation of pixels in the images. They supposed that steganography algorithm decreases the correlation that exists between pixels in an image. Therefore, they extracted a Co-Occurrence matrix for images. In the next step, feature vector was extracted from the Co-Occurrence matrix.

In [7], the authors used two-dimensional Moment of Characteristic Function (MCF). The MCFs were extracted from sub bands of DWT from image and prediction error image. In [8], they used the moment of DCT block histogram with different sizes. The authors of [9] used the extracted feature form DCT domain. First, they divided the image into blocks with 8×8 sizes and calculated the two dimensional DCT. The correlations between block-DCT in intra-block and inter-block were exploited as the features. The authors in [10] used information theory in active steganalysis domain. At first the cover was estimated with maximum a posteriori probability (MAP). Then they applied ICA to separate the message and cover form each other. Their experiment shows that 75% of embedded message can be extracted for large message sizes. A comparative survey on the different methods on steganalysis is given in [11].

In this paper, a new steganalysis method is introduced. At first, according to extracted features form sub-block of the images, we selected the best sub-blocks. Then wavelet feature are extracted form selected blocks. Finally the stego-images are detected with SVM classifier. Remainder of this paper is organized as follows: Sections 2 covers the description of our method that contains the segmentation method and clustering of the segments. We explain feature extraction and classification in Section 3 and 4, respectively. In Section 5, we demonstrate our experimental results. Conclusions are drawn in Section 6.

2 Segmentation and Clustering the Segments

Some of adaptive steganography methods use local information for embedding the data. They insert the data in selected regions. For selecting these regions, they use the places that Human Visual System (HVS) is less sensitive in them and therefore they use HVS model in their data insertion policy [12]. This idea inspires us to use the HVS model to break the steganography algorithm.

Based on pattern recognition techniques, image textures can be categorized in four groups: regular, near regular, irregular and stochastic. The difference between these groups is in the regularity level of the texture [13].

If we compare the stego-images with clean-images, we find that with inserting data in the image, its textures and edges are increased [5]. The feature based on texture of image has better results for breaking the steganography methods that work on selecting pixels for inserting data.

In this paper, we use HVS model for classifying the sub-blocks of the images under test. In our method, in the first step we classify the segments of an image into three classes. Every block shows a different level of image texture.

The important point in our proposed method is the segmentation. Classification of the images' blocks is done based on HVS model of the blocks. We use the HVS model described in [14] to determine maximum tolerance (MT) of DCT coefficients before the changes are sensible. The proposed method contains one function for sensitivity and two for masking components based on luminance. At first we break the image into disjoint blocks with 8×8 pixels. Then we calculate DCT coefficients for every block of the image. We show i, j-th pixel in k-th blocks DCT number with $C[i,j;k]$ (in blocks with 8×8 pixels, $1 \leq i$, $j \leq 8$). Every block contains 16 DCT coefficients.

For sensitivity function, we use a sensitivity table. This table shows the smallest value of DCT coefficients for a block that is discernible without masking noise. Therefore, the smaller values in Table 1 show that the eye is more sensitive to them.

Table 1. DCT frequency sensitivity table [15]

j \ i	1	2	3	4	5	6	7	8
1	1.4	1.01	1.16	1.65	2.4	3.43	4.79	6.55
2	1.01	1.45	1.32	1.52	2	2.71	3.67	4.93
3	1.16	1.32	2.24	2.59	2.98	3.64	4.6	5.88
4	1.66	1.52	2.59	3.77	4.55	5.3	6.28	7.6
5	2.4	2	2.98	4.55	6.15	7.46	8.71	10.17
6	3.43	2.71	3.64	5.3	7.46	9.62	11.58	13.51
7	4.79	3.67	4.6	6.28	8.71	11.58	14.5	17.29
8	6.56	4.93	5.88	7.6	10.17	13.51	17.29	21.15

Luminance adaptation indicates the boundary of DCT changing for denoting the increasing the intensity of the block. For each block (k), the luminance masked threshold can be calculated based on Eq. (1).

$$t_k[i,j,k] = t[i,j]\left(\frac{C_0[0,0,k]}{C_{0,0}}\right)^{a_T}, \quad C_{0,0} = \frac{1}{N}\sum_k C_0[0,0,k] \tag{1}$$

where $t[i,j]$ is the i, j-th element of sensitivity matrix, a_T is the DC coefficient of the k-th block in c (set to 0.649 [14]), $C_{0,0}$ is the mean DC coefficient in the image.

Another parameter that affects MT is contrast masking. This parameter determines the amount of modification in a frequency that can be invisible to eye by considering the energy presented in the frequency.

$$s(i,j,k) = \max\{t_L[i,j,k], C_0[i,j,k]^{0.3} \times t_L[i,j,k]^{0.3} \tag{2}$$

$s(i,j,k)$ determines the amount of acceptable change for $C[i,j;k]$ before breaking the JND threshold. For each block we use the mean of $s(i,j;k)$ and call them $s(k)$.

In the next step of our method we should assign every block into one of the classes. At first we should define the classes. It is an important phase, because in continuation of this method, we want to select the best class for feature extraction. To determine the classes, Fuzzy Clustering (FC) algorithm is used.

In the training step of the clustering, we calculate $s(k)$ for 20 randomly selected images of our database. After extracting a mean value of DCT coefficient for every block, we use FC for clustering the blocks into three clusters. The outputs of the clustering are the center of each class that is used in the labeling step. In labeling phase, we determine the label of input block (class-1, class-2 or class-3). Each block is assigned to the class with the minimum Euclidian distance to the feature of the block. The area with low texture is the best area for image steganalysis, therefore we select the block of class 3 for feature extraction.

3   Feature Extraction

The feature vector should be sensitive to the embedded data and not to the cover. The performance of wavelet transform for feature extraction is demonstrated in [1]. Therefore, in this paper we use wavelet domain features similar to the feature vector in [16].

In the feature extraction phase, at first, we apply a 3-level wavelet on the image. We denote the vertical, horizontal and diagonal of level n of the wavelet by $h_n(i,j)$, $v_n(i,j)$ and $d_n(i,j)$, that i and j are the index of the sub-band. At the next step, the sub-bands of the clean image are estimated with local variance and Wiener filer. We use the MAP to estimate the local variance of the clean image. The MAP estimation is applied for neighborhood of size $N \in \{3, 5, 7, 9\}$ with Eq. (3).

$$\sigma_N^2(i,j) = \max\left(0, \frac{1}{N^2}\sum_{i,j\in N}\left(w^2(i,j)-\sigma_0^2\right)\right),\tag{3}$$

$$\hat{\sigma}^2(i,j) = \min\left(\hat{\sigma}_3^2(i,j),\cdots,\hat{\sigma}_9^2(i,j)\right)$$

where σ_0^2 is a constant parameter and according to [16] set to 0.5, w_n denotes the wavelets' sub-bands in level n ($w_n \in \{h_n, v_n, d_n\}$) and N is the neighborhood size. After the estimation of the local variances of the every sub-band is done, the Wiener filter is applied to estimate the clean sub-band coefficients with Eq. (4).

$$w_{den}(i,j) = w(i,j)\frac{\hat{\sigma}^2(i,j)}{\hat{\sigma}^2(i,j)+\hat{\sigma}_0^2}\tag{4}$$

Then we should calculate the difference between the estimated and the original sub-band; we call them "estimation error" For each sub-band (v_n, h_n and d_n, $n \in \{1,2,3\}$), the three first absolute central moments of the estimation error is calculated using Eq. (5):

$$m_p = \frac{1}{\|e_n\|}\sum\left|e_n(i,j)-\overline{e}_n\right|^p\tag{5}$$

where

$$\overline{e}_n = \frac{1}{\|e_n\|}\sum e_n(i,j)\tag{6}$$

and e_n is the estimation error in level n and $\|e_n\|$ denotes the size of estimation error of sub-band. The Feature vector is extracted for each block of class-3 separately. Because of limitation classifier for input feature vector, the average of the feature vectors is used for classification.

4 Classification

After feature extraction, we should classify the signals into two classes. We need a classifier algorithm to assign a label to the features that are: clean or stego-image. Support Vector Machine (SVM) is a useful method for data classification. A classification problem involves two steps: training and testing. SVM at first learn and then use the learned data in tests phase. So the goal of train of SVM is preparing a model from the training set [17]. In the test phase, SVM uses the model for prediction. The SVM has two main categories: linear algorithm and non-linear algorithm. In linear algorithm we found an optimum hyper plane for data separation according to train data. Some training data are selected to show the hyper plane. The Selected data are called the support vectors. In non-linear algorithm, we map the dimension of data to a new space with higher dimensions. Data in the new space should be linearly separable; then we use linear SVM algorithm in this new space. To decrease the computation in the new space, kernel trick is used.

For every step, we should determine the data set (this is explained in the Experiment results Section, below). In our method, we use LibSVM with RBF kernel as the classifier. In the proposed method because of HVS model, we have some limitations (e.g. the size of base block). So if we use HVS model with less limitation, we could get more local information from image.

5 Experiment Result

We should verify our method with a database. We select the images of our database from two databases that can be downloaded from [18, 19]. According to [20], using multiple databases reduces the steganalysis performance. To generate stego-image, three methods of steganography have been used: Spread Spectrum (SS), Spread Spectrum Image Steganography (SSIS) and StegHide (downloaded from [21]).

For each steganography tool, 1000 random images are selected from 3714 images. After selection, the center block of the image with a size of 256×256 is selected and saved as a grayscale image with TIFF format. The tools are applied to 50% of the selected images (for every tool, we have 500 stego-images and 500 clean images). The remaining images were used as the clean images. 100% of the embedding capacity for each image has been used in the steganography methods. Therefore, for each image, every tool itself determines the full capacity of the image, and inserts that much data. We use random data with Gaussian distribution as the message.

To have a comparison, we selected three steganalysis methods (from [22, 16, and 1]). To have a fair comparison, the selected methods are applied to our database and the required feature vectors for each method are extracted.

After the feature extraction, we divide the feature vectors into training and test. 70% of the total feature vectors are used in the training phase of the classification, and the remaining 30% feature vectors are used in the test phase.

The training vectors for each method are applied to the classifier for extracting the model.

After training the classifier, we start the test phase. Table 2 shows the results of our method in breaking three different steganography methods. Fig. 1 shows the difference between Accuracy, True Positive (TP) and True Negative (TN) percentages of our method and three other steganalysis methods for breaking three steganography methods.

Table 2. The Accuracy, False Positive (FP), False Negative (FN), TP and True Negative (TN) percentage of our method for breaking 3 steganography methods

Steganography Method	Accuracy	FP	FN	TP	TN
SS (with c =1)	87.50	15.00	10.00	90.00	85.00
SS (with c =2)	94.67	9.33	1.33	98.67	90.67
SSIS	86.43	17.14	10.00	90.00	82.86
StegHide	68.75	43.75	18.75	81.25	56.25

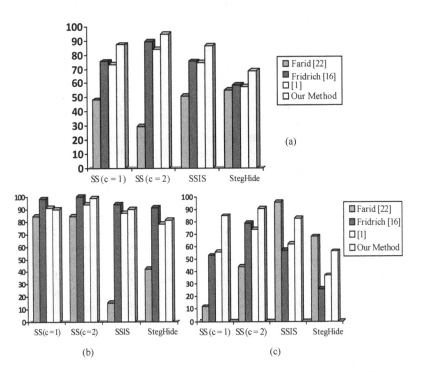

Fig. 1. (a), (b) and (c) are respectively the comparison of the Accuracy, TP and TN (percentage) of our method and the other three steganalysis methods

In the above table, the performance of our method applied on three steganography methods is showed. The best accuracy rate of our method is on SS method with $c=2$. The important point of this table is the balance results in different rates of our method under different steganography tools. In our method the difference between TP and TN is not significant. However, TP in our method is larger than TN and this is important in steganalysis algorithms.

In above figures, our method is compared with three different steganalysis methods. The experimental results show that the proposed method has outperformed the other methods. Because of different length of inserted message with StegHide, all methods have low accuracy. However, the smallest difference of our method and other method occur for method [1] in SSIS that is 10%.

6 Conclusions

Steganalysis and steganography are two methods that compete against each other. Steganography hides information and steganalysis tries to detect it. In blind steganalysis, we don't have any assumptions about the type of the steganography method. As a result, we should detect only based on the received signal.

We introduced a new blind steganalysis in this paper. In the proposed method more information is extracted from the image, in addition to the feature vector.

We segmented the image and only extracted the feature vector from the suspected segments of the image. Therefore, we expected more accuracy rate in comparison to the other methods. The results of our experiments confirm our prospect that if we extract the features considering the texture of each block, the performance of the method improves.

Acknowledgment

This research has been supported by Iran Telecommunication Research Center (ITRC) under a contract with Sharif University of Technology.

References

1. Shi, Y.Q., Xuan, G., Zou, D., Gao, J.: Steganalysis Based on Moments of Characteristic Functions Using Wavelet Decomposition, Prediction-Error Image, and Neural Network. In: International Conference on Multimedia and Expo (ICME), Amsterdam, The Netherlands (July 2005)
2. Ambalavanam, A.: Contributions in Mathematical Approaches to Steganalysis. In: International Symposium on Multimedia (ISM), San Diego, CA, USA, December 2006, pp. 371–375 (2006)
3. Chandramouli, R.: A Mathematical Approach to Steganalysis. In: Proc. SPIE Security and Watermarking of Multimedia Contents IV, San Jose, CA (January 2002)
4. Watters, P.A., Martin, F., Stripf, S.H.: Visual Steganalysis of LSB-encoded Natural Images. In: International Conference on Information Technology and Applications (ICITA), Sydney, Australia, July 2005, vol. 1, pp. 746–751 (2005)

5. Shuanghuan, Z., Hongbin, Z.: Image Texture Energy-Entropy-Based Blind Steganalysis. In: Workshop on Signal Processing Systems, Shanghai, China, October 2007, pp. 600–604 (2007)
6. Sullivan, K., Madhow, U., Chandrasekaran, S., Manjunath, B.S.: Steganalysis for Markov Cover Data with Applications to Images. IEEE Transactions on Information Forensics and Security 1, 275–287 (2006)
7. Chunhua, C., Shi, Y.Q., Chen, W., Xuan, G.: Statistical Moments Based Universal Steganalysis using JPEG 2-D Array and 2-D Characteristic Function. In: IEEE International Conference on Image Processing, Atlanta, GA, USA, October 2006, pp. 105–108 (2006)
8. Chen, C., Shi, Y.Q., Xuan, G.: Steganalyzing Texture Images. In: International Conference on Image Processing (ICIP), Antonio, Texas, September 2007, vol. 2, pp. 153–156 (2007)
9. Fu, D., Shi, Y.Q., Zou, D.K., Xuan, G.R.: JPEG Steganalysis Using Empirical Transition Matrix in Block DCT Domain. In: Workshop on Multimedia Signal Processing, Siena, Italy, October 2006, pp. 310–313 (2006)
10. Xu, B., Zhang, Z., Wang, J., Liu, X.: Improved BSS Based Schema for Active Steganalysis. In: International Conference on Software Engineering, Artificial Intelligence, Networking, and Parallel/Distributed Computing, Phuket, Thailand, August 2007, vol. 3, pp. 815–818 (2007)
11. Luo, X.-Y., Wang, D.-S., Wang, P., Liu, F.-L.: Review: A review on blind detection for image steganography. Signal Processing 88(9), 2138–2157 (2008)
12. Herrera-Moro, D.R., Rodríguez-Colín, R., Feregrino-Uribe, C.: Adaptive Steganography Based on Textures. In: International Conference on Electronics, Communications and Computers, Puebla, Mexico, December 2007, p. 34 (2007)
13. Wong, P.W., Dittmann, J., Memon, N.D.: Textural Features Based Universal Steganalysis. In: Security, Forensics, Steganography, and Watermarking of Multimedia Contents, vol. 6819 (January 2008)
14. Ahmidi, N., Lotfi Neyestanak, A.A.: A Human Visual Model for Steganography. In: Conference on Electrical and Computer Engineering, Niagara Falls, Ontario, Canada, May 2008, pp. 001077–001080 (2008)
15. Watson, A.B.: DCT Quantization Matrices Visually Optimized for Individual Images. In: Human Vision, Visual Processing, and Digital Display IV, Proc. SPIE 1913-14, San Jose, CA (1993)
16. Goljan, M., Fridrich, J., Holotyak, T.: New Blind Steganalysis and Its Implications. In: Proceedings of the Security, Steganography, and Watermarking of Multimedia Contents VIII, vol. 6072, pp. 1–13 (2006)
17. Hsu, C.W., Chang, C.C., Lin, C.J.: A Practical Guide to Support Vector Classification. Technical Report (May 2008)
18. http://vision.cs.aston.ac.uk/datasets/UCID/data/ ucid.v2.tar.gz
19. http://www.adastral.ucl.ac.uk/~gwendoer/?steganalysis/
20. Cancelli, G., Doerr, G., Barni, M., Cox, I.J.: A Comparative Study of ± Steganalyzers. In: Workshop on Multimedia Signal Processing, Cairns, Queensland, Australia, October 2008, pp. 791–796 (2008)
21. http://garr.dl.sourceforge.net/sourceforge/steghide/ steghide-0.5.1-win32.zip
22. Farid, H., Lyu, S.: Detecting Hidden Messages Using Higher-Order Statistics and Support Vector Machines. In: Workshop on Information Hiding, Noordwijkerhout, The Netherlands, October 2002, pp. 340–354 (2002)

An Effective Combination of MPP Contour-Based Features for Off-Line Text-Independent Arabic Writer Identification

Mohamed Nidhal Abdi, Maher Khemakhem, and Hanene Ben-Abdallah

Mir@cl Lab, FSEGS, University of Sfax,
BP 1088, Sfax, Tunisia
Nidhal.Abdi@gmail.com,
{Maher.Khemakhem,Hanene.BenAbdallah}@fsegs.rnu.tn

Abstract. This paper proposes an off-line, text-independent, Arabic writer identification approach, using a combination of probability distribution function (PDF) features. In writer identification, the success of PDFs in terms of homogeneity, classification and identification rates encouraged researchers to study them with different types of structural features. Intensive experiments achieved on 82 writers from the IFN/ENIT database show, in particular, that 6 simple feature vectors based on the length, direction, angle and curvature measurements, which are extracted from the minimum-perimeter polygon (MPP) contours of the pieces of Arabic words, can be used to reach promising Arabic writer identification rates. The results are obtained using a set of distance metrics and the Borda ranking algorithm for classification, and the best identification rates are 90.2% for top1, and 97.5% for top10, which confirm the consistency of the proposed approach.

Keywords: behavioural biometrics, OCR, text-independent Arabic writer identification, off-line handwriting recognition, structural features combination, classification.

1 Introduction

The handwriting-based identification of individuals is a behavioural recognition modality that is becoming an increasingly active research area of the pattern recognition domain [1]. It generally proceeds by matching unknown handwritings against a database of samples with known authorship. Moreover, current writer identification techniques can be classified into on-line *vs.* off-line, text-dependent *vs.* text-independent and structural *vs.* statistical [1],[2].

Handwriting-based writer identification concerns a broad range of real-world applications, ranging from forensic [3] and historical document analysis [4], to handwriting recognition system enhancement [5]. Its recent and good performances make it to a certain extent a comparable tool to the strong physiologic modalities of identification, such as DNA and fingerprints [2]. However, a number of difficulties and challenges remains an open study field. In particular, handwriting as neuro-motor behaviour is viewed as an inherently stochastic and therefore noisy process [6]. As a

D. Ślęzak et al. (Eds.): SIP 2009, CCIS 61, pp. 209–220, 2009.
© Springer-Verlag Berlin Heidelberg 2009

result, an important challenge consists in developing and enhancing techniques that capture writer uniqueness (*between-writer* variability) while being reasonably insensitive to same writer style variations (*within-writer* variability) [7]. In some studies, the within-writer variability is explained according to biophysical and biomechanical factors on the one hand, and to psychological factors on the other hand. The biophysical and biomechanical factors consist of the intrinsic motor system signal-to-noise ratio [8], and the kinematic adaptation of the neuro-motor system and the motor effectors to various writing settings [9]. And the psychological factors are mainly related to the writer's stress level and psychological attitude [6]. Also, the handwriting may spontaneously change during the writer's life time [10], not to mention the ability of writers to voluntarily alter their handwriting style.

Despite the growing interest in writer identification, Arabic writer identification has not been addressed as extensively as Latin or Chinese writer identification for example, until recently [11]. In fact, a number of new approaches has been proposed for Arabic in recent years. In particular, Faddaoui and Hamrouni opted for a set of 16 Gabor filters [12] for handwriting texture analysis. Also, Nejad and Rahmati used a Gabor multi-channel based method [13] for Persian writer identification. Similarly, Ubul *et al.* proposed another Gabor multi-channel wavelet for the Uyghur language in China [14], which is written using the Arabic and the Persian characters. On the other hand, Al-Dmour and Zitar addressed the problem of Arabic writer identification using a set of hybrid spectral-statistical measures [15]. In [16], Gazzeh and Ben Amara applied spatial-temporal textural analysis in the form of lifting scheme wavelet transforms. Angular features were considered as well in the task of Arabic writer identification. Indeed, Bulacu and Shoemaker [11] considered a set of edge-based joint directional probability distributions, like contour-direction probability distribution function (PDF), contour-hinge PDF and direction co-occurrence PDF. To improve performance, joint directional probability distributions were combined with grapheme-emission distribution. Similarly, Al'Maadid *et al.* [17] employed edge-based directional probability distributions, combined with moment invariants and structural word features, such as area, length, height, length from baseline to upper edge and length from base line to lower edge. On the other hand, Abdi *et al.* used stroke measurements of Arabic words, such as length, ratio and curvature, in the form of PDFs and cross-correlation transform of features [18]. Also, Rafiee and Motavalli [19] introduced a new Persian writer identification method, using baseline and width structural features, and relying on a feed forward neural network for the classification. In [20] as well, Ben Amara and Gazzeh used neural networks and support vector machines on Arabic handwriting for writer identification.

This paper presents a novel approach for off-line, text-independent Arabic writer identification. In this approach, six feature vectors are computed from the handwriting. The feature vectors are the following: weighted edge direction PDF (*f1*), edge length/direction PDF (*f2*), angle PDF (*f3*), angle co-occurrence PDF (*f4*), cross-correlation of angle co-occurrence distribution (*f5*) and curvature PDF (*f6*). In addition, classification is carried out using a set of distance metrics and the Borda ranking algorithm [21]. The remainder of this paper is organized as follows: Section 2 describes the proposed approach. Section 3 details and explains the feature vectors extraction. Experiments and obtained results are presented and discussed in section 4. Finally, the conclusion of this work is drawn in the last section.

2 Proposed Approach

This approach deals with free-style, unrestrained, cursive Arabic handwriting. In particular, it captures writer individuality using a set of writer-sensitive feature vectors. These feature vectors are in the form of PDFs and cross-correlation transforms of features. The choice of PDFs is justified by their reported discriminative capacity in the context of stochastic processes in general, and of text-independent writer identification specifically [2]. In addition, PDFs are homogenous and do not require complex classification schemes to give satisfactory identification rates. Indeed, the simple distance metrics for classification, and the Borda ranking algorithm for feature combination, led to promising results in our experiments. The feature vectors in this approach are computed from Arabic word contours: First, pieces of Arabic word (PAW) contours are obtained. Then, length, direction, angle and curvature measurements of edges are extracted from the minimum perimeter polygon (MPP) approximation of the obtained contours. Finally, the measurements are counted and normalised in PDFs and cross-correlation distributions and used in the classification step.

3 Feature Extraction

In this section, feature extraction is described, and feature vectors are explained. First, diacritic points are removed (Fig. 1 (a), (b)). Then, the handwriting image is morphologically dilated (Fig. 1 (c)). Finally, the 4-boundary image $\mathbf{b} \in \{0, 1\}^x$ of the

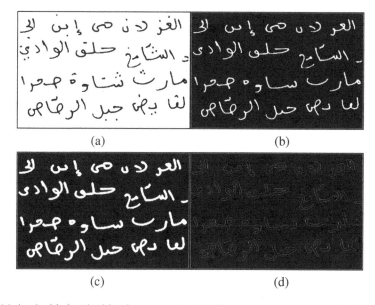

(a)   (b)

(c)   (d)

Fig. 1. (a) An Arabic handwriting image sample. (b) Diactric points are removed. (c) Handwriting image is dilated. (d) PAW contours obtained in the 4-boundary image of the sample. Generally, a PAW body corresponds to a connected-component in the handwriting after diacritic point removal.

handwriting is identified in order to extract PAW contours (Fig. 1 (d)): a pixel **x** belongs to the boundary if it is "on", and has at least one "off" pixel in its four neighborhood $N_4(\mathbf{x})$ [22],[23].

$$\mathbf{b(x)} = \begin{cases} 1 \text{ if } \mathbf{x} = 1 \text{ and } \sum N_4(\mathbf{x}) \in [0..3] \\ 0 \text{ otherwise .} \end{cases} \tag{1}$$

After contour extraction, contours are approximated. For this purpose, a minimum-perimeter polygon (MPP) algorithm with a 3×3 pixel grid is used [24]. Therefore, only the essential of the contour shape is retained as a *simple* polygon (Fig. 2).

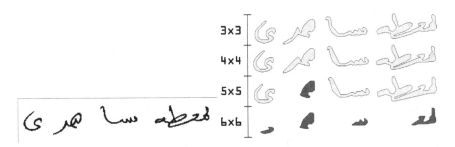

Fig. 2. An example of MPPs obtained using different grid sizes. Contour protrusions disappear with grid size increase. MPPs that no longer represent sensibly their respective PAWs outlines are filled in darker color.

Then, the following four edge measurements are taken:

- Length (*L*): represents the number of pixels in the edge. An edge is considered *upper* if it is located above the PAW central line, and *lower* otherwise,
- Direction (ϕ): denotes the edge angle relatively to the horizontal straight line (Fig. 3 (a)). ϕ is computed using the law of cosines. $0 < \phi \leq \pi$, an edge is considered *vertical* if $\pi/4 \leq \phi \leq 3\pi/4$, and *horizontal* otherwise,

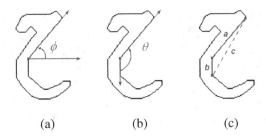

(a)　　　　　(b)　　　　　(c)

Fig. 3. (a) Edge direction (ϕ) relatively to the horizontal straight line. (b) Angle (θ) formed by two adjacent edges. (c) Curvature (*C*): $C = \dfrac{\|c\|}{\|a\| + \|b\|}$.

- Angle (θ): represents the angle formed by two adjacent edges (Fig. 3 (b)). Similarly to ϕ, θ is determined using the law of cosines. $0 < \theta \leq \pi$, no distinction is made between concave and convex angles [24]. An angle is considered *upper* if its vertex is located above the PAW central line, and *lower* otherwise,
- Curvature (C): obtained by dividing the distance between the extremity endpoints of two adjacent edges by the sum of the edge lengths (Fig. 3 (c)). $C \in [0, 1]$.

Next, the following six feature vectors are computed using the contour measurement information. These features fall in 3 categories: direction features, angle features and curvature features [17], [25], [26].

3.1 Weighted Edge Direction Probability Distribution Function (*f1*)

Direction features are strongly influenced by the handwriting slant, which is an essential and straightforward component of the individual handwriting style. Direction features are insensitive to scale. However, these features are particularly sensitive to writing settings (the wrist and writing surface positions, the pen-grip etc.) and are subject to the writer's psychological attitude or his/her voluntary control [6],[9].

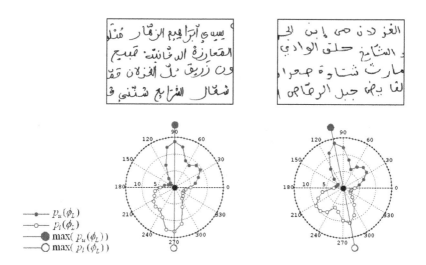

Fig. 4. The weighted edge direction PDF of two Arabic handwriting image samples, plotted in polar diagrams. max($p_u(\phi_L)$) and max($p_l(\phi_L)$) represent respectively the dominant upper and the dominant lower slants of the handwriting.

The weighted edge direction PDF, denoted $p(\phi_L)$, is the probability distribution of direction (ϕ) weighted by length (L) (Fig. 4). A total of 16 direction intervals is considered. First, the histograms of edge lengths sum per direction are computed separately for upper and lower edges. Then, these histograms are normalised, and interpreted respectively as weighted upper edge direction PDF, $p_u(\phi_L)$, and weighted lower edge direction PDF, $p_l(\phi_L)$. Finally, $p(\phi_L)$ is considered as the concatenation of the two PDFs. It represents a 32-dimensionnel feature vector. In literature, a similar feature vector is described in [11].

3.2 Edge Length/Direction Probability Distribution Function (*f2*)

The following 608-dimensionnel feature vector is the co-probability p of length (L) and direction (ϕ), denoted $p(L, \phi)$. L is considered from 1 to 76 by intervals of 2 pixels (Fig. 5 (a)). As for ϕ, a total of 16 direction intervals is retained. The edge length/direction PDF is computed separately for vertical edges, $p_v(L, \phi)$ (Fig. 5 (b)), and horizontal edges, $p_h(L, \phi)$ (Fig. 5 (c)). Similarly to (*f1*), $p(L, \phi)$ is formed by the concatenation of $p_v(L, \phi)$ and $p_h(L, \phi)$.

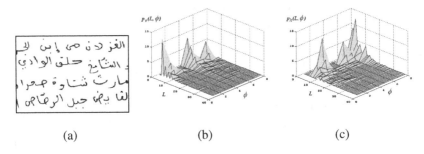

(a)   (b)   (c)

Fig. 5. (a) An Arabic handwriting image sample. (b) The vertical edge length/direction PDF of the sample, $p_v(L, \phi)$. (c) The horizontal edge length/direction PDF of the sample, $p_h(L, \phi)$.

This feature vector gave one of the best identification rates. However, it is the only considered feature vector in our approach which is scale sensitive [27].

3.3 Angle Probability Distribution Function (*f3*)

Angle measures are used to capture the handwriting roundness/sharpness by the mean of angle intervals, which are experimentally proven to reflect efficiently writer uniqueness (Fig. 6 (a),(b)). Angle features are described as more stable than features that mainly depend on the sole slant direction, and are largely insensitive to scale [17], [27].

In the angle PDF, angle instances are counted by intervals into a histogram having 24 bins. The histogram is normalized to a probability distribution, denoted $p(\theta)$, in the form of a 24-dimensionnel feature vector. Despite its low dimensionality compared to the rest of the described features, this feature vector performed well in terms of identification rates.

3.4 Angle Co-occurrence Probability Distribution Function (*f4*)

The angle co-occurrence PDF, denoted $p(\theta_1, \theta_2)$, is the co-probability distribution p of two consecutive angles θ_1 and θ_2 on the PAW MPP contour approximation. Both

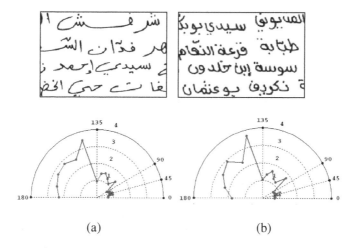

(a)   (b)

Fig. 6. The log-normalised polar diagram of the angle PDF, $p(\theta)$. Two Arabic handwriting image samples are considered: (a) a "rounded" handwriting style. (b) a "sharp" handwriting style.

angles, θ_1 and θ_2, are taken from 0 to π by intervals of $\pi/24$, which performed best and are retained for experimentation on handwriting image samples (Fig. 7 (a)). Indeed, the resulting normalised distribution is a 576-dimensionnel feature vector (Fig. 7 (b)). This feature vector gave the best results among the described features in this approach.

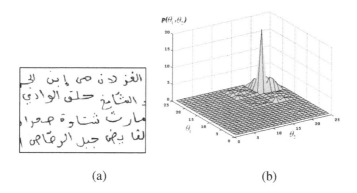

(a)   (b)

Fig. 7. (a) An Arabic handwriting image sample. (b) The angle co-occurrence PDF of the sample, $p(\theta_1, \theta_2)$.

3.5   Cross-Correlation of Angle Co-occurrence Distribution (*f5*)

This feature vector is justified by our experimental results confirming that cross-correlation between probability distributions efficiently reflects writer individuality [18]. In particular, it reflects how strong upper angle and lower angle co-occurrence

PDFs, respectively denoted $p_u(\theta_1, \theta_2)$ and $p_l(\theta_1, \theta_2)$, are related using the cross-correlation transformation.

$$f5 = (p_u(\theta_1, \theta_2) \star p_l(\theta_1, \theta_2)) \; . \tag{2}$$

(*f5*) represents the PDF of the difference between the random variables having the angle probability distributions p_u and p_l. The obtained feature vector is 1151 values in length.

3.6 Curvature Probability Distribution Function (*f6*)

Curvature is a scale-independent ratio measure of contours. It is also fairly discriminative according to our experiments. Besides, curvature is described as complementary to angle measure [18],[26].

The computation of this 32-dimensionnel feature vector, denoted $p(C)$, starts by counting curvature instances by intervals into a histogram having 32 bins. Then, the histogram is normalised to a probability distribution function, in order to finally obtain the feature vector.

4 Experimental Results

Training and testing are performed on 82 writers from the IFN/ENIT database (Fig. 8). The database consists of 26,459 images of 937 Tunisian villages and town names. It is widely considered as a major database for the evaluation of Arabic handwriting recognition systems. Indeed, it was used for the ICDAR 2005 Arabic OCR competition [28]. However, one difficulty that arises is the lack of data per writer. In fact, only an average of 60 words is available per writer. This constitutes a challenge, since dealing effectively with handwriting, as a stochastic process, depends on the amount of training data in use.

Fig. 8. Arabic handwriting image samples from the IFN/ENIT database, belonging to different writers

The effectiveness of our approach is tested via two types of experiments: the first type is designed to evaluate feature vectors individually, whereas the second type is aimed at testing feature vector combinations. The handwriting is considered at a resolution of 96dpi, in the form of b/w word padded image samples. Approximately, 66% of the data is used for training and 34% for testing, while keeping the training and the testing sets strictly disjoint. As for classification, several commonly used distance measures are tested, such as: χ^2, Euclidean, standardized Euclidean, Manhattan, Mahalanobis, Minkowski, Hamming and Chebechev. In Table 1, the best performing distance metrics for the contour-based features are presented.

Table 1. Overview of the contour-based feature vectors used for writer identification, their dimensionality and best forming distance

Feature	Explanation	PDF	Dim.	Distance
f1	Weighted edge direction	$p(\phi_L)$	32	χ^2
f2	Edge length/direction	$p(L, \phi)$	608	Manhattan
f3	Angle	$p(\theta)$	24	χ^2
f4	Angle co-occurrence	$p(\theta_1, \theta_2)$	576	Manhattan
f5	Cross-corr. of angle co-occ.	$(p_u(\theta_1, \theta_2) \star p_l(\theta_1, \theta_2))$	1151	S. Euclidean
f6	Curvature	$p(C)$	32	χ^2

For every feature, Table 1 shows its number, explanation, dimensionality and best performing distance metric between training and testing samples. On the other hand, Table 2 gives the individual feature vector identification rates. The feature (f4), which is an angle feature vector, shows the best identification rate of 71.9% for Top1, and 96.3% for Top10. Moreover, the direction feature vectors performed well, with (f2) giving 65.8% for Top1, and 95.1% for Top10. However, the curvature distribution gave one of the lowest identification rates. But despite its individual performance, it influenced positively the identification rate of the feature combination. This could be explained by the complementarity of the curvature feature with the angle and the direction features.

Table 2. The individual identification rates of the contour-based feature vectors, using their best performing distance metrics as shown in Table 1

Feature.	Identification (%)				
	Top1	Top2	Top3	Top5	Top10
f1	43.9	62.1	69.5	81.7	91.4
f2	65.8	78.0	87.8	90.2	95.1
f3	43.9	69.5	76.8	82.9	92.6
f4	71.9	81.7	84.1	93.9	96.3
f5	37.8	51.2	60.9	71.9	82.9
f6	39.0	53.6	59.7	68.2	86.5

Next, the feature vectors are combined, and the results are presented in Table 3. The Borda ranking algorithm is used for the classification task as following [21]: First, the rank of writer candidates is retained according to the features. Then, the Borda rank is computed iteratively by merging ranks from the first to the last feature vector for each writer candidate. The features order is also indicated in Table 3. In every iteration $t+1$, the new rank r is merged with the existing one $\tilde{r}$ of the previous iteration t.

$$\tilde{r}_{t+1} = \alpha r_{t+1} + (1-\alpha)\tilde{r}_t \tag{3}$$

Finally, writer candidates are classified according to the obtained ranks. Experimentally, $\alpha = 0.6$, which gave the best identification rate, is retained.

Table 3. The identification rates of the contour-based feature vector combinations

Feature combination	Identification (%)				
	Top1	Top2	Top3	Top5	Top10
f2 & f3 & f4	82.9	89.0	91.4	95.1	96.3
f4 & f3 & f2	84.1	89.0	93.9	95.1	96.3
f3 & f1 & f4 & f2	87.8	91.4	93.9	96.3	97.5
f3 & f1 & f2 & f4	87.8	92.6	93.9	95.1	96.3
f6 & f1 & f3 & f4 & f2	86.5	91.4	93.9	95.1	97.5
f4 & f6 & f1 & f3 & f2	86.5	91.4	93.9	96.3	97.5
f5 & f6 & f1 & f3 & f4 & f2	89.0	91.4	95.1	95.1	97.5
f5 & f6 & f1 & f3 & f2 & f4	**90.2**	**92.6**	**93.9**	**96.3**	**97.5**

As shown in Table 3, the best combinations are dominated by the best performing feature vectors (f4) and (f2). Indeed, these feature vectors are the last merged by the Borda ranking algorithm, giving them the biggest influence on the final ranks. The other feature vectors also enhanced the final combination in different proportions. The best feature vector combination involves all the feature vectors. In addition, it achieves 90.2% for Top1 and 97.5% for Top10. The results reflect the efficiency of our proposed approach, despite the reduced amount of training data.

The comparison of the final results with other Arabic writer identification approaches is difficult. The difficulty is due, in fact, to the diversity of experimentation conditions in terms of writers count, database type and experimentation methodology. Moreover, most of the approaches prototypes aren't available for a standard experimentation set.

5 Conclusion

We proposed in this paper a novel approach for off-line, text-independent, Arabic writer identification. What makes this approach attractive is the simplicity and the effectiveness of the six feature vectors used, which are extracted from the minimum-perimeter polygon contours of the pieces of Arabic words, based on length, direction, angle and curvature measurements. Intensive experiments were performed on 82 writers from the IFN/ENIT database, using a set of distance metrics and the Borda ranking algorithm for classification, and the best achieved results were 90.2% for top1, and 97.5% for top10. Several studies concerning especially the improvement of our approach and its adaptation to the writer verification are currently conducted.

References

1. Plamondon, R., Lorette, G.: Automatic Signature Verification and Writer Identification – The State of the Art. J. Pattern Recognition 22, 107–131 (1989)
2. Schomaker, L.: Advances in Writer Identification and Verification. In: 9th international Conference on Document Analysis and Recognition (ICDAR 2007), vol. 2, pp. 1268–1273 (2007)
3. Srihari, S.N., Lee, S.: Automatic Handwriting Recognition and Writer Matching on Anthrax-Related Handwritten Mail. In: 8th International Workshop on Frontiers in Handwriting Recognition (IWFHR 2002), pp. 280–284 (2002)
4. Fornes, A., Llados, J., Sanchez, G., Bunke, H.: Writer Identification in Old Handwritten Music Scores. In: 8th IAPR Workshop on Document Analysis Systems, pp. 347–353 (2008)
5. Sas, J.: Handwriting Recognition Accuracy Improvement by Author Identification. In: Rutkowski, L., Tadeusiewicz, R., Zadeh, L.A., Żurada, J.M. (eds.) ICAISC 2006. LNCS (LNAI), vol. 4029, pp. 682–691. Springer, Heidelberg (2006)
6. van Galen, G.P., van Huygevoort, M.: Error, Stress and the Role of Neuromotor Noise in Space-Oriented Behaviour. J. Biological Psychology 51, 151–171 (2000)
7. Djioua, M., Plamondon, R.: Studying the Variability of Handwriting Patterns using the Kinematic Theory. J. Human Movement Science (in press, 2009) (Corrected Proof)
8. Plamondon, R.: Looking at Handwriting Generation from a Velocity Control Perspective. J. Acta Psychologica 82, 89–101 (1993)
9. Ray, D.N.: Writing Position as a Factor in Handwriting Variability. Thesis (M.A. In: Police Science and Administration), State College of Washington (1959)
10. Seidler, R.D.: Differential Effects of Age on Sequence Learning and Sensorimotor Adaptation. Brain Research Bulletin 70, 337–346 (2006)
11. Bulacu, M., Schomaker, L., Brink, A.: Text-Independent Writer Identification and Verification on Off-Line Arabic Handwriting. In: 9th International Conference on Document Analysis and Recognition (ICDAR 2007), vol. 2, pp. 769–773 (2007)
12. Feddaoui, N., Hamrouni, K.: Personal Identification based on Texture Analysis of Arabic Handwriting Text. In: IEEE International Conference on Information and Communications Technologies (ICTTA 2006), vol. 1, pp. 1302–1307 (2007)
13. Nejad, F.S., Rahmati, M.: A New Method for Writer Identification and Verification based on Farsi/Arabic Handwritten Texts. In: 9th International Conference on Document Analysis and Recognition (ICDAR 2007), vol. 2, pp. 829–833 (2007)
14. Ubul, K., Hamdulla, A., Aysa, A., Raxidin, A., Mahmut, R.: Research on Uyghur Off-Line Handwriting-Based Writer Identification. In: 9th International Conference on Signal Processing (ICSP 2008), pp. 1656–1659 (2008)
15. AL-Dmour, A., Zitar, R.A.: Arabic Writer Identification based on Hybrid Spectral-Statistical Measures. J. Experimental and Theoretical Artificial Intelligence 19, 307–332 (2007)
16. Gazzah, S., Ben Amara, N.E.: Arabic Handwriting Texture Analysis for Writer Identification using the DWT-lifting Scheme. In: 9th International Conference on Document Analysis and Recognition (ICDAR 2007), vol. 2, pp. 1133–1137 (2007)

17. Al-Ma'adeed, S., Mohammed, E., Al Kassis, D., Al-Muslih, F.: Writer Identification using Edge-Based Directional Probability Distribution Features for Arabic Words. In: IEEE/ACS International Conference on Computer Systems and Applications (AICCSA 2008), pp. 582–590 (2008)
18. Abdi, M.N., Khemakhem, M., Ben-Abdallah, H.: A Novel Approach for Off-Line Arabic Writer Identification Based on Stroke Feature Combination. In: 24th IEEE International Symposium on Computer and Information Sciences, ISCIS 2009 (in press, 2009)
19. Rafiee, A., Motavalli, H.: Off-Line Writer Recognition for Farsi text. In: 6th Mexican International Conference on Artificial Intelligence (MICAI 2007), Special Session, pp. 193–197 (2008)
20. Ben Amara, N., Gazzah, S.: Neural Networks and Support Vector Machines for Writer Identification using Arabic Scripts. In: International Conference on Machine Intelligence (2005)
21. Jeong, S., Kim, K., Chun, B., Lee, J., Bae, Y.J.: An Effective Method for Combining Multiple Features of Image Retrieval. In: IEEE Region 10 Conference (TENCON 1999), vol. 2, pp. 982–985 (1999)
22. van den Boomgard, R., van Balen, R.: Methods for Fast Morphological Image Transforms using Bitmapped Images. J. Computer Vision, Graphics, and Image Processing: Graphical Models and Image Processing 54, 254–258 (1992)
23. Ritter, G.X., Wilson, J.N.: Handbook of Computer Vision Algorithms in Image Algebra. Technical Report. Centre for Computer Vision and Visualization, University of Florida (1996)
24. Kim, C.E., Sklansky, J.: Digital and Cellular Convexity. J. Pattern Recognition 15, 359–367 (1982)
25. Bulacu, M., Schomaker, L.: Text-Independent Writer Identification and Verification Using Textural and Allographic Features. IEEE Transactions on Pattern Analysis and Machine Intelligence 29, 701–717 (2007)
26. Khorsheed, M.S., Clocksin, W.F.: Structural Features of Cursive Arabic Script. In: Proceedings of the British Machine Vision Conference (1999)
27. Abdi, M.N.: Reconnaissance Automatique de l'Ecriture Arabe Imprimée. Thesis (M.S. in: Computer Science), Faculté des Sciences Economiques et de Gestion, Sfax, Tunisia (2005)
28. Margner, V., El-Abed, H.: Databases and Competitions: Strategies to Improve Arabic Recognition Systems. In: Doermann, D., Jaeger, S. (eds.) SACH 2006. LNCS, vol. 4768, pp. 82–103. Springer, Heidelberg (2008)

Ringing Artifact Removal in Digital Restored Images Using Multi-Resolution Edge Map

Sangjin Kim[1], Sinyoung Jun[1], Eunsung Lee[1], Jeongho Shin[2], and Joonki Paik[1]

[1] Image Processing and Intelligent Systems Laboratory, Graduate School of Advanced Imaging Science, Multimedia, and Film, Chung-Ang University, 221 Heuksuk-Dong, Dongjak-Ku, Seoul 156-756, Korea
layered372@wm.cau.ac.kr, jjun0427@wm.cau.ac.kr, lessel7@wm.cau.ac.kr, paikj@cau.ac.kr
[2] Department of Web Information Engineering, Hankyong National University, 67 Seokjeong-Dong, Anseong, Kyonggi-Do 456-749, Korea
shinj@hknu.ac.kr

Abstract. This paper presents a novel approach to reducing ringing artifact in digitally restored images by using multi-resolution edge map. The performance of reducing ringing artifacts depends on the accurate classification of the local features in the image. The discrete wavelet transform (DWT) provides effective insight into both spatial and frequency characteristics of an image. Through the DWT analysis, we show that ringing artifacts can be suppressed to a great extent by using multiple-level edge maps, which provide enhanced matching to local edges. Base on the experimental results, the proposed method can reduce ringing artifacts with minimized edge degradation by using DWT analysis.

Keywords: Ringing Artifact, Multi-Resolution Edge Map, Discrete Wavelet Transform.

1 Introduction

In recent times, digital image restoration techniques have attracted increasing interests in various applications such as super-resolution and digital auto-focusing, to name a few. Therefore, the quality of the image is an important factor. As the image restoration technique has wider application areas, the ringing artifacts removal technique, which is the side effects of the image restoration process, becomes more necessary.

Conventional image restoration algorithms, such as the constrained least squares (CLS) and the Wiener filter, try to reduce the processing time by using fast fourier transform (FFT). These approaches are often inappropriate for real applications because of linear space invariance assumption, which regularizes the restoration operation uniformly across the entire image. Such frequency-domain image restoration techniques result in ringing artifacts in the neighborhood of abrupt intensity transitions due to the uniform amplification of the high-frequency component [1]. Ringing

D. Ślęzak et al. (Eds.): SIP 2009, CCIS 61, pp. 221–227, 2009.

artifacts generally results from the poor match between the stationary image model and the actual image data [2]. There have been various researches for reducing ringing artifacts. The adaptive post-filtering approach extracts local statistics and constructs an edge map of the image [3]. Accurate localization of edge alleviates ringing artifacts by using quantized discrete cosine transform (DCT) coefficients of a block containing a straight edge [4]. In this paper, we present a multi-resolution, computationally simple post-filtering approach to reduce ringing artifacts in the restored images.

2   Ringing Artifacts Removal Using the DWT-Based Adaptive Edge Map

In this section we describe the proposed ringing reduction algorithm. The basic idea of the proposed algorithm is the application of DWT for making an edge map. Based on the human visual system, each pixel in an image plays a different role in describing the image. Since the human visual system is very sensitive to high frequency changes and especially to edges in the image, the edges are very important to the perception of the image [3]. Therefore, the proposed algorithm is to classify local features in the restored image, and adaptively apply the smoothness constraints. As mentioned in the previous section, the restored image has ringing artifacts. In order to reduce ringing artifacts in the restored Image, we present detailed descriptions for edge extraction and the post filtering in the following subsections, respectively.

2.1 Wavelet Analysis for Extracting Edge Region

We suppose to have a restored image with ringing artifacts, which is the output of the well-known constraint least square (CLS) filter [5]. We use the incorporation of wavelet decomposition for finding the edge.

Fig. 1. shows a 2D wavelet transform. The horizontal, vertical, and diagonal component of the wavelet transform using Daubechies D4 wavelet is shown in Fig. 1(a). Note, for example, that the horizontal edges of the original image are present in the horizontal detail coefficients of the upper-right quadrant of Fig. 1(a). The vertical edges of the image can be similarly identified in the vertical detail coefficients of the lower-left quadrant. To combine this information into a single edge image, we simply set the approximation coefficients to zero, compute the inverse transform, and take the edge image resulting from the absolute value. In the same way, we take the edge image resulting in second-level decomposition Fig. 1(b). Fig. 2(a) shows edge detection results by computing the absolute value of the inverse transform coefficients in the first level and Fig. 2(b) in the second level.

The main idea of the proposed method comes from the observation that the edge image resulting from the second level is clearer than the first level. A compromise between the two images obtains the more accurate edge map.

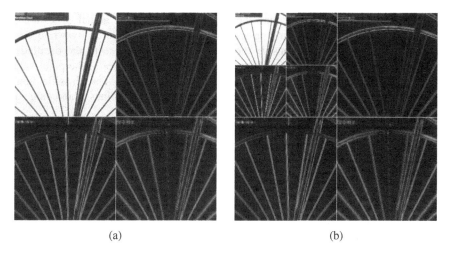

Fig. 1. First and second level decomposition (a) Degraded image of size 512X512 first-level decomposition (b) Degraded image of size 512X512 second-level decomposition

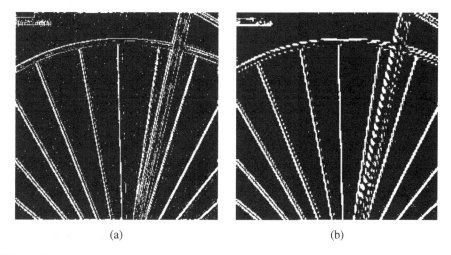

Fig. 2. Wavelets in edge detection (a) The edge resulting from computing the absolute value of the inverse transform in first-level, (b) The edge resulting from computing the absolute value of the inverse transform in second- level

2.2 Ringing Artifacts Removal

According to the previous analysis, we find the major edge map. Ringing artifacts generally occur in the neighborhood of sharp edges in the image [6]. Such regions are marked both by the occurrence of high frequency entailed in the description of the sharp edge together with smooth regions on both sides of the edge. Ringing–artifact-prone regions are detected by a series of morphological operations. The edge map is dilated to indicate the region around the major edges. Then the actual edges are excluded. Fig. 3 (b) shows the ringing-artifact-prone region rendered by white color. Variation of grayscale values inside the detected region are due to the oscillations introduced by the ringing artifacts. Therefore the average variance of the pixel values in the DWT domain can be used to quantify the degradation. The classification is obtained by applying a pre-specified threshold to the variance values. Candidate edge regions (e_i) are obtained as

$$e_i[m,n] = \begin{cases} 1, & if \ |y_i[m,n]| \ > \ T \\ 0, & otherwise \end{cases},$$

$$for \ i \in \{HL, LH, HH\}$$

(1)

where y_{HL}, y_{LH}, and y_{HH} respectively represent vertical, horizontal, and diagonal wavelet transform coefficients.

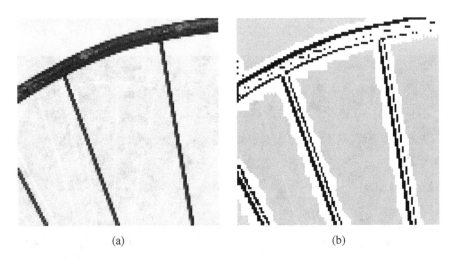

(a)   (b)

Fig. 3. Mask used to control filtering for alleviation of ringing artifacts. (a) Original image, (b) Mask generated by selecting the proposed edge map (black), the pixel surrounding the edges (white).

A smoothing operator, such as Gaussian filtering, alleviates ringing artifacts at the cost of blurring the sharp edges and other detail in the image. This trade-off has motivated adaptive methods. One approach is to locate edges in the image and to perform filtering operation on condition to the proximity to edges [4]. The regions are marked as smooth, edge, and texture regions. An adaptive 3X3 low pass filter is applied to the pixel surrounding the edges as shown with white color. Since ringing noise and texture pixels have similar variance values, the filter with the large central weight is applied to the texture pixels far away from the edges to preserve the details, and the smoothing filter is applied to the pixels close to edges to selectively filter out the ringing noise.

3   Experimental Results

In this section, we analyze the performance of the proposed ringing reduction algorithm using both objective and subjective measures. For the experiment, we used standard test images, 512X512, 8bit, Bike and Lena. Fig. 4(a) shows the restored Bike image by using the CLS filter. The input image was degraded by the 7X7 uniform blur and 40dB additive Gaussian noise, and Fig. 4(b) shows the filtered result by using the proposed ringing reduction algorithm. Figs. 4(c) and 4(d) show the magnified versions of Figs. 4(a) and 4(b), respectively.

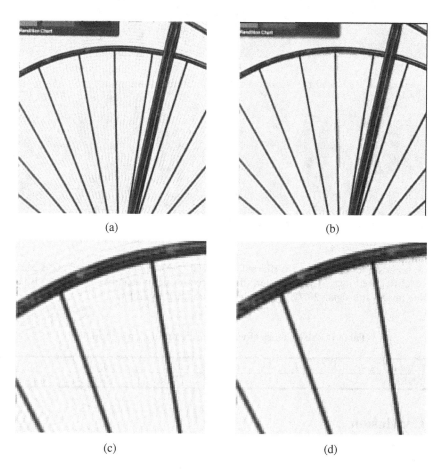

Fig. 4. (a) image restored by CLS filter with ringing artifacts (= 0.01, PSNR=21.17 dB), (b) result image using proposed algorithm (PSNR = 25.52 dB), (c) crop by 115X105 image from a, (d) crop by 115X105 image from b

The proposed algorithm is compared with various ringing reduction algorithms such as an iterative method [7] and bilateral filtering [8] in the sense of peak-to-peak signal-to-noise ratio (PSNR). As shown in Fig. 5, the proposed algorithm can effectively reduce ringing artifacts with preserving edge details near eyes and the hat and PSNR values are summarized in Table 1.

Fig. 5. (a) original Lena image, (b) Blurred image with 11X11 uniform blur and 50dB gaussian noise, (c) Restored image by CLS filter, (d) Restored image by iterative method, (e) Bilateral filtering image with sigma 20, (f) The result using proposed algorithm

Table 1. PSNR values of various image restoration algorithms

Image in Fig. 4.	(b)	(c)	(d)	(e)	(f)
PSNR(dB)	22.38	24.58	27.28	28.22	29.63

4   Conclusion

In this paper, we propose an adaptive ringing reduction algorithm using wavelet-based multi-resolution analysis. The proposed method shows significant improvement in accuracy over existing methods. The location of edges can be accurately estimated due to the wavelet representation. The proposed algorithm can also be applied to the ringing artifact removal in compressed images. Experimental results show effectively removed ringing artifacts with significant reduction in computational load. The simplified algorithm using the first and the second level of wavelet decomposition has the advantage over many post-processing algorithm.

Acknowledgement. This research was supported by Basic Science Research Program through National Research Foundation (NRF) of Korea funded by the Ministry of Education, Science and Technology (2009-0081059), by the MKE (The Ministry of Knowledge Economy), Korea, under the HNRC (Home Network Research Center) – ITRC (Information Technology Research Center) support program supervised by the NIPA (National IT Industry Promotion Agency) (NIPA-2009-C1090-0902-0035), and by Ministry of Culture, Sports and Tourism(MCST) and Korea Culture Content Agency (KOCCA) in the Culture Technology (CT) Research & Development Program 2009.

References

1. Lagendijk, R., Biemond, J., Boekee, D.: Regularized Iterative Image Restoration with Ringing Reduction . IEEE Trans. on Acoust., Speech, Signal Processing 36, 1874–1887 (1988)
2. Ritzgerrell, A., Dowski, E., Cathey, W.: Defocusing Transfer Function for Circularly Symmetric Pupils. Applied Optics 36, 5796–5804 (1997)
3. Kong, H., Vetro, A., Sun, H.: Edge Map Guided Adaptive Post-Filter for Blocking and Ringing Artifacts Removal. In: IEEE Int. Symp. Circuits and Systems (ISCAS) vol. 3, pp. 929–932 (2004)
4. Popovici, I., Withers, W.: Locating Edges and Removing Ringing Artifacts in JPEG Image by frequency-domain analyses. IEEE Trans. Image Processing 16, 1470–1474 (2007)
5. Gonzalez, R., Woods, R.: Digital Image Processing, 2nd edn. Prentice Hall, Englewood Cliffs (2001)
6. Yang, S., Hu, Y., Nguyen, T., Tull, D.: Maximum-Likelihood Parameter Estimation for Image Ringing-Artifact Removal. IEEE Trans. on Circuits and Systems for Video Technology 11, 963–973 (2001)
7. Yuan, L., Sun, J., Quan L., Shum H.: Progressive Inter-scale and Intra-scale Non-Blind Image Deconvolution. ACM Trans. on Graph 27, 74:1–74:10 (2008)
8. Zhang, M., Gunturk, B.: Multi-Resolution Bilateral Filtering for Image Denoising. IEEE Trans. on Image Processing 17, 2324–2333 (2008)

Upmixing Stereo Audio into 5.1 Channel Audio for Improving Audio Realism

Chan Jun Chun, Yong Guk Kim, Jong Yeol Yang, and Hong Kook Kim

Department of Information and Communications
Gwangju Institute of Science and Technology
1 Oryong-dong, Buk-gu, Gwangju 500-712, Korea
{cjchun,bestkyg,jyyang,hongkook}@gist.ac.kr

Abstract. In this paper, we address issues associated with upmixing stereo audio into 5.1 channel audio in order to improve audio realism. First, we review four different upmixing methods, including a passive surround decoding method, a least-mean-square based upmixing method, a principal component analysis based upmixing method, and an adaptive panning method. After that, we implement a simulator that includes the upmixing methods and audio controls to play both stereo and upmixed 5.1 channel audio signals. Finally, we carry out a MUSHRA test to compare the quality of the upmixed 5.1 channel audio signals to that of the original stereo audio signal. It is shown from the test that the upmixed 5.1 channel audio signals generated by the four different upmixing methods are preferred to the original stereo audio signals.

Keywords: Audio upmixing, stereo audio, multi-channel audio, passive surround decoding, least mean square, principal component analysis, adaptive panning.

1 Introduction

As technologies related to audio systems have advanced, the demand for multi-channel audio systems has increased too. Such audio systems not only provide more realistic sound, but also offer more ambient effects than standard stereo audio systems. For example, if audio content having fewer channels than can be provided by a target system is available, the target audio system cannot take full advantage of it. Therefore, in order to utilize such audio content, it is necessary to use an upmixing method that converts mono or stereo audio formats into a multi-channel audio format suitable for such a system.

Many multi-channel audio systems currently exist with a wide range in the number of channels available. Since 5.1 channels are better for creating the effects of ambience and spaciousness than stereo channels, we need to develop upmixing methods that convert audio from a stereo format to a 5.1 channel format. Since one of the typical approaches for creating additional channels is to use a correlation property between stereo channels, we first review and implement four correlation-based methods, including a passive surround decoding method [1], a least- mean-square based method [2], a principal component analysis based method [2], and an adaptive panning

D. Ślęzak et al. (Eds.): SIP 2009, CCIS 61, pp. 228–235, 2009.
© Springer-Verlag Berlin Heidelberg 2009

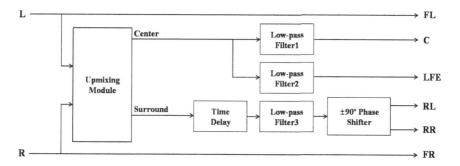

Fig. 1. Procedure for upmixing a stereo audio to a 5.1 channel audio format

method [3]. We then compare the upmixed audio contents obtained by each method to the original stereo audio contents.

The remainder of this paper is organized as follows. Following this introduction, we review four different upmixing methods for converting audio from a stereo format to a 5.1 channel format in Section 2. After that, we design a simulator for upmixing stereo contents using these methods in Section 3. In Section 4, the quality of the upmixed 5.1 channel audio is compared with that of the original stereo audio by a means of a multiple stimuli with hidden reference and anchor (MUSHRA) test [4]. Finally, we conclude this paper in Section 5.

2 Audio Upmixing Algorithm from Stereo to 5.1 Channel Audio

In this section, we describe four different upmixing methods for converting audio from a stereo format to a 5.1 channel format. Fig. 1 shows the upmixing procedure, where the channels are labeled FL (front left), FR (front right), C (center), LFE (low frequency enhancement), RL (rear left), and RR (rear right). As illustrated in the figure, the FL and the FR channels for the 5.1 channel audio format are directly obtained from the original stereo channels, while the remaining channels are generated from the center channel and the surround channels. Therefore, it is discussed in the following subsections how to derive the center channel and the surround channels by using each upmixing method.

2.1 Passive Surround Decoding Method

The passive surround decoding (PSD) method is an early passive version of the Dolby Surround Decoder [1]. In this method, the center channel is obtained by adding the original left and right channels. On the other hand, the surround channel can be derived by subtracting the right channel from the left channel. Note that in order to maintain a constant acoustic energy, the center and the surround channel are lowered by 3 dB, which is implemented by multiplying $1/\sqrt{2}$ to the center and the surround channels. That is, the center and the surround channels are obtained by using the equations of

$$Center\,(n) = (x_L(n) + x_R(n))/\sqrt{2}\,, \tag{1}$$

$$Surround\,(n) = (x_L(n) - x_R(n))/\sqrt{2}\,, \tag{2}$$

where $x_L(n)$ and $x_R(n)$ represent the left and the right samples at the time index n, respectively.

2.2 LMS-Based Upmixing Method

The least-mean-square (LMS)-based upmixing method creates the center and surround channels using the LMS algorithm [2][5]. In this method, one of the original stereo channels is taken as the desired signal, $d(n)$, and the other is considered as the input, $x(n)$, of the adaptive filter. The error signal, $e(n)$, is then the difference of the output, $y(n)$, of the filter and the desired signal, $d(n)$. The output, $y(n)$, is defined as a linear combination of the input signals by using the equation of,

$$y(n) = \mathbf{w}^T(n)\mathbf{x}(n) = \mathbf{w}(n)\mathbf{x}^T(n)\,, \tag{3}$$

where $\mathbf{x}(n) = [x(n)\ x(n-1)\ \cdots\ x(n-N+1)]^T$ and $\mathbf{w}(n) = [w_0\ w_1\ \cdots\ w_{N-1}]^T$. In Eq. (3), $\mathbf{w}(n)$ is a coefficient vector of the N-tapped adaptive filter and is obtained based on the LMS algorithm described as

$$\mathbf{w}(n+1) = \mathbf{w}(n) + 2\mu\,e(n)\mathbf{x}(n)\,, \tag{4}$$

where μ is a constant step size which is set to 10^{-4} in this paper. In this case, $y(n)$ and $e(n)$ are considered as the signals for the center channel and the surround channel, respectively.

2.3 PCA-Based Upmixing Method

The principal component analysis (PCA)-based upmixing method decomposes the original stereo channels into correlated and uncorrelated portions [2]. In order to derive the center and the surround channels, we first need to find a 2x2 covariance matrix, $\mathbf{A}$, such that

$$\mathbf{A} = \begin{bmatrix} \text{cov}(x_L, x_L) & \text{cov}(x_L, x_R) \\ \text{cov}(x_R, x_L) & \text{cov}(x_R, x_R) \end{bmatrix}, \tag{5}$$

where $\text{cov}(x_p, x_q)$ is the covariance of x_p and x_q, and that p and q represent the left and right channels, respectively. The covariance matrix, $\mathbf{A}$, gives two eigenvectors which are the basis vectors for a new coordinate system [6]. These eigenvectors are then used as weight vectors corresponding to the left and right channels to generate the center and the surround channels, such as

$$Center(n) = c_L x_L(n) + c_R x_R(n),$$ (6)

$$Surround(n) = s_L x_L(n) + s_R x_R(n).$$ (7)

The eigenvector $[c_L \ c_R]$ corresponding to the greatest eigenvalue becomes the weight vector for the center channel. Thus, the other eigenvector $[s_L \ s_R]$ becomes the weight vector for the surround channel. Typically, the PCA-based method is implemented as frame-based processing that may cause unwanted artifacts at the frame boundaries. These artifacts can be eliminated using an overlap-and-add technique [7]. In order to perform the overlap-and-add technique, analysis and synthesis windows that perfectly satisfy the overlap-and-add reconstruction conditions are required. In this paper, we chose the same window for the analysis and the synthesis, which is denoted as

$$w(n) = \begin{cases} \sin\left(\dfrac{\pi(n+\frac{1}{2})}{2(N-M)}\right), & 0 \le n \le N-M-1 \\ 1, & N-M \le n \le M-1, \\ \sin\left(\dfrac{\pi(N-n-\frac{1}{2})}{2(N-M)}\right), & M \le n \le N-1 \end{cases}$$ (8)

where N is the length of a frame, and M is the overlap region. Here, N and M are set to 1024 and 128, respectively.

2.4 Adaptive Panning Method

The adaptive panning (ADP) method proposed in [3] generates the center and the surround channels by panning the original stereo channels. The weight vector for ADP is recursively estimated using the LMS algorithm. Let us now define $y(n)$ to be a linear combination of the original stereo channels as

$$y(n) = \mathbf{w}^T(n)\mathbf{x}(n) = \mathbf{w}(n)\mathbf{x}^T(n),$$ (9)

where $\mathbf{x}(n) = [x_L(n) \ x_R(n)]^T$ and $\mathbf{w}(n) = [w_L(n) \ w_R(n)]^T$. Two coefficients, $w_L(n)$ and $w_R(n)$, which are the elements of the weight vector corresponding to the left and the right channels, respectively, are then estimated using the LMS algorithm defined as

$$w_L(n+1) = w_L(n) - \mu\, y(n)[x_L(n) - w_L(n) y(n)],$$ (10)

$$w_R(n+1) = w_R(n) - \mu\, y(n)[x_R(n) - w_R(n) y(n)],$$ (11)

where μ is a constant step size and is set to 10^{-10}. Finally, the center and surround channels can be determined as

$$Center(n) = w_L(n)x_L(n) + w_R(n)x_R(n),\qquad(12)$$

$$Surround(n) = w_R(n)x_L(n) - w_L(n)x_R(n).\qquad(13)$$

2.5 Low-Pass Filters

For 5.1 channel audio contents such as movie and live music, voice and dialog are usually emphasized when they are play through the center channel. Therefore, the center channel is further processed by a low-pass filter, where we design a finite-duration impulse response (FIR) low-pass whose length is 256 and cut-off frequency is 4 kHz, denoted as Low-pass Filter1 in Fig. 1. On one hand, the low frequency enhancement (LFE) channel is used to emphasize low frequency region ranged from 100 to 200 Hz. To this end, an FIR low-pass filter having a cut-off frequency of 200 Hz, denoted as Low-pass Filter2 in Fig. 1, is used here, where the number of taps for the filter is 256. In the surround channels, a low-pass filter (Low-pass Filter3 in Fig. 1) is also used to simulate a high-frequency absorption effect. An FIR low-pass filter with 256 taps and a cut-off frequency of 7 kHz is used in this paper.

2.6 Time Delay and ±90° Phase Shifter

The rear left and the rear right channels are intended to provide ambience and spaciousness effects. A time delay element is used to provide such ambience effects, and a

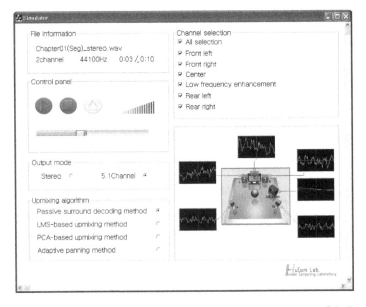

Fig. 2. A snapshot of the simulator for upmixing a stereo audio format to a 5.1 channel audio format in real-time

±90° phase shifter is needed to present spaciousness effects. Assuming that the distance from front loudspeakers to the wall is about 2 meters and the distance from rear loudspeakers to the wall is about 1 meter, a time delay of 12 ms is applied to the surround channels. In addition, a discrete Hilbert transform is used as a phase shifter [8]. By using FIR approximations having a constant group delay, we can implement a discrete Hilbert transform. In particular, the approximation is done using a Kaiser window which is defined as

$$h(n) = \begin{cases} \dfrac{I_0(\beta(1-[(n-n_d)/n_d]^2)^{1/2})}{I_0(\beta)} \cdot \dfrac{\sin(\pi(n-n_d)/2)}{\pi(n-n_d)/2}, & 0 \le n \le M \\ 0, & otherwise \end{cases}, \quad (14)$$

where M is the order of the FIR discrete Hilbert transform, and n_d is $M/2$. In this paper, M and β are set to 31 and 2.629, respectively.

3 Design and Implementation of Audio Upmixing Simulator

In this section, we present a simulator used to implement the four upmixing methods described in the previous section. Fig. 2 shows a snapshot of the simulator that is designed based on the procedure shown in Fig. 1. The simulator can play both stereo and 5.1 channel audio files. Furthermore, it enables us to have 5.1 channel audio files from stereo audio files by using one of the four different upmixing methods. The simulator mainly consists of five parts as follows.

1) File information
   This part offers information about the currently loaded audio file. It displays the filename, the number of channels, the sampling rate, etc.
2) Control panel
   The buttons in the control panel part control how to play the audio file. We can play, pause, stop, and open audio files. In addition, we can turn up and turn down the volume of the file being played.
3) Output mode
   Once the simulator has loaded a stereo audio file, the stereo file can be played in either a stereo audio format or an upmixed 5.1 channel format. In case of 5.1 channel audio files, the simulator only plays them in the 5.1 channel format.
4) Upmixing algorithm
   When an audio file is played by upmixing from stereo to 5.1 channel audio format, we can select one of the upmixing methods described in Section 3.
5) Channel selection
   We can select the activity of each speaker.

4 Performance Evaluation

In this section, we compared the quality of the upmixed 5.1 channel signals with that of the original stereo signals. First of all, we were in full compliance with the ITU multi-channel configuration standard defined by the ITU-R Recommendation

BS. 775-1 [9]. A multiple stimuli with hidden reference and anchor (MUSHRA) test [4] was conducted by using five music genres such as rock, ballad, hip-hop, classical, and heavy metal. Eight people with no auditory disease participated in this experiment. For the MUSHRA listening test, the audio contents to be compared in the test were listed as

- Hidden reference
- 3.5 kHz low-pass filtered anchor
- 7 kHz low-pass filtered anchor
- Stereo audio content containing the front left and the front right channels of the hidden reference
- Upmixed audio content by the passive surround decoding method
- Upmixed audio content by the LMS-based method
- Upmixed audio content by the PCA-based method, and
- Upmixed audio content by the adaptive panning method.

Fig. 3 shows the MUSHRA test result. As shown in the figure, the upmixed 5.1 channel audio contents by any of the upmixing methods were preferred to the stereo audio contents. This implies that 5.1 channel audio provides a better listening environment than stereo audio. Moreover, it was found out that the adaptive panning method using a robust tracking algorithm outperformed the other methods.

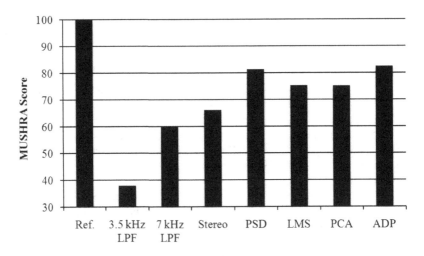

Fig. 3. Comparison of MUSHRA test scores for the audio signals upmixed by different methods

5   Conclusion

In this paper, we described four different upmixing methods for converting audio from a stereo format to a 5.1 channel format based on correlation techniques. After implementing these techniques, we then designed a simulator that was able to upmix stereo audio files and play them in real time to produce a better listening environment.

In order to evaluate the performance of the upmixing algorithms and compare the upmixed 5.1 channel signals with the original stereo signals, a MUSHRA test was conducted. It was shown from the test result that the 5.1 channel audio generated by any of the upmixing methods provided better audio quality than the original stereo audio and the adaptive panning method showed the best performance among all the methods.

Acknowledgments. This research was supported in part by the Ministry of Knowledge and Economy (MKE), Korea, under the Information Technology Research Center (ITRC) support program supervised by the National IT Industry Promotion Agency (NIPA) (NIPA-2009-C1090-0902-0017), and by a basic research project grant provided by GIST in 2009.

References

1. Dolby Laboratory: `http://www.dolby.com/professional/getting-dolby-technologies/index.html`
2. Bai, M.R., Shih, G.-Y., Hong, J.-R.: Upmixing and downmixing two-channel stereo audio for consumer electronics. IEEE Trans. on Consumer Electronics 53, 1011–1019 (2007)
3. Irwan, R., Aarts, R.M.: Two-to-five channel sound processing. J. Audio Eng. Soc. 50, 914–926 (2002)
4. ITU-R BS. 1534-1: Method for the Subjective Assessment of Intermediate Quality Levels of Coding System (2003)
5. Widrow, B., Stearns, S.D.: Adaptive Signal Processing. Prentice-Hall, NJ (1985)
6. Jolliffe, I.T.: Principal Component Analysis. Springer, Heidelberg (2002)
7. Bosi, M., Goldberg, R.E.: Introduction to Digital Audio Coding and Standards. Kluwer Academic Publishers, MA (2002)
8. Oppenheim, A.V., Schafer, R.W., Buck, J.R.: Discrete-time Signal Processing. Prentice-Hall, NJ (1989)
9. ITU-R BS.775-1: Multi-Channel Stereophonic Sound System with or without Accompanying Picture (1994)

Multiway Filtering Based on Multilinear Algebra Tools

Salah Bourennane and Caroline Fossati

Ecole Centrale Marseille,
Institut Fresnel Marseille cedex 20 France
salah.bourennane@fresnel.fr

Abstract. This paper presents some recent filtering methods based on the lower-rank tensor approximation approach for denoising tensor signals. In this approach, multicomponent data are represented by tensors, that is, multiway arrays, and the presented tensor filtering methods rely on multilinear algebra. First, the classical channel-by-channel SVD-based filtering method is overviewed. Then, an extension of the classical matrix filtering method is presented. It is based on the lower rank-$(K_1, \ldots, K_N)$ truncation of the HOSVD which performs a multimode Principal Component Analysis (PCA) and is implicitly developed for an additive white Gaussian noise. Two tensor filtering methods recently developed by the authors are also overviewed. The performances and comparative results between all these tensor filtering methods are presented for the cases of noise reduction in color images.

Keywords: Tensor, multilinear algebra, filtering, image.

1 Introduction

Tensor data modeling and tensor analysis have been improved and used in several application fields such as quantum physics, economy, chemometrics, psychology, data analysis, etc. Nevertheless, only recent studies focus their interest on tensor methods in signal processing applications. Tensor formulation in signal processing has received great attention since the recent development of multicomponent sensors, especially in imagery (color or multispectral images, video, etc.) and seismic fields (antenna of sensors recording waves with polarization properties). Indeed, the digital data obtained from these sensors are fundamentally higher order tensor objects, that is, multiway arrays whose elements are accessed via more than two indexes. Each index is associated with a dimension of the tensor generally called "nth-mode" [7, 8, 15, 16].

For the last decades, the classical algebraic processing methods have been specifically developed for vector and matrix representations. They are usually based on the covariance matrix, the cross-spectral matrix, or more recently, on the higher order statistics. Their overall aim is classically to determine a subspace

D. Ślęzak et al. (Eds.): SIP 2009, CCIS 61, pp. 236–249, 2009.

associated with the signal or the parameters to estimate. They mainly rely on three algebraic tools such as:

(1) The Singular Value Decomposition (SVD) [12], which is used in Principal Component Analysis (PCA).
(2) Penrose-Moore matrix inversion [12].
(3) The matrix lower rank approximation, which, according to Eckart-Young theorem [9], can be achieved thanks to a simple SVD truncation.

These methods have proven to be very efficient in several applications.

When dealing with multicomponent data represented as tensors, the classical processing techniques consist in rearranging or splitting the data set into matrices or vectors in order for the previously quoted classical algebraic processing methods to be applicable. The original data structure is then built anew, after processing.

In order to keep the data tensor as a whole entity, new signal processing methods have been proposed [19, 20]. Hence, instead of adapting the data tensor to the classical matrix-based algebraic techniques (by rearrangement or splitting), these new methods propose to adapt their processing to the tensor structure of the multicomponent data. This new approach implicitly implies the use of multilinear algebra and mathematical tools that extend the SVD to tensors.

Two main tensor decomposition methods that generalize the matrix SVD have been initially developed in order to achieve a multimode Principal Component Analysis and recently used in tensor signal processing. They rely on two models, which are the TUCKER3 model and the PARAFAC model.

These two decomposition methods differ in the tensor rank definition on which they are based. The HOSVD-$(K_1, \ldots, K_N)$ and the rank-$(K_1, \ldots, K_N)$ approximation rely on the nth-mode rank definition, that is, the rank of the tensor nth-mode flattening matrix [7, 8]. The rank-$(K_1, \ldots, K_N)$ approximation [8] relies on an optimization algorithm which is initialized by the HOSVD-$(K_1, \ldots, K_N)$ [7]. The rank-$(K_1, \ldots, K_N)$ approximation improves the approximation obtained with the HOSVD-$(K_1, \ldots, K_N)$. The goal of this paper is to present an overview of the principal results concerning this new approach of data tensor filtering. More details on the algorithms presented in this survey can be found in [19–22]. These algorithms are analogous to Multilinear ICA, but were developed independently for image filtering. The presented algorithms are based on a signal subspace approach, so they are efficient when the noise components are uncorrelated, the signal and the additive noise are uncorrelated, and when some rows or columns of the image are redundant. In this case it is possible to distinguish between a signal subspace and a noise subspace, as for the traditional SVD-based filtering and Wiener filtering algorithms. Wiener filtering requires prior knowledge on the expected noise-free signal or image. However, multiway filtering methods provide the following advantage over traditional filtering methods: by apprehending a multiway data set as a whole entity, they take into account the dependence between modes thanks to ALS algorithms. The goal of the paper is also to present some simulations and comparative results concerning color images and multicomponent seismic signal filtering.

The paper is organized as follows: Section 2 presents the tensor data and a short overview of its main properties. Section 3 introduces the tensor formulation of the classical noise-removal problem as well as some new tensor filtering notations. Firstly, we explain how the channel-by-channel SVD-based method processes successively each component of the data tensor. Secondly, we consider two methods that take into account the relationships between each component of the considered tensor. These two methods are based on the nth-mode signal subspace. The first method for signal tensor estimation is based on multimode PCA achieved by rank-$(K_1, \ldots, K_N)$ approximation. The second method is a new tensor version of Wiener filtering. Section 4 presents some comparative results where the overviewed multiway filtering methods are applied to noise reduction in color images. Section 5 concludes the paper.

The following notations are used in the rest of the paper: scalars are denoted by italic lowercase roman, like a; vectors by boldface lowercase roman, like $\mathbf{a}$; matrices by boldface uppercase roman, like $\mathbf{A}$; tensors by uppercase calligraphic, like $\mathcal{A}$. We distinguish a random vector, like $\mathbf{a}$, from one of its realizations, by using a supplementary index, like $\mathbf{a}_i$.

2   Tensor Representation and Properties

We define a tensor of order N as a multidimensional array whose entries are accessed via N indexes. A tensor is denoted by $\mathcal{A} \in \mathbb{R}^{I_1 \times \cdots \times I_N}$, where each element is denoted by $a_{i_1 \cdots i_N}$, and $\mathbb{R}$ is the real manifold. Each dimension of a tensor is called nth-mode, where n refers to the nth index. Fig. 1 shows how a color image can be represented by a third order tensor $\mathcal{A} \in \mathbb{R}^{I_1 \times I_2 \times I_3}$, where I_1 is the number of rows, I_2 is the number of columns, and I_3 is the number of color channels. In the case of a color image, we have $I_3 = 3$. Let us define $E^{(n)}$ as the nth-mode vector space of dimension I_n, associated with the nth-mode of

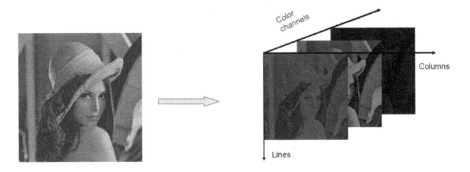

Fig. 1. 'Lena' standard color image and its tensor representation

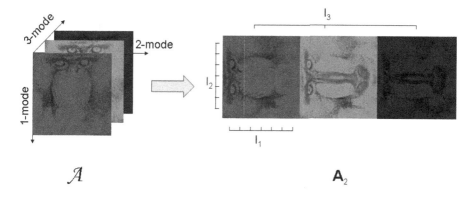

Fig. 2. $2nd$-mode flattening of tensor $\mathcal{A}$: $\mathbf{A}_2$

tensor $\mathcal{A}$. By definition, $E^{(n)}$ is generated by the column vectors of the nth-mode flattening matrix. The nth-mode flattening matrix $\mathbf{A}_n$ of tensor $\mathcal{A} \in \mathbb{R}^{I_1 \times \cdots \times I_N}$ is defined as a matrix from $\mathbb{R}^{I_n \times M_n}$, where:

$$M_n = I_{n+1} I_{n+2} \cdots I_N I_1 I_2 \cdots I_{n-1}. \tag{1}$$

For example, when we consider a third-order tensor, the definition of the matrix flattening involves the dimensions I_1, I_2, I_3 in a backward cyclic way [4, 7, 14]. When dealing with a 1st-mode flattening of dimensionality $I_1 \times (I_2 I_3)$, we formally assume that the index i_2 varies more slowly than i_3. For all $n = 1$ to 3, $\mathbf{A}_n$ columns are the I_n-dimensional vectors obtained from $\mathcal{A}$ by varying the index i_n from 1 to I_n and keeping the other indexes fixed. These vectors are called the nth-mode vectors of tensor $\mathcal{A}$. An illustration of the $2nd$-mode flattening of a color image is presented in Fig. 2.

In the following, we use the operator "$\times_n$" as the "nth-mode product", that generalizes the matrix product to tensors. Given $\mathcal{A} \in \mathbb{R}^{I_1 \times \cdots \times I_N}$ and a matrix $\mathbf{U} \in \mathbb{R}^{J_n \times I_n}$, the nth-mode product between tensor $\mathcal{A}$ and matrix $\mathbf{U}$ leads to the tensor $\mathcal{B} = \mathcal{A} \times_n \mathbf{U}$, which is a tensor of $\mathbb{R}^{I_1 \times \cdots I_{n-1} \times J_n \times I_{n+1} \times \cdots \times I_N}$, whose entries are given by:

$$b_{i_1 \cdots i_{n-1} j_n i_{n+1} \cdots i_N} = \sum_{i_n=1}^{I_n} a_{i_1 \cdots i_{n-1} i_n i_{n+1} \cdots i_N} u_{j_n i_n}. \tag{2}$$

Next section presents the recent filtering methods for tensor data.

3 Tensor Filtering Problem Formulation

The tensor data extend the classical vector data. The measurement of a multidimensional and multiway signal $\mathcal{X}$ by multicomponent sensors with additive noise $\mathcal{N}$, results in a data tensor $\mathcal{R}$ such that:

$$\mathcal{R} \mathcal{X} + \mathcal{N}. \tag{3}$$

$\mathcal{R}$, $\mathcal{X}$ and $\mathcal{N}$ are tensors of order N from $\mathbb{R}^{I_1 \times \cdots \times I_N}$. Tensors $\mathcal{N}$ and $\mathcal{X}$ represent noise and signal parts of the data respectively. The goal of this study is to estimate the expected signal $\mathcal{X}$ thanks to a multidimensional filtering of the data [19–22]:

$$\widehat{\mathcal{X}} = \mathcal{R} \times_1 \mathbf{H}^{(1)} \times_2 \mathbf{H}^{(2)} \times_3 \cdots \times_N \mathbf{H}^{(N)}, \tag{4}$$

From a signal processing point of view, the nth-mode product is a nth-mode filtering of data tensor $\mathcal{R}$ by nth-mode filter $\mathbf{H}^{(n)}$. Consequently, for all $n = 1$ to N, $\mathbf{H}^{(n)}$ is the nth-mode filter applied to the nth-mode of the data tensor $\mathcal{R}$.

In this paper we assume that the noise N is independent from the signal X, and that the nth-mode rank K_n is smaller than the nth-mode dimension I_n ($K_n < I_n$, for all $n = 1$ to N). Then it is possible to extend the classical subspace approach to tensors by assuming that, whatever the nth-mode, the vector space $E^{(n)}$ is the direct sum of two orthogonal subspaces, namely $E_1^{(n)}$ and $E_2^{(n)}$, which are defined as follows:

- $E_1^{(n)}$ is the subspace of dimension K_n, spanned by the K_n singular vectors associated with the K_n largest singular values of matrix $\mathbf{X}_n$; $E_1^{(n)}$ is called the signal subspace [1, 17, 26, 27].
- $E_2^{(n)}$ is the subspace of dimension $I_n - K_n$, spanned by the $I_n - K_n$ singular vectors associated with the $I_n - K_n$ smallest singular values of matrix $\mathbf{X}_n$; $E_2^{(n)}$ is called the noise subspace [1, 17, 26, 27].

The dimensions $K_1, K_2, \ldots, K_N$ can be estimated by means of the well-known AIC (Akaike Information Criterion) or MDL (Minimum Description Length) criteria [25], which are entropy-based information criteria. Hence, one way to estimate signal tensor X from noisy data tensor R is to estimate $E_1^{(n)}$ in every nth-mode of R. The following section presents three tensor filtering methods based on nth-mode signal subspaces. The first method is an extension of classical matrix filtering algorithms. It consists of a channel-by-channel SVD-based filtering.

The second filtering method is based on multimode PCA achieved by rank-$(K_1, \ldots, K_N)$ approximation.

3.1   Channel-by-Channel SVD-Based Filtering

The classical algebraical methods operate on two-dimensional data matrices and are based on the Singular Value Decomposition (SVD) [1–3], and on Eckart-Young theorem concerning the best lower rank approximation of a matrix [9] in the least-squares sense.

In the first method, a preprocessing is applied to the multidimensional and multiway data. It consists in splitting data tensor $\mathcal{R}$, representing the noisy multicomponent image into two-dimensional "slice matrices" of data, each representing a specific channel. According to the classical signal subspace methods [6], the left and right signal subspaces, corresponding to respectively the column and the row vectors of each slice matrix, are simultaneously determined by processing the SVD of the matrix associated with the data of the slice matrix.

Channel-by-channel SVD-based filtering is based on a common efficient method, but exhibits a major drawback: it does not take into account the relationships between the components of the processed tensor. Moreover, channel-by-channel SVD-based filtering is appropriate only on some conditions. For example, applying SVD-based filtering to an image is generally appropriate when the rows or columns of an image are redundant, that is, linearly dependent. In this case, the rank K of the image is equal to the number of linearly independent rows or columns. It is only in this case that it would be safe to throw out eigenvectors from $K+1$ on. It is only in this special case that the noise subspace is orthogonal to the signal subspace. Otherwise, the noise simply increases the variance of the signal subspace and underestimating the signal subspace dimension would result in throwing out both signal and noise information. Thus, one would lose spatial resolution.

The next subsection presents a multiway filtering method that processes jointly, and not successively, each component of the data tensor.

3.2   Tensor Filtering Based on Multimode PCA

Assuming that the dimension K_n of the signal subspace is known for all $n = 1$ to N, one way to estimate the expected signal tensor $\mathcal{X}$ from the noisy data tensor $\mathcal{RX} + \mathcal{N}$, is to orthogonally project, for every nth-mode, the vectors of tensor $\mathcal{R}$ on the nth-mode signal subspace $E_1^{(n)}$, for all $n = 1$ to N. This statement is equivalent to replace in (4) the filters $\mathbf{H}^{(n)}$ by the projectors $\mathbf{P}^{(n)}$ on the nth-mode signal subspace:

$$\widehat{\mathcal{X}} = \mathcal{R} \times_1 \mathbf{P}^{(1)} \times_2 \cdots \times_N \mathbf{P}^{(N)}. \tag{5}$$

In this last formulation, projectors $\mathbf{P}^{(n)}$ are estimated thanks to a multimode PCA applied to data tensor R. This multimode PCA-based filtering generalizes the classical matrix filtering methods [10, 11],[13], and implicitly supposes that the additive noise is *white* and *Gaussian*.

In the vector or matrix formulation, the definition of the projector on the signal subspace is based on the eigenvectors associated with the largest eigenvalues of the covariance matrix of the set of observation vectors. Hence, the determination of the signal subspace amounts to determine the best approximation (in the least-squares sense) of the observation matrix or the covariance matrix.

As an extension to the vector and matrix cases, in the tensor formulation, the projectors on the nth-mode vector spaces are determined by computing the rank-$(K_1, \dots, K_N)$ approximation of $\mathcal{R}$ in the least-squares sense. From a mathematical point of view, the rank-$(K_1, \dots, K_N)$ approximation of R is represented by tensor $R^{K_1, \dots, K_N}$ which minimizes the quadratic tensor Frobenius norm $\|R - B\|^2$ subject to the condition that $B \in \mathbb{R}^{I_1 \times \dots \times I_N}$ is a rank-$(K_1, \dots, K_N)$ tensor. The description of TUCKALS3 algorithm, used in rank-$(K_1, \dots, K_N)$ approximation is provided in the following.

Rank-$(K_1, \ldots, K_N)$ approximation - TUCKALS3 algorithm

1. **Input:** data tensor R, and dimensions $K_1, \ldots, K_N$ of all nth-mode signal subspaces.
2. **Initialization** $k = 0$: For $n = 1$ to N, calculate the projectors $\mathbf{P}_0^{(n)}$ given by HOSVD-$(K_1, \ldots, K_N)$:
   (a) nth-mode flatten R into matrix $\mathbf{R}_n$;
   (b) Compute the SVD of $\mathbf{R}_n$;
   (c) Compute matrix $\mathbf{U}_0^{(n)}$ formed by the K_n eigenvectors associated with the K_n largest singular values of $\mathbf{R}_n$. $\mathbf{U}_0^{(n)}$ is the initial matrix of the nth-mode signal subspace orthogonal basis vectors;
   (d) Form the initial orthogonal projector $\mathbf{P}_0^{(n)} = \mathbf{U}_0^{(n)} \mathbf{U}_0^{(n)^T}$ on the nth-mode signal subspace;
   (e) Compute the HOSVD-$(K_1, \ldots, K_N)$ of tensor R given by
   $$B_0 = R \times_1 \mathbf{P}_0^{(1)} \times_2 \cdots \times_N \mathbf{P}_0^{(N)};$$
3. **ALS loop**
   Repeat until convergence, that is, for example, while $\|\mathcal{B}_{k+1} - \mathcal{B}_k\|^2 > \epsilon$, $\epsilon > 0$ being a prior fixed threshold,
   (a) For $n = 1$ to N:
      i. Form $\mathcal{B}^{(n),k}$:
      $$\mathcal{B}^{(n),k} = \mathcal{R} \times_1 \mathbf{P}_{k+1}^{(1)} \times_2 \cdots \times_{n-1} \mathbf{P}_{k+1}^{(n-1)} \times_{n+1} \mathbf{P}_k^{(n+1)} \times_{n+2} \cdots \times_N \mathbf{P}_k^{(N)};$$
      ii. nth-mode flatten tensor $\mathcal{B}^{(n),k}$ into matrix $\mathbf{B}_n^{(n),k}$;
      iii. Compute matrix $\mathbf{C}^{(n),k} = \mathbf{B}_n^{(n),k} \mathbf{R}_n^T$;
      iv. Compute matrix $\mathbf{U}_{k+1}^{(n)}$ composed of the K_n eigenvectors associated with the K_n largest eigenvalues of $\mathbf{C}^{(n),k}$. $\mathbf{U}_k^{(n)}$ is the matrix of the nth-mode signal subspace orthogonal basis vectors at the kth iteration;
      v. Compute $\mathbf{P}_{k+1}^{(n)} = \mathbf{U}_{k+1}^{(n)} \mathbf{U}_{k+1}^{(n)^T}$;
   (b) Compute $\mathcal{B}_{k+1} = \mathcal{R} \times_1 \mathbf{P}_{k+1}^{(1)} \times_2 \cdots \times_N \mathbf{P}_{k+1}^{(N)}$;
   (c) Increment k.
4. **Output:** the estimated signal tensor is obtained through $\widehat{\mathcal{X}} = \mathcal{R} \times_1 \mathbf{P}_{k_{stop}}^{(1)} \times_2 \cdots \times_N \mathbf{P}_{k_{stop}}^{(N)}$. $\widehat{\mathcal{X}}$ is the rank-$(K_1, \ldots, K_N)$ approximation of $\mathcal{R}$, where k_{stop} is the index of the last iteration after the convergence of TUCKALS3 algorithm.

In this algorithm, the second order statistics comes from the SVD of matrix $\mathbf{R}_n$ at step 2b, which is equivalent, up to $\frac{1}{M_n}$ multiplicative factor, to the estimation of tensor R nth-mode vectors [22]. The definition of M_n is given in (1). In the same way, at step 3(a)iii, matrix $\mathbf{C}^{(n),k}$ is, up to $\frac{1}{M_n}$ multiplicative factor, the estimation of the covariance matrix between tensor R and tensor $\mathcal{B}^{(n),k}$ nth-mode vectors. According to step 3(a)i, $\mathcal{B}^{(n),k}$ represents data tensor R filtered in every mth-mode but the nth-mode, by projection-filters $\mathbf{P}_l^{(m)}$, with $m \neq n$, $l = k$ if $m > n$ and $l = k+1$ if $m < n$. TUCKALS3 algorithm has recently been used to process a multimode PCA in order to perform white noise removal in color images [20].

A good approximation of the rank-$(K_1, \ldots, K_N)$ approximation can simply be achieved by computing the HOSVD-$(K_1, \ldots, K_N)$ of tensor R [8, 18]. Indeed, the HOSVD-$(K_1, \ldots, K_N)$ of R consists of the initialization step of TUCKALS3 algorithm, and hence can be considered as a suboptimal solution for the rank-$(K_1, \ldots, K_N)$ approximation of tensor R [8].

3.3   Multiway Wiener Filtering

Let $\mathbf{R}_n$, $\mathbf{X}_n$ and $\mathbf{N}_n$ be the nth-mode flattening matrices of tensors $\mathcal{R}$, $\mathcal{X}$ and $\mathcal{N}$, respectively.

In the previous subsection, the estimation of signal tensor X has been performed by projecting noisy data tensor R on each nth-mode signal subspace. The nth-mode projectors have been estimated thanks to the use of multimode PCA achieved by rank-$(K_1, \ldots, K_N)$ approximation. In spite of the good results given by this method, it is possible to improve the tensor filtering quality by determining nth-mode filters $\mathbf{H}^{(n)}$, $n = 1$ to N, in (4), which optimize an estimation criterion. The most classical method is to minimize the mean squared error between the expected signal tensor X and the estimated signal tensor $\widehat{X}$ given in (4):

$$e(\mathbf{H}^{(1)}, \ldots, \mathbf{H}^{(N)}) = E[\|X - R \times_1 \mathbf{H}^{(1)} \times_2 \cdots \times_N \mathbf{H}^{(N)}\|^2]. \tag{6}$$

Due to the criterion which is minimized, filters $\mathbf{H}^{(n)}$, $n = 1$ to N, can be called "nth-mode Wiener filters" [21].

According to the calculations presented in [21], the minimization of (6) with respect to filter $\mathbf{H}^{(n)}$, for fixed $\mathbf{H}^{(m)}$, $m \neq n$, leads to the following expression of nth-mode Wiener filter:

$$\mathbf{H}^{(n)} = \gamma_{\mathbf{XR}}^{(n)} \mathbf{\Gamma}_{\mathbf{RR}}^{(n)}{}^{-1}, \tag{7}$$

where

$$\gamma_{\mathbf{XR}}^{(n)} = E\left[\mathbf{X}_n \mathbf{T}^{(n)} \mathbf{R}_n^T\right] \tag{8}$$

is the $\mathbf{T}^{(n)}$-weighted covariance matrix between the random column vectors of signal $\mathbf{X}_n$ and data $\mathbf{R}_n$, with:

$$\mathbf{T}^{(n)} = \mathbf{H}^{(1)} \otimes \cdots \otimes \mathbf{H}^{(n-1)} \otimes \mathbf{H}^{(n+1)} \otimes \cdots \otimes \mathbf{H}^{(N)}, \tag{9}$$

where $\otimes$ stands for Kronecker product, and:

$$\mathbf{\Gamma}_{\mathbf{RR}}^{(n)} = E\left[\mathbf{R}_n \mathbf{Q}^{(n)} \mathbf{R}_n^T\right], \tag{10}$$

is the $\mathbf{Q}^{(n)}$-weighted covariance matrix of the data $\mathbf{R}_n$, with:

$$\mathbf{Q}^{(n)} = \mathbf{T}^{(n)}{}^T \mathbf{T}^{(n)}. \tag{11}$$

In order to obtain $\mathbf{H}^{(n)}$ through (7), we suppose that the filters $\{\mathbf{H}^{(m)}, m = 1 \text{ to } N, m \neq n\}$ are known. Data tensor R is available, but signal tensor X is unknown. So, only the term $\Gamma_{\mathbf{RR}}^{(n)}$ can be derived, and not the term $\gamma_{\mathbf{XR}}^{(n)}$. Hence, some more assumptions on $\mathcal{X}$ have to be made in order to overcome the indetermination over $\gamma_{\mathbf{XR}}^{(n)}$ [19, 21]. In the one-dimensional case, a classical assumption is to consider that a signal vector is a weighted combination of the signal subspace basis vectors. In extension to the tensor case, [19, 21] have proposed to consider that the nth-mode flattening matrix $\mathbf{X}_n$ can be expressed as a weighted combination of K_n vectors from the nth-mode signal subspace $E_1^{(n)}$:

$$\mathbf{X}_n \mathbf{V}_s^{(n)} \mathbf{O}^{(n)}, \tag{12}$$

with $\mathbf{X}_n \in \mathbb{R}^{I_n \times M_n}$, and $\mathbf{V}_s^{(n)} \in \mathbb{R}^{I_n \times K_n}$ being the matrix containing the K_n orthonormal basis vectors of nth-mode signal subspace $E_1^{(n)}$. Matrix $\mathbf{O}^{(n)} \in \mathbb{R}^{K_n \times M_n}$ is a weight matrix and contains the whole information on expected signal tensor X. This model implies that signal nth-mode flattening matrix $\mathbf{X}_n$ is orthogonal to nth-mode noise flattening matrix $\mathbf{N}_n$, since signal subspace $E_1^{(n)}$ and noise subspace $E_2^{(n)}$ are supposed mutually orthogonal.

Supposing that noise N in (3) is white, Gaussian and independent from signal X, and introducing the signal model (12) in (7) leads to a computable expression of nth-mode Wiener filter $\mathbf{H}^{(n)}$:

$$\mathbf{H}^{(n)} = \mathbf{V}_s^{(n)} \gamma_{\mathbf{OO}}^{(n)} \mathbf{\Lambda}_{\mathbf{\Gamma s}}^{(n)^{-1}} \mathbf{V}_s^{(n)^T}, \tag{13}$$

where $\gamma_{\mathbf{OO}}^{(n)} \mathbf{\Lambda}_{\mathbf{\Gamma s}}^{(n)^{-1}}$ is a diagonal weight matrix given by:

$$\gamma_{\mathbf{OO}}^{(n)} \mathbf{\Lambda}_{\mathbf{\Gamma s}}^{(n)^{-1}} = diag \left[\frac{\beta_1}{\lambda_1^\Gamma}, \cdots, \frac{\beta_{K_n}}{\lambda_{K_n}^\Gamma} \right], \tag{14}$$

where $\lambda_1^\Gamma, \ldots, \lambda_{K_n}^\Gamma$ are the K_n largest eigenvalues of $\mathbf{Q}^{(n)}$-weighted covariance matrix $\Gamma_{\mathbf{RR}}^{(n)}$ (see (10)). Parameters $\beta_1, \ldots, \beta_{K_n}$ depend on $\lambda_1^\gamma, \ldots, \lambda_{K_n}^\gamma$ which are the K_n largest eigenvalues of $\mathbf{T}^{(n)}$-weighted covariance matrix $\gamma_{\mathbf{RR}}^{(n)} = E[\mathbf{R}_n \mathbf{T}^{(n)} \mathbf{R}_n^T]$, according to the following relation:

$$\beta_{k_n} = \lambda_{k_n}^\gamma - \sigma_\Gamma^{(n)^2}, \ \forall \ k_n = 1, \ldots, K_n \tag{15}$$

Superscript γ refers to the $\mathbf{T}^{(n)}$-weighted covariance, and subscript Γ to the $\mathbf{Q}^{(n)}$-weighted covariance. $\sigma_\Gamma^{(n)^2}$ is the degenerated eigenvalue of noise $\mathbf{T}^{(n)}$-weighted covariance matrix $\gamma_{\mathbf{NN}}^{(n)} = E\left[\mathbf{N}_n \mathbf{T}^{(n)} \mathbf{N}_n^T\right]$. Thanks to the additive noise and the signal independence assumptions, the $I_n - K_n$ smallest eigenvalues of $\gamma_{\mathbf{RR}}^{(n)}$ are equal to $\sigma_\Gamma^{(n)^2}$, and thus, can be estimated by the following relation:

$$\widehat{\sigma}_\Gamma^{(n)^2} = \frac{1}{I_n - K_n} \sum_{k_n = K_n+1}^{I_n} \lambda_{k_n}^\gamma. \tag{16}$$

In order to determine the nth-mode Wiener filters $\mathbf{H}^{(n)}$ that minimize the mean squared error (6), the Alternating Least Squares (ALS) algorithm has been proposed in [19, 21]. It can be summarized in the following steps:

1. **Initialization** $k = 0$: $\mathcal{R}^0 = \mathcal{R} \Leftrightarrow \mathbf{H}_0^{(n)} = \mathbf{I}_{I_n}$, Identity matrix, $\forall\, n = 1 \ldots N$.
2. **ALS loop**

   Repeat until convergence, that is, $\left\| \mathcal{R}^{k+1} - \mathcal{R}^k \right\|^2 < \epsilon$, with $\epsilon > 0$ prior fixed threshold,

   (a) for $n = 1$ to N:

      i. Form $\mathcal{R}^{(n),k}$: $\mathcal{R}^{(n),k} = \mathcal{R} \times_1 \mathbf{H}_{k+1}^{(1)} \times_2 \cdots \times_{n-1} \mathbf{H}_{k+1}^{(n-1)} \times_{n+1}$
   $\mathbf{H}_k^{(n+1)} \times_{n+2} \cdots \times_N \mathbf{H}_k^{(N)}$;

      ii. Determine $\mathbf{H}_{k+1}^{(n)} = \mathbf{Z}^{(n)} \arg\min \left\| \mathcal{X} - \mathcal{R}^{(n),k} \times_n \mathbf{Z}^{(n)} \right\|^2$
   subject to $\mathbf{Z}^{(n)} \in \mathbb{R}^{I_n \times I_n}$ thanks to the following procedure:

       A. nth-mode flatten $\mathcal{R}^{(n),k}$ into $\mathbf{R}_n^{(n),k} = \mathbf{R}_n (\mathbf{H}_{k+1}^{(1)} \otimes \cdots \otimes \mathbf{H}_{k+1}^{(n-1)} \otimes$
   $\mathbf{H}_k^{(n+1)} \otimes \cdots \otimes \mathbf{H}_k^{(N)})^T$, and $\mathcal{R}$ into $\mathbf{R}_n$;

       B. Compute $\gamma_{\mathbf{RR}}^{(n)} = E[\mathbf{R}_n \mathbf{R}_n^{(n),k^T}]$,

       C. Determine $\lambda_1^\gamma, \ldots, \lambda_{K_n}^\gamma$, the K_n largest eigenvalues of $\gamma_{\mathbf{RR}}^{(n)}$;

       D. For $k_n = 1$ to I_n, estimate $\sigma_\Gamma^{(n)^2}$ thanks to (16) and for $k_n = 1$
   to K_n, estimate β_{k_n} thanks to (15);

       E. Compute $\mathbf{\Gamma}_{\mathbf{RR}}^{(n)} = E[\mathbf{R}_n^{(n),k} \mathbf{R}_n^{(n),k^T}]$;

       F. Determine $\lambda_1^\Gamma, \ldots, \lambda_{K_n}^\Gamma$, the K_n largest eigenvalues of $\mathbf{\Gamma}_{\mathbf{RR}}^{(n)}$;

       G. Determine $\mathbf{V}_s^{(n)}$, the matrix of the K_n eigenvectors associated
   with the K_n largest eigenvalues of $\mathbf{\Gamma}_{\mathbf{RR}}^{(n)}$;

       H. Compute the weight matrix $\gamma_{\mathbf{OO}}^{(n)} \mathbf{\Lambda}_{\mathbf{\Gamma s}}^{(n)^{-1}}$ given in (14);

       I. Compute $\mathbf{H}_{k+1}^{(n)}$, the nth-mode Wiener filter at the $(k+1)^{\text{th}}$ iteration, using (13);

   (b) Form $\mathcal{R}^{k+1} = \mathcal{R} \times_1 \mathbf{H}_{k+1}^{(1)} \times_2 \cdots \times_N \mathbf{H}_{k+1}^{(N)}$;

   (c) Increment k;

3. **output:** $\widehat{\mathcal{X}} = \mathcal{R} \times_1 \mathbf{H}_{k_{stop}}^{(1)} \times_2 \cdots \times_N \mathbf{H}_{k_{stop}}^{(N)}$, with k_{stop} being the last iteration after convergence of the algorithm.

4 Simulation Results

In the following simulations, the channel-by-channel SVD-based filtering defined in subsection 3.1 and the rank-$(K_1, \ldots, K_N)$ approximation based multiway and multidimensional filtering are applied to the denoising of color images and multispectral images, and to the denoising of seismic signals. Color images, multispectral images, and seismic signals can be represented by a third order tensor from $\mathbb{R}^{I_1 \times I_2 \times I_3}$, where I_1, I_2, and I_3 take different values. In all these applications, the efficiency of denoising is tested in the presence of an additive Gaussian noise.

A multidimensional and multiway white Gaussian noise $\mathcal{N}$ which is added to signal tensor $\mathcal{X}$ can be expressed as:

$$\mathcal{N}\alpha \cdot \mathcal{G}, \tag{17}$$

where every element of $\mathcal{G} \in \mathbb{R}^{I_1 \times I_2 \times I_3}$ is an independent realization of a normalized centered Gaussian law, and where α is a coefficient that permits to set the SNR in noisy data tensor $\mathcal{R}$.

4.1 Performance Criterion

Following the representation of (3), the multiway noisy data tensor is expressed as $\mathcal{R}\mathcal{X} + \mathcal{N}$, where $\mathcal{X}$ is the expected signal tensor and $\mathcal{N}$ is the additive noise tensor. Let us define the Signal to Noise Ratio (SNR, in dB) in the noisy data tensor by:

$$SNR = 10log(\frac{\|\mathcal{X}\|^2}{\|\mathcal{N}\|^2}). \tag{18}$$

4.2 Denoising of Color Images

Denoising of color images has already been studied in several works [5, 23, 24]. Some solutions have been brought from the field of wavelet processing, exhibiting good results in terms of output SNR. These studies only concern bidimensional data, whereas the methods that we compare are adapted to the processing of third order tensors as a whole, and in particular to three-channel images. We focus on subspace-based methods. We first consider the channel-by-channel SVD-based filtering, the rank-(K_1, K_2, K_3) approximation and multiway Wiener filtering (Wmm-(K_1, K_2, K_3)), applied to images impaired by an additive white Gaussian noise.

The channel-by-channel SVD-based filtering of noisy image $\mathcal{R}$ (see Fig. 3) yields the image of Fig. 3c, and rank-$(30, 30, 2)$ approximation of noisy data tensor $\mathcal{R}$ yields the image of Fig. 3d. From the resulting image, presented on Fig.3, we note that dimension reduction leads to a loss of spatial resolution. However the choice of a set of values K_1, K_2, K_3 which are small enough is the condition for an efficient noise reduction effect.

Concerning the qualitative results obtained with this color image, we note that the intra-class variance of the pixel values of each component (or color mode) of the resulting image is lower for the image obtained with Wmm-$(30, 30, 2)$ (see Fig. 3e) than for those images obtained with other methods applied in this subsection. This permits, for example, to apply after denoising a high level classification method with a higher efficiency than when classification is applied after channel-by-channel SVD-based filtering or HOSVD-$(30, 30, 2)$.

For the $256 \times 256 \times 3$ Sailboat image of Fig. 3, the computational times needed when $Matlab^{\circledR}$ programs are used on a 3Ghz Pentium 4 processor running Windows are as follows. HOSVD-$(30, 30, 2)$ lasts 1.61 sec., the channel-by-channel SVD-based filtering lasts 1.94 sec., the rank-$(30, 30, 2)$ approximation run with

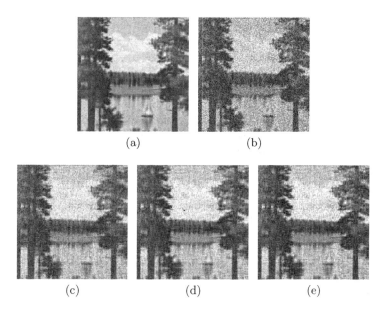

(a)   (b)

(c)   (d)   (e)

Fig. 3. (a) non-noisy image. (b) image to be processed, impaired by an additive white Gaussian noise, with SNR=8.1dB. (c) channel-by-channel SVD-based filtering of parameter K=30. (d) rank-$(30, 30, 2)$ approximation. (e) Wmm-$(30, 30, 2)$ filtering.

25 iterations lasts 54.1 sec. and Wmm-$(30, 30, 2)$ run with 25 iterations lasts 40.0 sec.

According to the simulations performed on a color image, it is possible to conclude that the more channels the image is composed of, the better the denoising.

5 Conclusion

In this paper, an overview on new mathematical methods dedicated to multi-component data is presented. Multicomponent data are represented as tensors, that is, multiway arrays, and the tensor filtering methods that are presented rely on multilinear algebra. First we present how to perform channel-by-channel SVD-based filtering. Then we review three methods that take into account the relationships between each component of a processed tensor. The first method consists of an extension of the classical SVD-based filtering method. In the case of an additive white Gaussian noise, the signal tensor is estimated thanks to a multimode PCA achieved by applying a lower rank-$(K_1, \ldots, K_N)$ approximation to the noisy data tensor, or a lower rank-$(K_1, \ldots, K_N)$ truncation of its HOSVD. This method is implicitly based on second order statistics and relies on the orthogonality between nth-mode noise and signal subspaces.

Finally, the reviewed method is a multiway version of the classical Wiener filtering. In extension to the one-dimensional case, the nth-mode Wiener filters are estimated by minimizing the mean squared error between the expected signal tensor and the estimated signal tensor obtained by applying the nth-mode Wiener filters to the noisy data tensor thanks to the nth-mode product operator. An Alternating Least Squares algorithm has been presented to determine the optimal nth-mode Wiener filters. The performances of this multiway Wiener filtering and comparative results with multimode PCA have been presented in the case of additive white noise reduction in a color image.

References

1. Abed-Meraim, K., Maître, H., Duhamel, P.: Blind multichannel image restoration using subspace based method. In: IEEE International Conf. on Accoustics Systems and Signal Processing, Hong Kong, China, April 6-10 (2003)
2. Andrews, H.C., Patterson, C.L.: Singular value decomposition and digital image processing. IEEE Trans. on Acoustics, Speech, and Signal Processing 24(1), 26–53 (1976)
3. Andrews, H.C., Patterson, C.L.: Singular Value Decomposition (SVD) image coding. IEEE Trans. on Communications, 425–432 (April 1976)
4. Bader, B.W., Kolda, T.G.: Algorithm 862: Matlab tensor classes for fast algorithm prototyping. ACM Transactions on Mathematical Software 32(4) (December 2006)
5. Bala, E., Ertüzün, A.: A multivariate thresholding technique for image denoising using multiwavelets. EURASIP Journal on ASP, 1205–1211 (August 2005)
6. Benjama, A., Bourennane, S., Frikel, M.: Seismic wave separation based on higher order statistics. In: IEEE International Conf. on Digital Signal Processing and its Applications, Moscow, Russia (1998)
7. De Lathauwer, L., De Moor, B., Vandewalle, J.: A multilinear singular value decomposition. SIAM Journal on Matrix Analysis and Applications 21, 1253–1278 (2000)
8. De Lathauwer, L., De Moor, B., Vandewalle, J.: On the best rank-(1) and rank-$(r_1, \ldots, r_N)$ approximation of higher-order tensors. SIAM Journal on Matrix Analysis and Applications 21(4), 1324–1342 (2000)
9. Eckart, Y., Young, G.: The approximation of a matrix by another of lower rank. Psychometrika 1, 211–218 (1936)
10. Freire, F.L.M., Ulrych, T.J.: Application of SVD to vertical seismic profiling. Geophysics 53, 778–785 (1988)
11. Glangeaud, F., Mari, J.-L.: Wave separation. Technip IFP edition (1994)
12. Golub, G.H., Van Loan, C.F.: Matrix computations, 3rd edn. The John Hopkins University Press edition, Baltimore (1996)
13. Hemon, M., Mace, D.: The use of Karhunen-Loeve transform in seismic data prospecting. Geophysical Prospecting 26, 600–626 (1978)
14. Kiers, H.: Towards a standardized notation and terminology in multiway analysis. Journal of Chemometrics 14, 105–122 (2000)
15. Kroonenberg, P.M.: Three-mode principal component analysis. DSWO press, Leiden (1983)
16. Kroonenberg, P.M., De Leeuw, J.: Principal component analysis of three-mode data by means of alternating least squares algorithms. Psychometrika 45(1), 69–97 (1980)

17. Mendel, J.M.: Tutorial on higher order statistics (spectra) in signal processing and system theory: theoretical results and some applications. Proc. of the IEEE 79, 278–305 (1991)
18. Muti, D.: Traitement du signal tensoriel. Application aux images en couleurs et aux signaux sismiques. Phd thesis, Institut Fresnel, Université Paul Cézanne, Aix-Marseille III, Marseille, France, December 2 (2004)
19. Muti, D., Bourennane, S.: Multidimensional estimation based on a tensor decomposition. In: IEEE Workshop on Statistical Signal Processing, St Louis, USA, September 28 - October 1 (2003)
20. Muti, D., Bourennane, S.: Multidimensional signal processing using lower rank tensor approximation. In: IEEE Int. Conf. on Accoustics, Systems and Signal Processing, Hong Kong, China, April 6-10 (2003)
21. Muti, D., Bourennane, S.: Multidimensional filtering based on a tensor approach. Signal Processing Journal 85, 2338–2353 (2005)
22. Muti, D., Bourennane, S.: Multiway filtering based on fourth order cumulants. Applied Signal Processing, EURASIP 7, 1147–1159 (2005)
23. Neelamani, R., Choi, H., Baraniuk, R.: Forward: Fourier-wavelet regularized deconvolution for ill-conditioned systems. IEEE Transactions on SP 52(2), 418–433 (2004)
24. She, L.W., Zheng, B.: Multiwavelets based denoising of sar images. In: Proceedings 5th international Conference on Signal Processing, Beijing, China, August 2008, pp. 321–324 (2000)
25. Wax, M., Kailath, T.: Detection of signals information theoretic criteria. IEEE Trans. on Acoust., Speech, Signal Processing 33(2), 387–392 (1985)
26. Yuen, N., Friedlander, B.: DOA estimation in multipath: an approach using fourth order cumulant. IEEE Trans. on Signal Processing 45(5), 1253–1263 (1997)
27. Yuen, N., Friedlander, B.: Asymptotic performance analysis of blind signal copy using fourth order cumulant. International Jour. Adaptative Contr. Signal Proc. 48, 239–265 (1996)

LaMOC – A Location Aware Mobile Cooperative System

Junzhong Gu, Liang He, and Jing Yang

Institute of Computer Applications, East China Normal University (ICA-ECNU), 200062
Shanghai, China
{jzgu,lhe,jyang}@cs.ecnu.edu.cn

Abstract. In this paper LaMOC- a Location Aware Mobile Cooperative System developed by ICA-ECNU is presented. LaMOC is a Location Based System, a Mobile Cooperative System and a Spatial Group Decision Support System. A map based browser is designed for clients. It's a map based co-browsing user interface. The scenario based developing of LaMOC, its system architecture, user interface, and key issues are discussed here.

Keywords: LBS, mobile Collaboration, map based browser, context awareness, Spatial Group Decision Support.

1 Introduction

Now it is a fast developing era. ICT evolves rapidly. Especially, fixed network based communication is more and more replaced by mobile communication. Geographic Information Systems (GIS) are now evolved to Web GIS. Mobile Internet attracts more users than traditional Internet. Such techniques converge to a new area-Location based service (LBS),[1,2] as shown in Fig. 1.

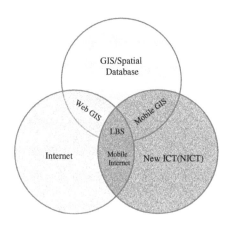

Fig. 1. Location based service

D. Ślęzak et al. (Eds.): SIP 2009, CCIS 61, pp. 250–258, 2009.

When designing a wireless infrastructure which has to support LBS, a number of mobile environment constraints must be taken into account:

- The service has to provide information of different types, covering different geographic regions, and coming from different sources;
- The service has to provide dynamic information;
- Interoperability of different LBS components must be managed;
- Mobile terminals have limited memory, limited computational power, limited screen size and resolution;
- Mobile networks have high cost, limited bandwidth, high latency, low connection stability and low availability.

ExPo2010 will be held in Shanghai from 1. May to 31. Oct. 2010. Nearly 70 million (everyday 400 thousands) visitors will come to the 5.28 km^2 Expo village. They are often grouped, in family, pair, as well as in tourist group. How to efficiently guiding such a huge number of visitors according to their actual location, distribution, visit plan as well as exhibition program is a big challenge. This is a sample scenario of LaMOC.

While this challenge appears, Mobile Internet, Web GIS and Mobile GIS, each respectively meets new challenges. As their harmonizing LBS is a severe test. A lot of problems should be solved, such as how to realize adaptive positioning, how to converge the actual position with GIS, how to atone for the inherent shortcomings of mobile computing, how to use back ground system to counteract the weaknesses of mobile terminals, etc. Furthermore, location based mobile cooperation is demanded here. Mobile cooperation of a group results in some new problems such as: collaborative model changes dynamically accompanied with the change of participants' location, in moving situation which is hard for participants to estimate, determine and make decisions, and so on. In this way, the system should be context aware, and adaptive reaction to the change of context. [3]

Therefore, LaMOC is focused on: minimizing user's efforts to access information in moving, being aware of environmental conditions and user's activities, offering natural interaction techniques, visualizing relevant information in 2D/3D maps on small screen effectively, and mobile cooperation of group users. In short, main challenges met by us: mobile computing and context awareness, on demand mapping and visualizing geographic (in short, Geo) data on small screen, spatial decision support, mobile cooperation, etc.

The paper is organized as follows: Section 2 is about the scenario based design and development; Section 3 presents LaMOC system; some key issues solved in LaMOC are discussed in Section 4.

2 Scenario Based Design and Development of LaMOC

For an application system as LaMOC, usage scenario plays an important role. Therefore a scenario-based approach is used by LaMOC. The scenario oriented design and development can be illustrated as in Fig.2.

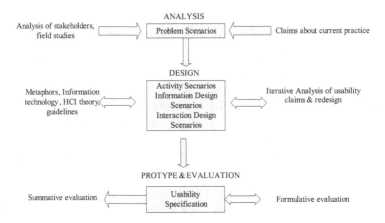

Fig. 2. Scenario-based frameworks [4]

The first phase (ANALYSIS Phase) in Fig. 2 is scenario analysis. Scenarios of LaMOC can be classified into 5 categories.

Scenario 1. User sessions begin and end, such as user Login/Logout. LaMOC user should log in before using the system, and log out when the end of the session. A personalized map will appear in the mobile device as soon as user logged in.

Scenario 2. Information Retrieval- LaMOC has facilities for Yellow Page retrieval; Location based information retrieval (IR); Internet based information retrieval and Intelligent (context aware) IR.

Location based IR in LaMOC means that it can answer user's questions based his/her actual location, such as:

Where am I now?
Who are near me now?
Is there any friend just near me?

LaMOC support context aware IR, i.e. the answer gotten from LaMOC system is personalized, environment adaptive and user-oriented.

Scenario 3. Group application

A group of members distributed in different places work together, to achieve a common target, such as to muster at a determined time, to select an optimal mustering site, to select an optimal route to arrive there in time, etc. Or a group of people communicates on line to share information. Participants here are distributed and mobile, i.e. their locations change during a session.

Scenario 4. Map based browsing

A map based browser is a main user interface of LaMOC. It's a map based co-browser, i.e. users in same group share information using the browser. Each user navigates on a personalized map, where his/her location is centralized.

Users can click an icon on the map to dial (make a phone call), click an icon to SMS (send a short message), as well as click an icon to IM (Instant Message).

Scenario 5. Location based application

In scenario 5, a user can based on his/her location find: a favorite restaurant near himself/herself just now; the nearest metro station; the best route to arrive the target place.

And the answer can be extended to the group, e.g. to find a favorite restaurant for the 5 distributed members of a group, or to find the nearest appropriate service zone on the highway for 3 cars just on the way.

In the DESIGN Phase Problem scenarios are studied.

The main problem scenarios met by us are: On-demand mapping, Context awareness, Mobile collaboration and so on. That is:

On-demand mapping: we focus on dynamic location based mapping, narrow band map data transmission, smart small display screen display mechanism, etc. On-demand mapping enable users to define by themselves the content, coverage, scale and visual appearance of the requested products. Methods such as on-demand mapping, real-time and object-oriented generalization of the content of geo-databases may be used to address the challenge of better delivering the location-related information to mobile users.

Context awareness is trying to answer: how to capture position information of clients? How to get parameters of mobile devices automatically? How to get user profile as well as group profile? How to get behavior information of individual user and group? What is the meaning of the captured context?

Mobile collaboration: the challenge is the dynamic change of the group organization structure, communication model, as well as collaborative mechanism. And it should answer how to react to the change.

In the DESIGN Phase, Activity scenarios analyzed are for example, registering, create group, active group, run a session, etc.

Then, a prototype of LaMOC is developed, as follows.

3 LaMOC System

3.1 Architecture of LaMOC

LaMOC is a mobile computing system, actually a hybrid environment, as shown in Fig. 3. Three layers appear in LaMOC: fixed hosts (FH); mobile support stations (MSS); and Mobile hosts (MH). Here, FH contains the computing devices, such as PC desktops, servers, etc. Databases are active on servers, to supply reliable data services. MH means laptops, mobile phones, etc. Huge differences appear in such mobile hosts, either in computing power, or the duration of the battery. The layer connecting FH and MH is MSS. In LaMOC the IP channel based on GPRS/CDMA/WiFi or 3G is used. A web gateway works as a portal to link mobile systems and fixed systems.

Even though it's layered, functionally LaMOC is a global system as shown in Fig. 4.

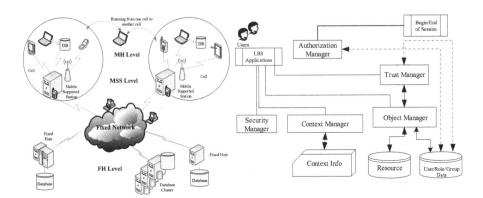

Fig. 3. LaMOC mobile computing referential model

Fig. 4. Software architecture of LaMOC

Some main components of LaMOC are:

- Object manager, manages objects, such as Yellow Page Object, Geo Object, etc;
- Trust manager-Trust manager is in charge of ensuring LaMOC running in a trusted environment. It should care the trust in person-thing/object, trust in person-person, trust in person-group, trust in person-system, and trust in group-group.
- Security manager-The security manager is in charge of protecting the safety of LaMOC. The portal of LaMOC is the only one channel to connect mobile devices and the back ground server group behind the web portal. To access servers the Firewall/NAT is used as protective screen. For the application security some safety measures are necessary, such as anti-pretend, strict certificate authority, trusted client and server connection, etc.

3.2 Map Based Browser and LaMOC User Interface

Mobile devices in LaMOC are smart phones. As a prototype, mobile devices in LaMOC are based on Microsoft Windows Mobile v6.0. GPS (Global Positioning System) is used for positioning. Because that GPS works not so good in-door, Pseudo-GPS is designed and developed in LaMOC project to support in-door positioning. A Pseudo-GPS station is actually a GPS signal sender located at ground, especially in buildings. In this way position info can be captured either out-door or in-door.

A location based personalized user interface is demanded. It's a map based browser.

Map Based Browser

Different from a text/picture based browser, such as Microsoft IE (Internet Explorer) or Google Chrome, a map based browser is designed as a main user interface of LaMOC. A map based browser is a Geo info based browser, which is user oriented and user favorite object oriented. It means, as soon as a user has registered (logged in), a user-centralized map will be presented at his/her terminal. The user preferred objects such as favorite restaurants near him/her and favorite stores near him/her, are illustrated as icons on the map. Actually, it's a map based co-browser. It has ability to co-navigate the map with other people at-a-distance. As soon as user A logs in LaMOC, a map based interface where his/her location shown in the center, will be presented. The other members in a same group will also be illustrated as icons in the map exactly corresponding to their location. Accompanied with it, A's information with his/her location will be automatically illustrated in the partners' map browsers.

The map based browser of LaMOC can be illustrated as in Fig. 5.

LaMOC Client

As explained, clients in LaMOC are based on smart phones. The software architecture in clients can be illustrated as in Fig. 6.

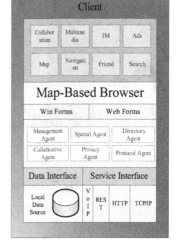

Fig. 5. Map based browser

Fig. 6. Client software architecture

As shown in Fig. 6, a map based browser is a main component in clients. A small mobile database is used to keep temporary data as well as context data. Soft agents, such as spatial agent, collaborative agent, and privacy agent, are running on clients. They are in charge of capturing context info, adapting the change of environment, and making intelligent reaction. Two internal interfaces, data interface and service interface, are used to connect agents and local data source/communication services.

3.3 Fixed Host Layer in LaMOC

The fixed host layer of LaMOC is composed of a server group and workstations.

Because of the inherent shortages of mobile devices, such as lower computing power, and limited battery capacity, a background server system remedies congenital deficiency of mobile devices. Some servers can be deployed remotely. LaMOC accesses them via Internet.

The FH Layer in LaMOC mainly is a group of servers, containing:

- Web Server- Web Server is used as a gateway to connect MH with FH ;
- Application Server-The kernel of Application Server is the LaMOC Engine. The engine analyses and dispatches the user commands to produce a corresponding running plan.
- GeoData Server- It manages Geo data, and carries out basic Geo computing;
- Database Server- It is in charge of fundamental data management, and managing yellow page data, user data, group data, and application data.
- Collaboration Server- It is used to coordinate participants of group and their activities.
- Context Server- The context server is used to store and manage context data, as well as process context data.

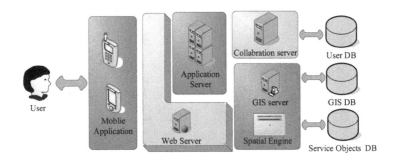

Fig. 7. The back ground system of LaMOC

4   Some Key Issues

4.1   Context Awareness

The word context means here physical context and logical context. That is:

- Physical context-location, terminal/device context, etc.
- Logical context-user profile, usage preferring, user/group behaviors, etc.

LaMOC has (hard and soft) sensors to capture context. Hard sensors such as GPS receivers are used to capture location data. Soft sensors are actually soft agents that can capture logical context.

There are different kinds of smart phones in market. Especially, their display screen size differs from one another. The way to solve adaptive problem in LaMOC is by device sensing and device awareness. An on-demanded map can be adapted automatically to suit different display screens of various smart phones.

GPS is used for positioning. But GPS has its inherent shortage. Mainly GPS signal is poor in some places, especially in tunnels, or in a building, where the signal is lost.

In LaMOC project a pseudo GPS system is developed by SITP[1]. A pseudo GPS station is actually a GPS signal sender, following the signal standard of GPS. It can be installed in buildings, tunnels, and so on. Thus, LaMOC sensor can capture position information covered nearly anywhere.

Logical context awareness in LaMOC means, automatically user/group profile capturing, user/group usage awareness, etc.

User and group profiles are defined and stored in the context database of LaMOC. They are static context of user and group. Besides, user/group usage as dynamic context is captured during the running of LaMOC. Soft sensors (agents) are used for usage awareness. The agents are engaged in:

- Collaborative awareness- e.g. where is my friends? Is there any of my friends near me?
- Group feature awareness – e.g. in the case of the system recommendation, decisions often depend on the group's historical activities. Group's historical activities are captured via group feature awareness. Such that, it can be answered: what are the common favorites of the group users?
- Partner behaviors awareness – e.g. which friends of mine is going to dinning ? What is the planed program of my partner next hour? Can I meet him on next hour?

4.2 Mobile Cooperation

Mobile collaboration is an imported feature of our scenarios. Mobile users of same group cooperate with each other to achieve common target. For example, a co-visitor in Expo Village recommends a common interested exhibition pavilion to his/her partners of the same group. Or a group of distributed tourists try to find each other and determine a muster place. Some things concerned mainly by us are:

- Spatial Group Decision Support (SGDS).
- Social network and group application
- Co-browsing-user interface

Let's look at some scenarios:

Scenario 1: A group of distributed users are on the way, someone occasionally supposes to take a group meeting as soon as possible. Other participants agree with him/her. Where is the suitable mustering place to the group?

Scenario 2: Five participants in a group are driving 5 cars on highway to a city at the same time. One suggests taking a rest, and others agree with him/her. Which rest service zone is optimal for such 5 cars?

Such scenarios involve Spatial Group Decision Support (SGDS). LaMOC is focused on supporting such applications. Geo Server and Collaboration Server are used to support SGDS. The Geo Server has the functionality of spatial modeling and Geo computing. The commercial GIS products are lack of the functionality for wide do-

---

[1] The Shanghai Institute of Technical Physics of the Chinese Academy of Sciences, one partner of LaMOC project.

main spatial modeling and intelligent computing. Therefore, a self developed Geo data management system, LaMOC spatial engine as the kernel, with spatial modeling and intelligent computing toolkits are developed in LaMOC project. PostgreSQL2 is used as the base system for LaMOC. The spatial engine has the Geo computing capability, such as finding the optimal object related to the user's location according to his/her favors, or finding the optimal route to the target place. LaMOC is focused on the group application. The spatial engine can compute to find the favorite and optimal restaurant for a group of distributed people, to find the optimal route for each participant to arrive there, and so on.

LaMOC supports group decision, such as finding an optimal mustering place for a group of Expo visitors distributed in an area such as Expo village.

Because LaMOC works in a mobile environment, compared with traditional cooperative environment, user mobility is a big challenge.

For example, five members take part in the cooperation, locating just in a restricted area, e.g. in Expo Village. In moving, 10 minutes later one body is out of the area, so that he/she leaves the cooperation activity automatically. Maybe other 2 person belonging to the same group now enter the area. Then these two persons join the collaborative task automatically.

A Collaboration Server is designed to support cooperation. Some algorithms and mechanisms have be worked out in LaMOC to suit such situation.

5 Conclusion

In this paper LaMOC- a Location Aware Mobile Cooperative System developed by ICA-ECNU is presented. Actually, it is a join work developed by East China Normal University, Fudan University, SITP, and some other partners. The work is supported by Science & Technology Commission Shanghai Municipality with the Grant No. 075107006. Restricted by the length, the details, e.g. some algorithms proposed, have not been presented here. We will report our works continually.

References

1. Wolfson, O.: Moving Objects Information Management: The Database Challenge. In: Halevy, A.Y., Gal, A. (eds.) NGITS 2002. LNCS, vol. 2382, p. 75. Springer, Heidelberg (2002)
2. Beaubrun, R., Moulin, B., Jabeur, N.: An Architecture for Delivering Location-Based Services. International Journal of Computer Science and Network Security 7(7) (July 2007)
3. Ward, A.M.R.: Sensor-driven Computing. PhD Thesis, University of Cambridge (1998)
4. Rosson, M.B., Carroll, J.M.: Usability Engineering. Morgan Kaufmann, San Francisco (2001)

---

2 http://www.postgresql.org/

Modelling of Camera Phone Capture Channel for JPEG Colour Barcode Images

Keng T. Tan, Siong Khai Ong, and Douglas Chai

Faculty of Computing, Health and Science, Edith Cowan University,
270 Joondalup Drive, Joondalup 6027, Australia
a.tan@ecu.edu.au,
siongo@student.ecu.edu.au,
d.chai@ecu.edu.au
http://www.chs.ecu.edu.au

Abstract. As camera phones have permeated into our everyday lives, two dimensional (2D) barcode has attracted researchers and developers as a cost-effective ubiquitous computing tool. A variety of 2D barcodes and their applications have been developed. Often, only monochrome 2D barcodes are used due to their robustness in an uncontrolled operating environment of camera phones. However, we are seeing an emerging use of colour 2D barcodes for camera phones. Nonetheless, using a greater multitude of colours introduces errors that can negatively affect the robustness of barcode reading. This is especially true when developing a 2D barcode for camera phones which capture and store these barcode images in the baseline JPEG format. This paper present one aspect of the errors introduced by such camera phones by modelling the camera phone capture channel for JPEG colour barcode images.

Keywords: Image-based modelling, Image processing & understanding.

1 Introduction

Two dimensional (2D) barcodes are becoming a pervasive interface for mobile devices, such as camera phones. Often, only monochrome 2D barcodes are used due to their robustness in an uncontrolled operating environment of camera phones. Recently we are seeing an emerging use of colour 2D barcodes for camera phones. Nonetheless, using a greater multitude of colours introduces errors that can negatively affect the robustness of barcode reading. Futhermore, most camera phones capture and store such colour 2D barcode images in the Joint Photographic Experts Group (JPEG) format [1]. As a lossy compression technique, JPEG also introduces a fair amount of error in the decoding of captured 2D barcode images. This JPEG quantization error is further compounded by the error introduced by the camera phone capture channel. Thus, robust decoding of colour 2D barcode JPEG images is a very challenging task. When possible, 2D barcode developers would rather not work with such lossy JPEG images [2]. Unfortunately, the most often image format supported on camera phones is the

D. Ślęzak et al. (Eds.): SIP 2009, CCIS 61, pp. 259–266, 2009.

baseline JPEG format. Moreover, there is an increasing use of colour 2D barcodes for camera phones. Examples of these colour 2D barcodes are the coloured ColorCode [3], High Capacity Colour Barcode (HCCB^{TM}) [4] and the Mobile Multi-Colour Composite (MMCC^{TM}) [5] 2D barcode. Figures 1 and 2 presents examples of such colour 2D barcodes for camera phones.

Fig. 1. Examples of ColorCode (left) and HCCB^{TM}(right)

Fig. 2. An example of MMCC^{TM}

1.1 Errors Pertinent to JPEG Images Captured on Camera Phones

When a 2D colour barcode image is captured by a camera phone and stored in JPEG format, there is essentially two major components of introduced errors, other than errors due to corruption of the 2D barcode symbol itself, such as smudges, tear and wear, or errors due to bad illumination of the barcode symbol being captured. These two major components of introduced errors are:

1. Errors due to the JPEG compression technique;
2. Errors due to the camera phone capture channel.

For errors introduced by the JPEG compression, there is essentially two major elements that introduced these errors, namely, the errors due to the quantisation process in the JPEG compression, and the errors due to the use of sub-optimal dequantisation matrix at the JPEG decoder. There is nothing much that can be done about the former but with the latter, we have found that a JPEG decoder based on optimised Discrete Cosine Transform (DCT) co-efficients distribution can result in higher quality decompressed images [6].

In this paper, we analysed the other major component of introduced error, the errors due to the camera capture channel. From our research [5] on the the MMCC^{TM}, we have observed that even under controlled laboratory conditions (i.e. perfect 2D barcode symbol and good barcode symbol illumination) there is still signification errors in the captured image such that the colours in the captured image shift significantly from their respective reference colour. An example of this observation is presented in Figure 3, where the colour of the data symbol, Symbol 7, has shifted from its origin at $RGB = (0, 255, 255)$ to a cluster of colour co-ordinates within the RGB colour space.

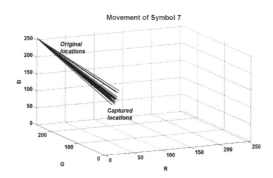

Fig. 3. An example of colour movement in RGB colour space for a capture JPEG colour barcode image

Section 2 presents our modelling of the camera phone capture channel. Section 3 presents the results of our simulations. Section 4 presents our conclusion.

2   Camera Phone Capture Channel Modelling

To model our colour 2D barcode, we have used the colour scheme from our previous work [5] and constructed a data cells only colour 2D barcode. Such a data cells only model is valid because in the decoding process for 2D barcodes, the data cells area of the barcode symbol is usually segmented from the located

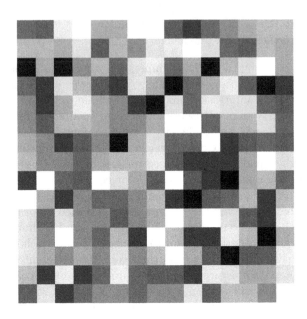

Fig. 4. An example of our colour 2D barcode model

barcode image prior to the colour decoding process [5]. An example of our colour 2D barcode is depicted in Figure 4.

For modelling the camera phone capture channel, we have used a Palm Treo 750 smart camera phone. This is a camera phone equipped with a 1.3 megapixel Charged Couple Devices (CCD) camera, with a Windows Mobile Ver.6 operating system. To model the camera capture channel for this camera phone, we have generated 20 random samples of the colour barcode model deoicted in Figure 4, where for each sample, we have captured 20 JPEG images of the colour 2D barcode under controlled laboratory conditions. Figure 5 illustrates a similar setting used in our experiment to model the camera phone capture channel.

From the experiment, after capturing the JPEG images of the colour 2D barcode model with our camera phone, we have found that the average observed mean values for the C_b and C_r colour components of the barcode images do not

Fig. 5. An experiment setting similar to that used in our modelling

change but its Y luminance component is, on the average, reduced by about 40%. We have analysed the captured JPEG images in the YC_bC_r domain because the JPEG compression manipulates colour images in this domain [1]. While this observation can be attributed to the camera phone capture channel error, it can also be caused by quantisation errors in the JPEG compression process of the captured images. Hence, we further compared this observation with JPEG compressed images of the same 2D barcodes samples under ideal conditions (i.e. directly compressing the original barcode images). What we have found is that there is definitely another error component contributing to the changes in the YC_bC_r components of our sample 2D barcode images. Figure 6 presents an example of such an observation.

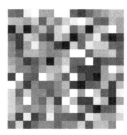 

Fig. 6. Comparing between a JPEG compressed 2D barcode sample image (left) and its camera phone captured equivalent (right)

To further model for this error component, we have taken steps to reproduce the same colour movements using image pre-processing techniques to model the noise in the colour 2D barcode sample images prior to the JPEG compression under ideal conditions. Figure 7 illustrates the steps we have taken in this comparative study.

From our study, we have found that we can reproduce, within limits, the error introduced by the camera capture channel. These results are presented in the next Section.

3   Simulation Results

From our simulations, we have observed that the camera capture channel produces 'darker' images of our sample colour 2D barcodes. This is consistant across all our randomised model samples. To reproduce such effects, we have experimented with various image processing algorithms, such as image blurring, image softening, manipulating the image contrast and adding noise to the image. We have also experimented with noise with different distributions, such as Gaussian, Laplacian and Pisson distributed noises.

From our experiments, we have found that adding Gaussian distributed noise gives us the closest effect to the colour movements observed in our camera channel modelling experiment. For Additive White Gaussian Noise (AWGN), we have

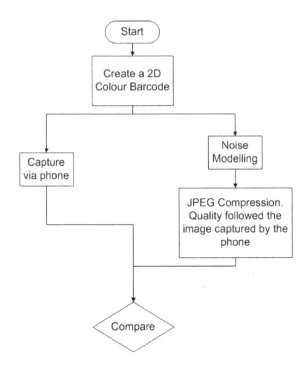

Fig. 7. A comparative study between the captured and the pre-processed colour 2D barcode images

experimented by adding AWGN with a distribution mean ranging from 0.1 to 1. For each distribution mean, we have calculated the differences between the colour shift vectors of the pre-processed colour 2D barcode sample images versus those of the camera phone captured images. The average differences in the YC_bC_r movements between our pre-processed sample images and our camera phone captured sample images is presented in Figure 8. The maximum differences in YC_bC_r movements between our pre-processed sample images and our camera phone captured sample images is presented in Figure 9.

From Figures 8 and 9, we have found that by pre-processing our 2D barcode sample images with an AWGN with a mean of 0.3 give us resultant JPEG sample images that is closest to our camera phone captured sample images. Thus, for the experiments in this paper, an AWGN channel with a distribution mean of 0.3 can closely model the camera phone capture channel for our colour 2D barcode. A visual comparison between this pre-processed sample image against its camera phone captured equivalent is presented in Figure 10.

4   Conclusion

Colour 2D barcodes are new emerging pervasive interfaces for camera phones. Nonetheless, using a greater multitude of colours introduces errors that can

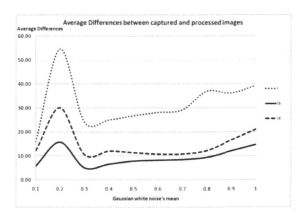

Fig. 8. Average differences in YC_bC_r values between our pre-processed sample images and our camera phone captured sample images

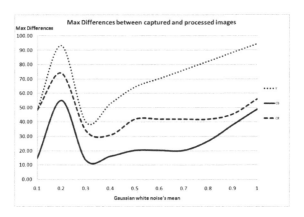

Fig. 9. Maximum differences in YC_bC_r values between our pre-processed sample images and our camera phone captured sample images

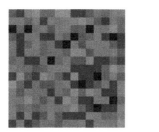

Fig. 10. Comparing between a pre-processed 2D barcode sample image (left) and its camera phone captured equivalent (right)

negatively affect the robustness of barcode reading. Futhermore, most camera phones capture and store such colour 2D barcode images in the JPEG format. As a lossy compression technique, JPEG also introduces a fair amount of error in the decoding of captured 2D barcode images. This JPEG quantization error is further compounded by the error introduced by the camera phone capture channel. Thus, robust decoding of colour 2D barcode JPEG images is a very challenging task. Hopefully, the work presented in this paper can help to address some of these challenges. From our results herein, we have concluded that a most likely model to model the camera phone capture channel errors for colour 2D barcodes captured by camera phones and stored in JPEG format is the AWGN channel. With this knowledge, coupled with our work [6] in improving the decompression of JPEG barcode images, it is likely that we will be able to improve the decoding of colour 2D barcode images for camera phones.

References

1. Pennebaker, W., Micthell, J.: JPEG Still Image Data Compression Standard. Van Nostrand Reinhold, New York (1992)
2. Private correspondance with inventor of mCode
3. Han, T.D., Cheong, C.H., Lee, N.K., Shin, E.D.: Machine readable code image and method of encoding and decoding the same. United States Patent, 7, 020, 327 (March 2006)
4. Gavin, J.: System and method for encoding high density geometric symbol set. United States Patent Application, 20, 050, 285, 761 (December 2005)
5. Kato, H., Tan, K.T., Chai, D.: MMCC^{TM}: Color 2D Bar code for True Ubiquitous Computing. Submitted to IEEE Transaction on Mobile Computing
6. Tan, K.T., Chai, D.: Improving Mobile 2D-Barcode Images Compression Using DCT Coefficient Distributions. Submitted to Journal of Electronic Imaging

Semantic Network Adaptation Based on QoS Pattern Recognition for Multimedia Streams

Ernesto Exposito[1,2], Mathieu Gineste[1,2], Myriam Lamolle[3], and Jorge Gomez[1,2,4]

[1] CNRS , LAAS , 7 av. du Colonel Roche, F-31077 Toulouse, France
[2] Université de Toulouse, UPS, INSA, INP, ISAE, LAAS, F-31077 Toulouse, France
{ernesto.exposito,mgineste,jorge.gomez}@laas.fr
[3] LIASD/LINC, IUT de Montreuil, Université Paris VIII
m.lamolle@iut.univ-paris8.fr
[4] Facultad de Matemáticas, Universidad Autónoma de Yucatán, México

Abstract. This article proposes an ontology based pattern recognition methodology to compute and represent common QoS properties of the Application Data Units (ADU) of multimedia streams. The use of this ontology by mechanisms located at different layers of the communication architecture will allow implementing fine per-packet self-optimization of communication services regarding the actual application requirements. A case study showing how this methodology is used by error control mechanisms in the context of wireless networks is presented in order to demonstrate the feasibility and advantages of this approach.

Keywords: self-optimizing, semantic pattern recognition, QoS, multimedia.

1 Introduction

Distributed applications were originally characterized by very basic communication requirements related to reliability and order. Today, multimedia applications demanding more complex requirements in terms of delay and bandwidth have been developed. Moreover, some of these applications are able to tolerate a non-perfect communication service (e.g. a partially ordered and partially reliable service). These applications have generally been designed in the framework of the Application Level Framing (ALF) approach [1]. In this approach, Application Data Units (ADU) can be processed independently by the various components of the multimedia system. The resulting QoS is strongly associated to the way this processing has been achieved by the application and networking components. Generally, components participating in the end-to-end communication path configure and provide their services using only a session level approach (e.g. each packet of the multimedia stream receives the same service). However, a more accurate knowledge of individual ADU requirements could largely help these components to improve the QoS provided and perceived by end-users. This paper introduces an ontology for the standard representation of ADU properties in order to allow any of the underlying communication mechanisms to access and use these properties. This approach is not intended to add new headers in

D. Ślęzak et al. (Eds.): SIP 2009, CCIS 61, pp. 267–274, 2009.

the data packets but to use existing descriptor headers to make their QoS properties publicly available to any underlying communication mechanisms, according to a well-defined set of rules. This ontology is called the Implicit Packet Meta Header (IPMH) ontology. The use of this ontology by mechanisms situated at different levels of the communication architecture will allow a global optimization of the communication processing according to the actual ADU requirements.

The rest of the paper is structured as follow. The section 2 presents the multimedia network context where IPMH can be employed. Section 3 introduces the IPMH and presents an abstract scheme for its representation and computation. Section 4 describes a study case based on the use of the IPMH to optimize error control mechanisms. Concluding remarks are finally proposed.

2 Multimedia Network Context: RTP-Based Systems

The Real-time Transport Protocol (RTP) proposed by the IETF [2] has become the standard for time constrained multimedia applications such as video on demand (VoD), audio and video conferencing, voice over IP (VoIP), television over IP (IPTV), etc. RTP has been proposed by the Audio/Video Transport workgroup (AVT) of IETF and an important number of RFCs have been defined in order to describe standards for streaming multimedia content using RTP.

RTP follows the principles of Application Level Framing (ALF) proposed by Clark and Tennenhouse [1]. ALF principle claims for breaking media data (i.e. audio or video content) into suitable aggregates. The frame boundaries of these aggregates are preserved by lower layers of the communication system (i.e. transport, network and data link layers). These aggregates are called Application Data Units (ADU) and are intended to be used as the minimal processing unit.

RTP standards follow a header definition approach to describe every ADUs aimed at providing information such as payload identification, sequence numbering and timestamps. RTP, however, does not provide any guarantees concerning the QoS and real-time constraints of the data transported. Lower layers should provide these guarantees. However, lower layers generally not used this important QoS information when delivering communication services. Next paragraphs introduce the standards describing the common header of the RTP protocol as well as the specific headers used to describe legacy multimedia streams.

2.1 Fixed RTP Header

In this section the fields of RTP fixed headers are detailed. These fields need to be specified for every RTP packet composing any RTP media stream.

- Version (2 bits) identifies the version of RTP (currently version 2).
- P: Padding (1 bit) indicates if the packet contains one or more additional padding octets.
- X: extension (1 bit) indicates if the fixed header is followed by exactly one header extension.

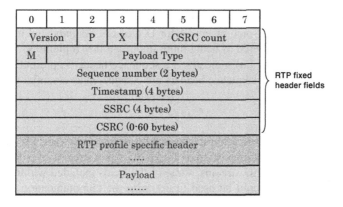

Fig. 1. Fixed RTP header

- CSRC count (4 bits) contains the number of CSRC or contributing sources for the payload that follow the fixed header.
- M: marker (1 bit) depends of the RTP profile, but generally is intended to specify if the ADU has been segmented in several RTP packets.
- Payload type (7 bits) is used to identify the format of the RTP payload. A set of default format types for audio and video streams is specified in [3].
- Sequence number (16 bits) indentifies the order of every RTP packet within the stream.
- Timestamp (32 bits) reflects the sampling instant of the first octet in the RTP data packet.
- SSRC (32 bits) uniquely identifies the synchronization source of the stream (e.g. the sender of a stream of packets derived from a media capture source such as a microphone or a camera).
- CSRC list (60 bytes) contains the contributing sources for the current payload if it is the result of the combination of multiple streams performed by an intermediate entity.

2.2 Specific RTP Profiles

The fixed RTP header can be enhanced with specific headers including additional fields required for more specialized RTP streams such as MPEG1, MPEG2, MPEG4, H.263, H.264, etc.

2.2.1 MPEG

The Moving Picture Experts Group or MPEG, is a working group of ISO/IEC responsible for the development of the widely used video and audio encoding standards. The fields included within the specific RTP/MPEG [4] video header are intended to be used by the receiving application in order to decode the MPEG video streams. However, some of these fields could be used by the communication system in order to differentiate the priority of RTP/MPEG packets and optimize the utilisation of constrained communication resources:

- Picture type (3 bits) indicates the type of picture being transported, 1 for I-pictures, 2 for P-pictures, 3 for B-pictures and 4 for D-pictures. As indicated in [4], I and P pictures could be considered as being more important that B pictures.
- Sequence header flag (1 bit) indicates if a sequence-header is present (value of 1). RTP packets containing sequence headers are considered as essential to decode the following packets.

2.2.2 H.263

H.263 is a video codec for low-bitrate, originally designed by the ITU-T. In [5] the standard for transporting various versions for H.263 video streams over RTP has been proposed. As for RTP/MPEG streams, some of the fields included within the specific RTP/H263 video stream could be used to optimize the communication resources, in particular:

- compression mode (1 bit) indicates the video mode used for video compression. P=0 implies normal I and P frames and P=1, indicates I, P and B frames. As for MPEG codec, I pictures are more important than P and B pictures.
- intra or inter-coded picture (1 bit) is used to identify the type of picture. I=0 for intra-coded picture and I=1 for inter-coded pictures. Inter-coded pictures are dependent of intra-coded pictures.

2.2.3 H.264

The H.264 [6] video codec has a very broad application range that covers all forms of digital compressed video (e.g. low bit-rate Internet streaming applications, HDTV broadcast or Digital Cinema applications) with nearly lossless coding. The codec specification distinguishes between a video coding layer (VCL) and a network abstraction layer (NAL). The VCL contains the signal processing functionality of the codec and a loop filter. The Network Abstraction Layer (NAL) encoder encapsulates the slice output of the VCL encoder into Network Abstraction Layer Units (NAL units), which are suitable for transmission over packet networks.

- NAL reference identificator (2 bits) a value of 00 indicates that the content of the NAL unit is not used to reconstruct reference pictures for inter picture prediction. Such NAL units can be discarded without risking the integrity of the reference pictures. Values greater than 00, indicate that the decoding of the NAL unit is required to maintain the integrity of the reference pictures. In addition to the specification above, according to this RTP payload specification, values greater than 00 indicate the relative transport priority, as determined by the encoder. Intermediate communication components can use this information to protect more important NAL units better than they do for less important NAL units. The highest transport priority is 11, followed by 10, and then by 01; finally, 00 is the lowest.

3 Implicit Packet Meta Header Ontology

The Implicit Packet Meta Header or IPMH is intended to provide information about the ADU properties of ALF-based multimedia streams. The attributes contained in this descriptor are aimed at helping the different entities participating in the end-to-end path transmission to optimize the real-time operations performed over every ADU. The overall IPMH architecture is illustrated in figure 2.

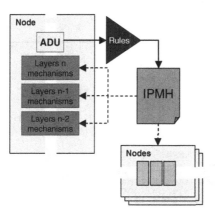

Fig. 2. IPMH Architecture

In this architecture, the IPHM is built from a specific set of rules associated to the actual multimedia stream. This header will be visible for any of the communication components present all along the transmission path, either from the vertical point of view (i.e., cross layer) or from the horizontal point of view (i.e., set of nodes crossed by the ADU).

3.1 IPMH Structure

The IPMH includes the following information semantic elements per ADU:

- Unique Identifier, in order to express reliability and order constraints;
- Class of ADU (e.g. I and P pictures for H.263 streams);
- Priority Class (e.g. I pictures are "more important" than P pictures);
- Tolerated Delay;
- Intra-Dependency (e.g. dependency among segmented I or P pictures);
- Inter-Dependency (e.g. P pictures depend on I pictures);

3.2 Mapping Rules

In order to allow any communication mechanisms to identify or deduce the IPMH for a particular packet, a set of mapping rules must be proposed. These rules should specify all the necessary information in order to deduce the IPMH components: type of attribute (e.g. Integer, Boolean, String, etc), position of attributes within the ADU (i.e. offset and length) and optionally conditions to be verified and values to be set if conditions are true. Of course, this set of rules depends on the implemented codec.

In order to facilitate the deployment of this approach, the XML language has been used to define a standard specification of the IPMH. Figure 3 illustrates the XSD schema for the IPMH. This schema can be used to map standard streams (e.g. RTP streams) or any proprietary codec. Proprietary applications could then use cross layer optimizations without publishing the codec format.

```xml
<?xml version="1.0"?>
<xsd:schema xmlns:xsd="http://www.w3.org/2001/XMLSchema">
        <xsd:element name="IPMH">
                <xsd:complexType mixed="false">
                        <xsd:choice minOccurs="0" maxOccurs="unbounded">
                                <xsd:group ref="IPMHGroup"/>
                        </xsd:choice>
                        <xsd:attribute name="description" type="xsd:string"/>
                </xsd:complexType>
        </xsd:element>

        <xsd:group name="IPMHGroup">
                <xsd:choice>
                        <xsd:element name="uniqueID" type="IPMHAttribute"/>
                        <xsd:element name="streamClass" type="IPMHAttribute"/>
                        <xsd:element name="ADUClass" type="IPMHAttribute"/>
                        <xsd:element name="classPriority" type="IPMHAttribute"/>
                        <xsd:element name="toleratedDelay" type="IPMHAttribute"/>
                        <xsd:element name="intraDep" type="IPMHAttribute"/>
                        <xsd:element name="interDep" type="IPMHAttribute"/>
                </xsd:choice>
        </xsd:group>

        <xsd:complexType name="IPMHAttribute" mixed="true">
                <xsd:attribute name="typeAttribute" type="xsd:integer"/>
                <xsd:attribute name="offset" type="xsd:integer"/>
                <xsd:attribute name="length" type="xsd:integer"/>
                <xsd:attribute name="condition" type="xsd:string"/>
                <xsd:attribute name="value" type="xsd:string"/>
        </xsd:complexType>

</xsd:schema>
```

Fig. 3. XML schema definition for IPMH ontology

3.3 Using the IPMH

QoS functions such as packet scheduling or error control could use the IPMH approach to optimize their operations. Next paragraphs present a non-exhaustive list of functions and possible per-packet attributes to perform this optimization:

- Flow scheduling (forwarding of packets between end-system and networks): tolerated delay, classes and priorities.
- Flow shaping (regulation of flow scheduling based on the flow requirements and underlying resources): tolerated delay.
- Flow policing (actions to be taken when the flow specification is violated): tolerated delay, classes, priorities and inter and intra dependencies.
- Flow synchronization (control of order and time requirements for the delivery of multiple streams): tolerated delay and inter and intra dependencies.
- Error control (detection, correction of errors): unique identifier, tolerated delay, classes, priorities and inter and intra dependencies.

4 Case Study

Packet losses are generally the consequence of network congestion in traditional IP networks. During data transmission, packets are temporally stored in intermediate nodes (e.g., proxies, routers, access points) while they are forwarded to their final destination. When storage capacities are exceeded, then packets are dropped. Moreover, in wireless networks, unreliability is also originated from bit-errors due to channel fades and interference.

In order to recover packet losses, various error control mechanisms have been implemented at different layers of the communication system. Losses of packets are usually recovered by using Automatic Repeat reQuest (ARQ) techniques. However, these mechanisms do not always take into account the actual requirements of the applications. Indeed, some applications are able to tolerate a certain amount of packet losses over the transmitted data (e.g., audio, video, images, etc.), while for other applications packet loss is unacceptable (i.e. binary or data files, text, etc.). Currently, most of the error control mechanisms offer only a fully reliable service for all the applications. These mechanisms could be specialized to provide a partially reliable service in order to be more compliant with time-constrained applications.

In order to evaluate the feasibility of the IPMH implementation as well as the benefits of using this approach in the layered communication mechanisms, an experimental study case including RTP-based applications and error control mechanisms has been carried out. This type of application presents some preference for timeliness over reliability. This case study involve the achievement of various experiences, all of them using real RTP traffic (i.e., a VoD application using a H263 codec) and real implementations of different error control mechanisms. This software has been evaluated using a network emulator providing a real-time dynamic evolution of network conditions. High, medium and low load network conditions are emulated by varying the available bandwidth and the packet loss rates. Results of these experiments are illustrated in Figure 4.

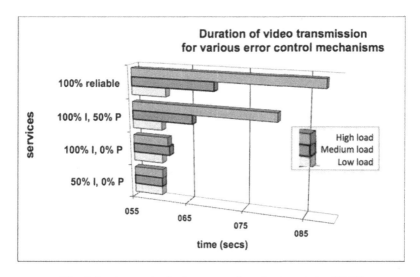

Fig. 4. Service results for error control mechanisms using IPMH

The performance results in terms of total duration of data transfer of a video stream have been compared for four kinds of error control mechanisms. Considered error control mechanisms are:

- 100 reliable: full reliable service (no losses are tolerated).
- 100% I, 50 % P: full reliability is required for I Pictures and 50 % for P pictures.
- 100% I, 0 % P: full reliability is required for I Pictures and 0 % for P pictures.
- 50% I, 0 % P: 50% reliability is required for I Pictures and 0 % for P pictures.

Results demonstrates that the delay accumulated for retransmission based error control mechanisms are higher when more reliability is requested for a video stream of 60 seconds. Using IPMH properties such as ADU type, priority or dependency allows implementing partial reliable services, which can largely reduce the transmission delay while respecting the reliability constraints indicated by the applications. While for fully reliable services the transmission delay for the complete video transmission was more than 85 seconds for a high loaded network, for partial reliable services using IPMH, the transmission delay can be controlled when the adequate reliability constraints are adjusted to the network conditions. Indeed, the worst case for the partial reliable service of (100%I, 50%P) was of 65 seconds (20 seconds less than the fully reliable service) for a high loaded network.

5 Conclusions

This paper presents the IPMH, a standardized way for the packet QoS properties to be computed and represented. The semantic offered by IPMH allows any of the underlying communication mechanisms to perform pattern recognition and use the computed QoS packet properties. It is build according to a well-defined set of rules applied onto the application data units. The use of the IPMH by several mechanisms situated at different levels of the communication architecture allows the development of optimized communication mechanisms regarding the real ADU requirements. This approach has been successfully implemented and evaluated in the context of various error control mechanisms using the QoS Information in order to optimize the reliability and time constraints.

References

1. Clark, D.D., Tennenhouse, D.L.: Architectural considerations for a new generation of protocols. In: Proceedings of IEEE SIGCOMM (Symposium on Communications Architectures and Protocols), Philadelphia, USA, September 1990, pp. 200–208 (1990)
2. Schulzrinne, H., Casner, S., Frederick, R., Jacobson, V.: RTP: A Transport Protocol for Real-Time Applications, IETF Request for Comments N°1889 (January 1996)
3. Schulzrinne, H.: RTP Profile for Audio and Video Conferences with Minimal Control, IETF RFC 3551 (July 2003)
4. Hoffman, D., et al.: RTP Payload Format for MPEG1/MPEG2 Video, IETF RFC 2250 (January 1998)
5. Ott, J., et al.: RTP Payload Format for ITU-T Rec. H.263 Video, IETF RFC 4629 (January 2007)
6. Wenger, S., Stockhammer, T., Hannuksela, M., Westerlund, M., Singer, D.: RTP payload format for H.264 video. Internet Engineering Task Force (IETF), RFC 3984 (February 2005)

Active Shape Model-Based Gait Recognition Using Infrared Images

Daehee Kim, Seungwon Lee, and Joonki Paik

Image Processing and Intelligent Systems Laboratory, Graduate School of Advanced Imaging Science, Multimedia, and Film, Chung-Ang University, 221 Heuksuk-Dong, Dongjak-Ku, Seoul 156-756, Korea
Wangcho100@wm.cau.ac.kr

Abstract. We present a gait recognition system using infra-red (IR) images. Since an IR camera is not affected by the intensity of illumination, it is able to provide constant recognition performance regardless of the amount of illumination. Model-based object tracking algorithms enable robust tracking with partial occlusions or dynamic illumination. However, this algorithm often fails in tracking objects if strong edge exists near the object. Replacement of the input image by an IR image guarantees robust object region extraction because background edges do not affect the IR image. In conclusion, the proposed gait recognition algorithm improves accuracy in object extraction by using IR images and the improvements finally increase the recognition rate of gaits.

Keywords: Active shape model, gait recognition.

1 Introduction

Recently, human recognition techniques have attracted increasing attention in intelligent surveillance applications. For recognition systems using Physiological feature-based human recognition systems use individual, unique characteristics, such as fingerprints, iris, and face. Because of the unique nature, the physiological feature-based system provides very high recognition rate up to 99%. In spite of the high recognition rate, the physiological features are not easily acquired due to the reluctance of participants. Alternatively, human behavioral features gains increasing attractions in applications for in conscious recognition of humans using gaits. [1][2][3]

Accurate infrared target tracking is critical in many military weapons systems where common knowledge indicates that improving infrared target detection and tracking has the potential to simultaneously minimize unwanted collateral damage and maximize the probability of successful target elimination. [4]

Gait recognition is a technology to identify a human based on the difference in a cycle of steps representing a behavioral feature. Most existing gait recognition techniques use CCD camera images, which are sensitive to illumination changes. The proposed gait recognition method is able to extract the object region using infrared (IR) images. Even under row-level illumination or complex background, an IR image contains the easily extracting object region.

D. Ślęzak et al. (Eds.): SIP 2009, CCIS 61, pp. 275–281, 2009.
© Springer-Verlag Berlin Heidelberg 2009

This paper is organized as follows. In section 2, we present the ASM algorithm for infrared images. Extraction method of the gait data is presented in section 3. Section 4 presents the experimental results and section 5 concludes the paper.

2 Active Shape Model for Object Extraction in Infrared Images

The proposed algorithm deals with several basic computer vision problems such as matching and comparing temporal signatures object and background, modeling human motion and dynamics, and shape recovery from partial occlusions. It is well known that the extraction of object area is very difficult with complex background or partial occlusions.

In this section we present a novel gait recognition system by deploying the ASM framework using the infrared video. The advantage of the proposed system is the robust recognition with complicated background, poor illumination, and occlusions.

We briefly revisit the ASM theory divided by including three steps: (a) shape variation modeling, (b) model fitting, and (c) local structure modeling.

2.1 Shape Variation Modeling

Given a frame of input video, initial landmark points should be assigned on the contour of the object either manually or automatically. Good landmark points should be at or close to the desired boundary of each object. A particular shape X is represented by a set of n landmark points which approximate its outline as

$$X = [x_1, x_2, \cdots, x_n, y_1, y_2, \cdots, y_n]^T, \tag{1}$$

Different sets of such landmark points make a training set. A shape in the training set is normalized in scale, and aligned with respect to a common frame. Although each aligned shape is in the $2n$-dimensional space, we can model the shape with a reduced number of modes using the principal component analysis (PCA) analysis. The main modes of the template model, X, are then described by the eigenvectors φ of the covariance matrix c, with the largest eigen-values.

2.2 Model Fitting

We can find the best shape and pose parameters to match a shape in the model coordinate frame, x, to a new shape in the image coordinate frame, y, by minimizing the following error function

$$E = (y - Mx)^T W^T (y - Mx), \tag{2}$$

where M represents the geometric transformation of scaling (s), translation (t), and rotation (θ). For instance if we apply the transformation to a single point, denoted by $[p, q]^T$, we have

$$M \begin{bmatrix} p \\ q \end{bmatrix} = s \begin{bmatrix} \cos\theta & \sin\theta \\ -\sin\theta & \cos\theta \end{bmatrix} \begin{bmatrix} p \\ q \end{bmatrix} + \begin{bmatrix} t_x \\ t_y \end{bmatrix}, \tag{3}$$

After a set of pose parameters, $\{\theta, t, s\}$, is obtained, the projection of y on to the model coordinate frame is given as

$$x_p = M^{-1}y. \tag{4}$$

Finally, the parameters are updated as

$$b = \phi^T (x_p - \bar{x}). \tag{5}$$

2.3 Local Structure Modeling

In order to interpret a given shape in the input image based on ASM, we must find a set of parameters that best match the model to the input shape. If we assume that the shape model represents boundaries and strong edges of the object, a profile across each landmark point has an edge like local structure. The nearest profile can be obtained by minimizing the following Mahalanobis distance between the sample and mean of the model as

$$f(g_{i,m}) = (g_{i,m} - \bar{g})^T S_g^T (g_{i,m} - \bar{g}), \tag{6}$$

where $g_{i,m}$ represents the shifted version of g_i by m samples along the normal direction of the corresponding boundary.

3 Extraction of Gait Data

In this section we present an object region extraction method based on the analysis of the object in section 2. For the analysis of human gait it is very important to extract the gait period. Most gait period extracting methods measure object features in the image on the basis of temporally proceeding frames. The proposed gait recognition method analyzes the shape of an object in the form of a four-dimensional (4D) vector as shown in Fig.1 in order to acquisition 4D data.

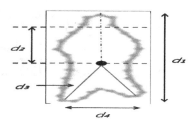

Fig. 1. A 4D vector with four gait parameters: $W = [d_1, d_2, d_3, d_4]^T$

The 4D vector consists of the height of the object shape (d1), the length of the upper torso (d2), from the pelvis to the foot length (d3), and the distance between the left and right foot (d4). We can detect the eigen gait by using the extracted 4D gait vectors. For normally obtained gait data, d4 data shows the periodic property and has the unique feature. Extracted 4D gait vectors are store in the database for future comparison with an arbitrary input data.

In the recognition process the cross-correlation of the 4D gait vectors extracted from the input image and from the stored database as,

$$c(n) = \Sigma \, d_i^{DB}(m) \cdot d_i^{input}(n+m), \text{ for } i = 1, 2, 3, 4. \tag{7}$$

The stored gait data with the maximum correlation is selected as the best matching candidate. Where d_i^{DB} represents the element of the i-th gait vector in the stored database, d_i^{input} the element of the i-th vector extracted from the input image, and $c(n)$ the cross-correlation shifted by n frames.

4 Experiment Results

We first compare object extraction performance using a CCD and an IR images. While the object was extracted by subtraction from background in the CCD image, the object in the IR image was extracted by using an experimentally chosen threshold as shown in Fig. 2.

(a)   (b)

(c)   (d)

Fig. 2. (a) Input CCD image, (b) extracted object in the CCD image, (c) input IR image, and (d) extracted object in the IR image

Fig.3 show results of the ASM in the CCD and IR images. For the CCD image, a number of mismatches are observed near the contour of the object. However the IR image provides the significantly improved shape of the object.

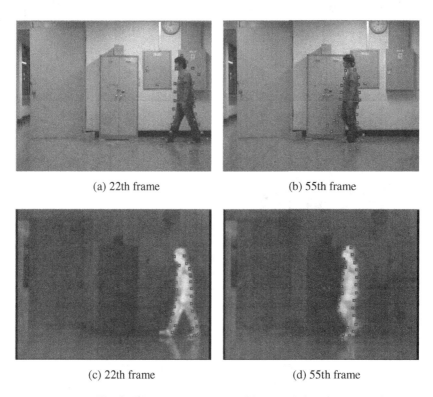

(a) 22th frame   (b) 55th frame

(c) 22th frame   (d) 55th frame

Fig. 3. The ASM results in the CCD and IR images

Table 1 summarizes the value of shape-accuracy of CCD and IR images using the ASM, respectively. The matching-ratio is achieved by using manually-assigned ground truth data.

Table 1. Font sizes of headings. Table captions should always be positioned *above* the tables.

Camera	object	Suitability	Disagreement	Matching ratio(%)
CCD	Object 1	21	11	62.6
	Object 2	18	14	56.3
IR	Object 1	30	3	93.8
	Object 2	29	3	90.6

Fig.4 shows the ASM results and gait data in the row-level illumination images acquired by both CCD and IR cameras. The region of object and gait data are clearly extracted in the IR image. On the other hand the appearance of the object is not clear in the CCD image.

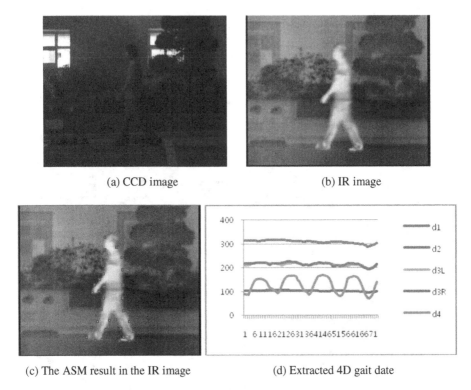

(a) CCD image   (b) IR image

(c) The ASM result in the IR image   (d) Extracted 4D gait date

Fig. 4. The ASM results in the row-level illumination images

Fig.5 (a) shows the measured gait data using cross-correlation when the same object walks in the different areas, and Fig.5 (b) shows the obtained gait data using cross-correlation when the different object walks in the same areas.

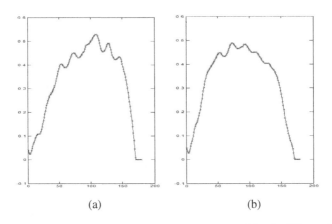

(a)   (b)

Fig. 5. The results of cross-correlations: (a) The same person walking in different locations, and (b) two different people walking in the same location

5   Conclusion

In this paper we obtain the gait data using the ASM in the IR image. Experiment results show that the proposed system can robustly extract the object region at the row-level illumination or with partial occlusion. For the future development of the gait recognition system, we need to consider conditions, to acquire input images to accurately simulate more practical situation.

Acknowledgement. This research was supported by Basic Science Research Program through National Research Foundation (NRF) of Korea funded by the Ministry of Education, Science and Technology (2009-0081059), and the Medium-term Strategic Technology Development Program funded by the Ministry of Knowledge Economy (MKE, Korea), and by Ministry of Culture, Sports and Tourism (MCST) and Korea Culture Content Agency (KOCCA) in the Culture Technology (CT) Research & Development Program 2009.

References

1. Jain, A.H.L., Pankanti, S., Bolle, R.: An identity verification system using fingerprints. Proc. IEEE 85(9), 1365–1388 (1999)
2. Turk, M., Pentland, A.: Face recognition using eigenfaces. In: Proc. IEEE Conf. Computer Vision and Pattern Recognition, pp. 586–591 (1991)
3. Cho, W., Kim, T., Paik, J.: Gait recognition using active shape models. In: Proc. Int. Conf. Advanced Concepts for Intelligent Vision Systems, pp. 384–394 (2007)
4. Colin, M.J., Nick, M., Joseph, P.H.: Dual Domain Auxiliary Particle Filter with Integrated Target Signature Update. In: Proc. IEEE Conf. Computer Vision and Pattern Recognition, pp. 54–59 (2009)

About Classification Methods Based on Tensor Modelling for Hyperspectral Images

Salah Bourennane and Caroline Fossati

Ecole Centrale Marseille,
Institut Fresnel Marseille cedex 20 France
salah.bourennane@fresnel.fr

Abstract. Denoising and Dimensionality reduction (DR) are key issue to improve the classifiers efficiency for Hyperspectral images (HSI). The multi-way Wiener filtering recently developed is used, Principal and independent component analysis (PCA, ICA) and projection pursuit (PP) approaches to DR have been investigated. These matrix algebra methods are applied on vectorized images. Thereof, the spatial rearrangement is lost. To jointly take advantage of the spatial and spectral information, HSI has been recently represented as tensor. Offering multiple ways to decompose data orthogonally, we introduced filtering and DR methods based on multilinear algebra tools. The DR is performed on spectral way using PCA, or PP joint to an orthogonal projection onto a lower subspace dimension of the spatial ways. We show the classification improvement using the introduced methods in function to existing methods. This experiment is exemplified using real-world HYDICE data.

Keywords: Multi-way filtering,Dimensionality reduction, matrix and multilinear algebra tools, tensor processing.

1 Introduction

Detection and classification are key issues in processing HyperSpectral Images (HSI). Spectral identification-based algorithms are sensitive to spectral variability and noise in acquisition. In recently, we have proposed multi-way Wiener algorithms that is robust to noise [1].

The emergence of hyperspectral images (HSI) implies the exploration and the collection of a huge amount of data. Imaging sensors provide typically up to several hundreds of spectral bands. This unreasonably large dimension not only increases computational complexity but also degrades the classification accuracy [2]. Thereof, dimensionality reduction (DR) is often employed. Due to its simplicity and ease of use, the most popular DR is the principal component analysis (PCA), referred to as PCA_{dr}. A refinement of PCA_{dr} is the independent component analysis (ICA_{dr}) [3]. While PCA_{dr} maximizes the amount of data variance by orthogonal projection, ICA_{dr} uses higher order statistics which characterize many subtle materials. On the other hand, projection pursuit (PP_{dr}) method [4, 5, 6, 7, 8] uses an index projection, to be defined, to find *interesting*

D. Ślęzak et al. (Eds.): SIP 2009, CCIS 61, pp. 282–296, 2009.

projections. Originally, the projections revealing the least gaussian distributions have been shown to be the most interesting. Regarding its index projections, which can be set, PP_{dr} method generalizes the matrix algebra based methods. In opposition with the previous DR methods, PP_{dr} uses a deflective rather than a global procedure by sequentially searching the projection. But all these matrix algebra methods require a preliminary step which consists in vectorizing the images. Therefore, they rely on spectral properties only, neglecting the spatial rearrangement. To overcome these disadvantages, [9,10,11] recently introduced a new HSI representation based on tensor. This representation involves a powerful mathematical framework for analyzing jointly the spatial and spectral structure of data. For this investigation, several tensor decompositions have been introduced [12,13,14]. In particular, the Tucker3 tensor decomposition [15,16] yields the generalization of the PCA to multi-way data. An extension of this decomposition can generalize the lower rank matrix approximation to tensors. This multilinear algebra tool is known as lower rank-(K_1, K_2, K_3) tensor approximation [17], denoted by $LRTA$-(K_1, K_2, K_3). In [1] multi-way filtering algorithms are developed and the shown results demonstrate that this new way to built the Wiener filter is very efficient for the multidimensional data such as HSI.

Based on this tensor approximation, this paper introduces two multilinear algebra methods for DR approach. These two methods yield a multi-way decorrelation and jointly perform a spatial-spectral processing: p spectral components are extracted, jointly with a lower spatial rank-(K_1, K_2) approximation. The latter spatial processing yields a projection onto a lower dimensional subspace that permits to spatially whiten the data. For the first method, the PCA_{dr} is considered for the extraction of the p spectral components, this multiway method has been introduced in [10]. And for the second one, the PP_{dr} is investigated. The two introduced multilinear algebra-based methods are referred to as $LRTA_{dr}$-(K_1, K_2, p) and as hybrid $LRTA$-PP_{dr}-(K_1, K_2, p). We show that the spatial projection onto a lower orthogonal subspace joint to spectral dimension reduction improves the classification efficiency compared to those obtained when matrix algebra-based DR is used.

The remainder of the paper is organized as follows: Section 2 introduces matrix algebra approaches to DR. Then, Section 3 reviews some tensor properties. Section 4 introduces the multilinear algebra method before drawing comparative classification in Section 5.

2   Matrix Algebra-Based DR Method

2.1   HSI Representation

To apply these matrix algebra-based methods, the HSI data are considered as a sampling of spectrum. Suppose that I_3 is the number of spectral bands and $I_1 \times I_2$ is the size of each spectral band image (*i.e.* $I_1 \cdot I_2$ is the number of samples). Each image pixel vector is an I_3-dimensional random variable. Those pixel vectors, referred to as spectral signatures, can be concatenated to yield a matrix $\mathbf{R} = [\mathbf{r}_1 \; \mathbf{r}_2 \; \cdots \; \mathbf{r}_{I_1 I_2}]$ of size $I_3 \times I_1 I_2$. In other words, the ith row in $\mathbf{R}$ is specified by the ith spectral band.

The DR approach extracts a lower number p of features called components, with $p < I_3$, such as

$$\mathbf{Y} = \mathbf{TR}, \tag{1}$$

where $\mathbf{T}$ is a linear transformation matrix of size $p \times I_3$, and $\mathbf{Y}$ the reduced matrix of size $p \times I_1 I_2$.

2.2 Principal Component Analysis Based DR Approach

In PCA_{dr} context, the extracted components are called principal components (PCs). Each PC is generated by projecting the data spaced onto the ith eigenvector associated with the ith largest eigenvalue of the covariance matrix. This orthogonal projection maximizes the amount of data variance. Therefore the p spectral PCs generate a reduced matrix $\mathbf{Y}_{PCs}$ of size $p \times I_1 I_2$. Let $\mathbf{U}$ the $p \times p$ matrix holding the p eigenvectors associated to the p first largest eigenvalues, the PCs are given by:

$$\mathbf{Y}_{PC} = \mathbf{U}^T \mathbf{X}. \tag{2}$$

In this case the matrix $\mathbf{U}^T$ is the transformation matrix referred to as $\mathbf{T}$ in Eq. (1). Sometimes, this transformation is associated to the sphering, given by:

$$\mathbf{Y} = \mathbf{\Lambda}^{-1/2} \mathbf{U}^T \mathbf{X}.$$

with, $\mathbf{\Lambda}$ the $p \times p$ eigenvalue diagonal matrix of the covariance matrix.

2.3 Independent Component Analysis Based DR Approach

ICA [18] is an unsupervised source separation process, that has been applied to linear blind separation problem [19]. Its application to linear mixture analysis for HSI has been found in [20] and to DR approach in [3]. ICA assumes that data are linearly mixed and separates them into a set of statistically independent components (ICs). Since ICA requires higher-order statistics, many subtle materials or rare targets are more easily characterized. ICA finds a $p \times I_3$ separating matrix $\mathbf{W}$ to generate p ICs (with $p < I_3$) such that:

$$\mathbf{Y}_{IC} = \mathbf{WR}. \tag{3}$$

Commonly, a pre-processing step performs a PCA to sphere and reduce the samples. To generate p ICs, $FastICA$ algorithm is selected using the absolute value of kurtosis as a measure of non-gaussianity. But, while PCs are generated by PCA_{dr} in accordance with decreasing magnitude of eigenvalues, the $FastICA$ algorithm does not necessarily generate ICs in order of information significance. This is emphasized since this method estimate simultaneously all the $\mathbf{w}$ vectors of the matrix $\mathbf{W}$ (see Eq. (3)).

2.4   Projection Pursuit Based DR Approach

Projection pursuit is a statistical analysis developed by [6] that searches the direction of the *interesting* projections. The *interesting* concept is defined by a projection index. Therefore, projection pursuit-based DR method (PP_{dr}) selects a lower dimensional projection from high dimensional data by maximizing (or minimizing) the index projection. This technique is the generalization of the DR approach: if the index projection is the variance, PP_{dr} is similar to the PCA_{dr}; if the projection index is the kurtosis, PP_{dr} is similar to the ICA_{dr}. But, instead of being a global method like PCA_{dr} and ICA_{dr}, the PP_{dr} is a deflective method. Thus, after finding the first direction which maximizes the projection index, the data are projected onto the orthogonal subspace of this direction and so forth.

In this paper, we select the non-gaussianity index. For this investigation, the *FastICA* algorithm in deflective mode estimates only one projection, $\mathbf{w}^T \mathbf{r}$ at a time. When i vectors $\mathbf{w}_1 \cdots \mathbf{w}_i$ components are estimated, *FastICA* is used to estimate the $\mathbf{w}_{i+1}$ vector and projection contribution of the i previous vectors are deducted, such as:

$$\mathbf{w}_{i+1} \longleftarrow \mathbf{w}_{i+1} - \sum_{j=1}^{i} \left(\mathbf{w}_{i+1}^T \mathbf{w}_j \right) \mathbf{w}_j. \tag{4}$$

The reduced matrix $\mathbf{Y}_P$, holding the p projections, is equivalent to $\mathbf{Y}_{IC}$ (Eq. (3)) using the non-gaussianity index but generated the vectors $\mathbf{w}$ using the previous process. PP_{dr} approach finds the p first projection pursuit directions to attain a lower p-dimensional space.

3   Tensor Representation and Some Properties

PCA_{dr}, ICA_{dr} and PP_{dr} are matrix algebra-based methods which require data rearrangement. In this paper, we use multilinear algebra tools to consider the whole data. Our methods are based on tensor representation to keep the initial spatial structure and insure the neighborhood effects. Tensor processing have proven its efficiency in several domains, telecommunications [21], image processing [1, 22, 23] and, more recently in hyperspectral image analysis [9, 10, 11].

In this representation, the whole HSI data is considered as a *third*-order tensor, the entries of which are accessed via three indices. It is denoted by $\mathcal{R} \in \mathbb{R}^{I_1 \times I_2 \times I_3}$, with elements arranged as $r_{i_1 i_2 i_3}$, $i_1 = 1, \ldots, I_1$; $i_2 = 1, \ldots, I_2$; $i_3 = 1, \ldots, I_3$ and $\mathbb{R}$ is the real manifold. Each index is called mode: two spatial and one spectral modes characterize the HSI tensor.

Tensor representation is mathematically grounded in multilinear algebra which studies the properties of data tensor $\mathcal{R}$ in a given n-mode. Let us define $E^{(n)}$, the n-mode vector space of dimension I_n, associated with the n-mode of tensor $\mathcal{R}$. By definition, $E^{(n)}$ is generated by the column vectors of the n-mode flattened matrix. The n-mode flattened matrix $\mathbf{R}_n$ of tensor $\mathcal{R} \in \mathbb{R}^{I_1 \times I_2 \times I_3}$ is defined as a matrix from $\mathbb{R}^{I_n \times M_n}$, with: $M_n = I_p I_q$ and $p, q \neq n$. An illustration of the n-mode flattening of a *third*-order tensor is represented in Fig. 1.

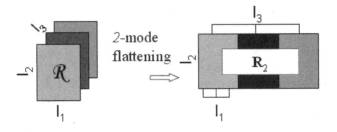

Fig. 1. 2-mode flattening of tensor $\mathcal{R}$

Algorithm 1. $LRTA_{dr}$-(K_1, K_2, p)

---

input : $\mathcal{R} \in \mathbb{R}^{I_1 \times I_2 \times I_3}$, K_1, K_2, p-values
output: $\mathcal{Y}_{PC\text{-}(K_1,K_2,p)} \in \mathbb{R}^{I_1 \times I_2 \times p}$
initialization: $i = 0, \forall n = 1, 2$
$\mathbf{U}^{(n),0} \leftarrow$ first K_n eigenvectors of matrix $\mathrm{E}\left[\mathbf{R}_n \mathbf{R}_n^T\right]$
repeat
    ALS loop:
    forall $n = 1, 2, 3$ **do**
        $\widehat{\mathcal{R}}^i \leftarrow \mathcal{R} \times_q \mathbf{U}^{(q),i^T} \times_r \mathbf{U}^{(r),i^T}$ with $q, r \neq n$
        $\mathrm{E}\left[\widehat{\mathbf{R}}_n^i \widehat{\mathbf{R}}_n^{i^T}\right]$ eigenvalue decomposition
        $\mathbf{U}^{(n),i+1} \leftarrow$ first K_n or p eigenvectors
        $i \leftarrow i + 1$
    end
until *convergence of Eq. (9)*
$\mathbf{P}^{(n)} \leftarrow \mathbf{U}^{(n)} \mathbf{U}^{(n)^T}$
$\mathcal{Y}_{PC\text{-}(K_1,K_2,p)} \leftarrow \mathcal{R} \times_1 \mathbf{P}^{(1)} \times_2 \mathbf{P}^{(2)} \times_3 \mathbf{\Lambda}^{-1/2} \mathbf{U}^{(3)^T}$

Tensor representation yields to process the whole data from spatial and spectral perspectives.

This representation naturally implies the use of multilinear algebraic tools and especially tensor decomposition and approximation methods. The most commonly used is the Tucker3 [15] decomposition, which generalizes to higher order, the singular value decomposition. This tensor decomposition is expressed as follows:

$$\mathcal{R} = \mathcal{C} \times_1 \mathbf{U}^{(1)} \times_2 \mathbf{U}^{(2)} \times_3 \mathbf{U}^{(3)}, \tag{5}$$

where $\mathcal{C}$ is the core tensor, $\mathbf{U}^{(n)}$ is the matrix of eigen vectors associated with the n-mode covariance matrix $\mathbf{R}_n \mathbf{R}_n^T$.

4 Multilinear Algebra-Based DR Method

4.1 Tensor Formulation of PCA_{dr} and PP_{dr}

Here, we write the equivalent of the Eqs. (1)-(3) in tensor formulation [17, 12, 13, 14]. Let $\mathcal{R} \in \mathbb{R}^{I_1 \times I_2 \times I_3}$ be the HSI data. The previously obtained matrix

Algorithm 2. hybrid $LRTA\text{-}PP_{dr}\text{-}(K_1, K_2, p)$

input : $\mathcal{R} \in \mathbb{R}^{I_1 \times I_2 \times I_3}$, K_1, K_2, p-values
output: $\mathcal{Y}_{P\text{-}(K_1,K_2,p)} \in \mathbb{R}^{I_1 \times I_2 \times p}$
initialization: $i = 0$, $\forall n = 1, 2$
$\mathbf{U}^{(n),0} \leftarrow$ first K_n eigenvectors of matrix $\mathrm{E}\left[\mathbf{R}_n \mathbf{R}_n^T\right]$
while $\left\| \mathcal{Y}_{P\text{-}(K_1,K_2,p)}^{i+1} - \mathcal{Y}_{P\text{-}(K_1,K_2,p)}^{i} \right\|_F^2 >$ *threshold* **do**
   ALS loop:
   for $n = 3$ **do**
      $\widehat{\mathcal{R}}^i \leftarrow \mathcal{R} \times_1 \mathbf{U}^{(1),i+1^T} \times_2 \mathbf{U}^{(2),i+1^T}$
      $\mathrm{E}\left[\widehat{\mathbf{R}}_n^i \widehat{\mathbf{R}}_n^{i^T}\right]$ eigenvalue decomposition
      $\mathbf{U}^{(3),i+1} \leftarrow p$ first eigenvectors
      $\widehat{\mathcal{R}}^i \leftarrow \widehat{\mathcal{R}}^i \times_3 \mathbf{U}^{(3),i+1^T}$
      $\mathbf{W} \leftarrow$ *fastica*, deflective approach from $\widehat{\mathcal{R}}^i$
   end
   for $n = 1, 2$ **do**
      $\widehat{\mathcal{R}}^i \leftarrow \mathcal{R} \times_q \mathbf{U}^{(q),i^T} \times_3 \mathbf{W}^i$ with $q, r \neq n$
      $\mathrm{E}\left[\widehat{\mathbf{R}}_n^i \widehat{\mathbf{R}}_n^{i^T}\right]$ eigenvalue decomposition
      $\mathbf{U}^{(n),i+1} \leftarrow$ first K_n
   end
   $\mathbf{P}^{(n)} \leftarrow \mathbf{U}^{(n),i+1} \mathbf{U}^{(n),i+1^T}$
   $\mathcal{Y}_{P\text{-}(K_1,K_2,p)}^{i+1} \leftarrow \mathcal{R} \times_1 \mathbf{P}^{(1)} \times_2 \mathbf{P}^{(2)} \times_3 \mathbf{W}$
   $i \leftarrow i + 1$

$\mathbf{R}$ (Section 2.1) is equivalent to the 3-mode flattened matrix of $\mathcal{R}$ denoted by $\mathbf{R}_3$. Thus in tensor formulation, the DR approach (Eq. (1)) including the image reshaping can be written as follows:

$$\mathcal{Y} = \mathcal{R} \times_3 \mathbf{T}, \tag{6}$$

where $\mathcal{Y}$ is the reduced *three*-order tensor $\in \mathbb{R}^{I_1 \times I_2 \times p}$ holding the p components. $\times_n$ is the n-mode product [17,12] generalizing the product between a tensor and a matrix along an n-mode.

In the same way, the equivalent to Eq. (2) including the image reshaping is formulated:

$$\mathcal{Y}_{PC} = \mathcal{R} \times_3 \mathbf{U}^{(3)^T}, \tag{7}$$

where $\mathcal{Y}_{PC}$ is a *three*-order tensor $\in \mathbb{R}^{I_1 \times I_2 \times p}$ holding the p PCs and $\mathbf{\Lambda}, \mathbf{U}^{(3)}$ the eigenvalues and eigenvectors of the covariance matrix of $\mathbf{R}_3$. $\mathbf{U}^{(3)}$ is the equivalent to $\mathbf{U}$ defined in Eq. (2).

And the equivalent to Eq. (3) including the image reshaping is formulated:

$$\mathcal{Y}_P = \mathcal{R} \times_3 \mathbf{W}, \tag{8}$$

where, $\mathcal{Y}_P$ is a *three*-order tensor $\in \mathbb{R}^{I_1 \times I_2 \times p}$ holding the p projections.

Eqs. (6)-(8) highlight the spectral or 3-mode processing in the traditional DR method, neglecting the spatial modes.

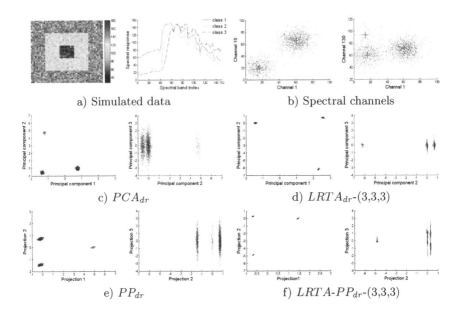

a) Simulated data   b) Spectral channels

c) PCA_{dr} d) $LRTA_{dr}$-(3,3,3)

e) PP_{dr} f) $LRTA$-PP_{dr}-(3,3,3)

Fig. 2. a) simulated tensor and scatter plots of b) close and farther spectral channels (red crosses are endmembers), c) PCs, d) PCs obtained using $LRTA_{dr}$-(3,3,3), e) projections, f) projections obtained using hybrid $LRTA$-PP_{dr}-(3,3,3)

4.2   Multilinear Algebra and PCA-Based DR Method

The multilinear algebra-based DR method has two objectives: *(i)* estimate the matrix $\mathbf{U}^{(3)}$ of Eq. (7) using spatial information; *(ii)* make a joint spatial-spectral processing with the aim of whitening and compressing the spatial and spectral modes. Consequently, the proposed method performs simultaneously a dimensionality reduction of the spectral mode ($p < I_3$), and a projection onto a lower (K_1, K_2)-dimensional subspace of the two spatial modes.

To attain these two objectives, the PCA-based DR proposed multilinear algebra tool has two approaches [10]: *(i)* find the lower rank-(K_1, K_2, K_3) approximation tensor of $\mathcal{R}$, this method is referred to as the $LRTA$-(K_1, K_2, K_3) and the resulting tensor as $\mathcal{Y}_{(K_1, K_2, K_3)}$; *(ii)* keep only the p first principal spectral components. Thereof, the proposed method that is the association of these two approaches, is referred to as the $LRTA_{dr}$-(K_1, K_2, p) and the resulting reduced and approximated tensor to as $\mathcal{Y}_{PC\text{-}(K_1, K_2, p)}$. Contrary to the $LRTA$-(K_1, K_2, K_3), the $LRTA_{dr}$-(K_1, K_2, p) extracts p principal components in the spectral mode. The spatial subspace dimension is the K_n-value, for $n = 1, 2$ and p ($< I_3$) the number of retained features. For convenience, the K_n-value is denoted by $K_{1,2}$ in the following.

The $LRTA$-(K_1, K_2, K_3) minimizes the following quadratic Frobenius norm: $\|\mathcal{R} - \mathcal{Y}\|_F^2$. Finding the lower rank approximation, using the Tucker3 decomposition, consists in estimating the $\mathbf{U}^{(n)}$ orthogonal matrix. [17] shows that minimizing the quadratic Frobenius norm with respect to $\mathcal{Y}$ amount to maximize with respect to $\mathbf{U}^{(n)}$ matrix the quadratic function:

$$g\left(\mathbf{U}^{(1)}, \mathbf{U}^{(2)}, \mathbf{U}^{(3)}\right) = \left\|\mathcal{R} \times_1 \mathbf{U}^{(1)T} \times_2 \mathbf{U}^{(2)T} \times_3 \mathbf{U}^{(3)T}\right\|_F^2 \qquad (9)$$

The least square solution involves the $LRTA\text{-}(K_1, K_2, K_3)$ expression:

$$\mathcal{Y}_{(K_1, K_2, K_3)} = \mathcal{R} \times_1 \mathbf{P}^{(1)} \times_2 \mathbf{P}^{(2)} \times_3 \mathbf{P}^{(3)}, \qquad (10)$$

where $\mathbf{P}^{(n)} = \mathbf{U}_{1\cdots K_n}^{(n)} \mathbf{U}_{1\cdots K_n}^{(n)T}$, $n = 1, 2, 3$.

The reduced tensor, $\mathcal{Y}_{PC\text{-}(K_1, K_2, p)}$, is obtained by simultaneously performing the previous approximation and by extracting the p PCs using PCA_{dr}. This statement can be written as follows [10]:

$$\mathcal{Y}_{PC\text{-}(K_1, K_2, p)} = \mathcal{R} \times_1 \mathbf{P}^{(1)} \times_2 \mathbf{P}^{(2)} \times_3 \mathbf{\Lambda}^{-1/2} \mathbf{U}^{(3)T}. \qquad (11)$$

The joint estimation of the orthogonal matrix $\mathbf{U}^{(n)}$, $\forall n$, which solve the Eq. (9), is feasible using an alternating least squares algorithm (ALS). The algorithm 1 summarizes the $LRTA_{dr}\text{-}(K_1, K_2, p)$ method.

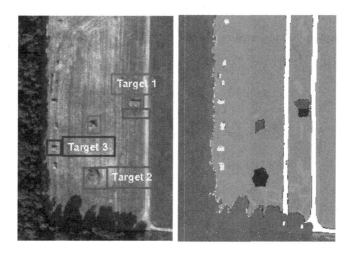

Fig. 3. Classes in the HYDICE image HSI02 and its ground truth

4.3 Multilinear Algebra and PP-Based DR Method

To pursue this previous work, we introduce a generalized multilinear algebra approach to DR using projection pursuit. The same objectives, presented in the previous section, are considered. This multilinear algebra-based method is referred to as the hybrid $LRTA\text{-}PP_{dr}\text{-}(K_1, K_2, p)$ and the resulting approximated and reduced tensor as $\mathcal{Y}_{P\text{-}(K_1, K_2, p)}$. Finding $\mathcal{Y}_{P\text{-}(K_1, K_2, p)}$ consists in minimizing the frobenius norm: $\left\|\mathcal{Y}_P - \mathcal{Y}_{P\text{-}(K_1, K_2, p)}\right\|_F^2$, with $\mathcal{Y}_P$ defined in Eq. (8).

Fig. 4. Classes in the HYDICE image HSI03 and its ground truth

Using the Tucker3 representation, the reduced tensor $\mathcal{Y}_P$ (Eq. (8)) can be decomposed as follows:

$$\mathcal{Y}_P = \mathcal{C} \times_1 \mathbf{U}^{(1)} \times_2 \mathbf{U}^{(2)} \times_3 \mathbf{U}^{(3)} \tag{12}$$

The lower rank-(K_1, K_2) approximated tensor of $\mathcal{Y}_P$ to be estimated, can be also decomposed following the Tucker3 representation as:

$$\mathcal{Y}_{P\text{-}(K_1,K_2,p)} = \mathcal{D} \times_1 \mathbf{U}^{(1)} \times_2 \mathbf{U}^{(2)} \times_3 \mathbf{U}^{(3)} \tag{13}$$

Following the minimizing criterion, we can express the core tensor $\mathcal{D}$ such as:

$$\mathcal{D} = \mathcal{Y}_P \times_1 \mathbf{U}^{(1)T} \times_2 \mathbf{U}^{(2)T} \times_3 \mathbf{U}^{(3)T}$$

As a result:

$$\mathcal{D} = \mathcal{R} \times_1 \mathbf{U}^{(1)T} \times_2 \mathbf{U}^{(2)T} \times_3 \mathbf{U}^{(3)T}\mathbf{W} \tag{14}$$

Finally, the $LRTA\text{-}PP_{dr}\text{-}(K_1, K_2, p)$ method is expressed as:

$$\mathcal{Y}_{P\text{-}(K_1,K_2,p)} = \mathcal{R} \times_1 \mathbf{P}^{(1)} \times_2 \mathbf{P}^{(2)} \times_3 \mathbf{W} \tag{15}$$

As previouly, the joint estimation of the orthogonal matrix $\mathbf{U}^{(n)}$, $\forall n$, and the demixing matrix $\mathbf{W}$ is feasible using an alternating least squares algorithm (ALS). The algorithm 2 summarizes the hybrid $LRTA\text{-}PP_{dr}\text{-}(K_1, K_2, p)$ method.

5   Experimental Results

In this section, two experiments are introduced. First, the interest of the multi-linear algebra-based filtering and DR methods are highlighted using simulated data and showing the scatter plots. Secondly, the classification improvement is exemplified using two real-world HYDICE data.

5.1   Experiment on Simulated Data

In order to show the interest of multi-way filtering on the classification, we consider a simulated data shown Fig. 5, where for the original image a gaussian

Table 1. Information classes and samples

HSI02			HSI03		
Classes	Training samples	Test samples	Classes	Training samples	Test samples
field	1 002	40 811	house	65	613
trees	1 367	5 537	road	210	2364
road	139	3 226	trees	148	11482
shadow	372	5 036	field	325	5353
target 1	128	519			
target 2	78	285			
target 3	37	223			

noise with variable variance is added. To quantify the benefits of our filter, for each SNR we calculate the overall(OA) classification rate exhibits after multi-way Wiener filtering. for example, the Spectral-Angle Mapper (SAM) is investigated.

For this experiment, a simulated data is considered (see Fig. 2a). The spatial size of this simulated data is 50×50 with 150 spectral channels. Three endmembers are present associated with a each class. The endmembers spectral signature are introduced in Fig. 2a to appreciate the similarity. A Gaussian random number generator simulated 2100 samples associated with the first class, 300 with the second class and 100 with the third one. The scatter plot in Figs. 2b show that close spectral channels are more correlated as farther ones. The set of dots, that portray the samples, highlights that the three classes are sometimes not distinct enough.

DR methods are applied to extract three components. Figures 2c and 2e respectively show the scatter plot of the spectral components obtained using PCA_{dr} and ICA_{dr}. Two scatter plots are represented: first and second components then the second and third components. This very basic classification context does not allow to compare the PCA_{dr} and ICA_{dr}-based spectral DR, but yields to show the interest of the multilinear algebra based methods. Then, the scatter plot obtained using PCA_{dr} and ICA_{dr} can be compared with those obtained using the $LRTA_{dr}$-(K_1, K_2, p) and hybrid $LRTA$-PP_{dr}-(K_1, K_2, p) in Figs. 2d and 2f respectively. For this experiment, the $K_{1,2}$-value are set to 3.Those comparisons denoted that the sets of dots are more concentrated and more distinct in Figs. 2d and 2f than those in Figs. 2c and 2e respectively. This is due to the spatial decorrelation of the components by projecting the data onto an orthogonal lower subspace of the $LRTA_{dr}$-(K_1, K_2, p) and the hybrid $LRTA$-PP_{dr}-(K_1, K_2, p) DR methods. This interesting remark forecasts classification improvement, that is exemplified, in the next section, using real-world HYDICE data.

5.2   Experiment on Real-World Data

Two real-world images collected by HYDICE [24] imaging are considered for this investigation, with a 1.5 m spatial and 10 nm spectral resolution. The first one, referred to as HSI02, has 150 spectral bands (from 435 to 2326 nm), 310 rows and 220 columns. HSI02 can be represented as a *three*-order tensor, referred to as $\mathcal{R} \in \mathbb{R}^{310 \times 220 \times 150}$. The second one, referred to as HSI03, has 164 spectral bands (from 412 to 2390 nm), 170 rows and 250 columns. HSI03 can be represented as a *three*-order tensor, referred to as $\mathcal{R} \in \mathbb{R}^{170 \times 250 \times 164}$.

Two experiments are introduced in this section. The first one shows the efficiency of the ALS algorithm and, the $K_{1,2}$ value influence is highlighted in the second experiment. To quantify the improvement, we focus on the overall classification (OA) rate exhibits after DR methods. Three classifiers are investigated: *(i)* maximum likelihood (ML), *(ii)* Mahalanobis distance (MD) and *(iii)* spectral angle mapper (SAM).

Efficiency of the ALS algorithm. The ALS algorithm guarantees the cross-dependency between the spatial and spectral processing realizing by the proposed multilinear algebra-based methods. To exemplified the ALS efficiency, Fig. 5 shows the OA evolution in function of the iteration index of the ALS

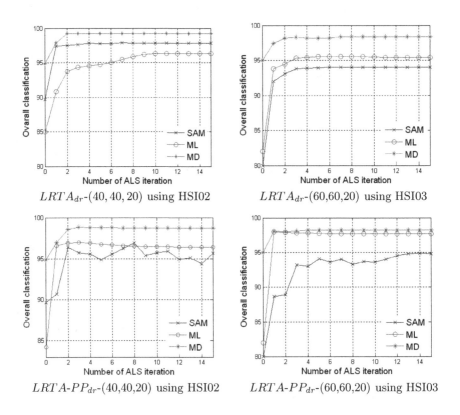

Fig. 5. OA value evolution in function of the number of ALS iteration

algorithm, when ML, SAM and, MD classifiers are considered. For this simulation the $K_{1,2}$ and p values are set to (40,40,20) and (60,60,20) according to the use of HSI02 and HSI03 respectively. It is worth noting that the OA collected values at the iteration 0 are the classification results obtained on the real-world image without DR processing. This reading highlights the noteworthy improvement from the first iteration of the ALS when $LRTA_{dr}$-(K_1, K_2, p) and hybrid $LRTA$-PP_{dr}-(K_1, K_2, p) are used. Figure 5 shows that the classification result is iteratively improved until a maximum value is reached whatever the classifier and HYDICE data used. This state is attained after few iterations, less than 10. However, when hybrid $LRTA$-PP_{dr}-(K_1, K_2, p) DR method is considered, the OA value obtained from the SAM classifier, after notable increase, oscillate around a value. The SAM classifier results seem to have more variations than the others classifiers in function of the hybrid $LRTA$-PP_{dr}-(K_1, K_2, p) iteration index. This is due probably to the fact that the hybrid $LRTA$-PP_{dr}-(K_1, K_2, p) performs a demixing matrix ($\mathbf{W}$) estimation at each iteration. The estimation of $\mathbf{W}$ using *FastICA* algorithm is not unique since it crucially depends of the initialization step. Moreover, the SAM classifier has an unclassified class for the pixel which has a spectral angle measure, with all classes, superior to a threshold (fixed to 0.15 in the experiment). Then, from one iteration to another, some samples with spectral angle measure close to the threshold can be unclassified or well-classified, this can explain the iterative fluctuation of the SAM classifier. Note that these classification results are obtained for a specific $K_{1,2}$ and p values and, the maximum attained OA value depends on these values.

Influence of the ($K_{1,2}$)-value. The influence of the number p of retained components is conceded. In this section, we assess the influence of the spatial dimension ($K_{1,2}$) of the projection subspace which is required by the proposed multilinear algebra-based DR methods. The $K_{1,2}$ value is the number of first eigenvalues of the $\mathbf{R}_n$ covariance matrix, kept in the Algorithms 1 and 2, for $n = 1,2$.

In this experiment, the classification results are assessed as a function of the ($K_{1,2}$)-value. For sake of clarity, only the OA exhibits by the Mahalanobis distance classifier are illustrated.

Figure 6 specifies the cumulative proportion variance in function of the number of eigenvalues kept. It concerns the eigenvalue of the covariance matrix of flattened $\mathbf{R}_n$, for $n = 1, 2$. Thereby, we appreciate at what time this gain is negligible and, we consider that the others eigenvalues providing no more information, can be suppressed. For example, concerning HSI02, we consider that from $K_{1,2}$ value equal to 50, the gain is less than 1% and becomes insignificant. Concerning HSI03, it is reached when $K_{1,2}$ value equal to 90. At the following, the $K_{1,2}$ is set to 50 when HSI02 is used and 90 when HSI03 is used.

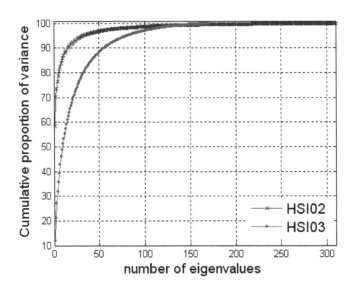

Fig. 6. Cumulative proportion of the variance in function of the eigenvalue number $K_{1,2}$ of the 1-mode covariance matrix

6   Conclusion

In this paper, we show the intrest of multi-way filtering in terms of classification. We also introduced two multilinear algebra-based DR methods, the $LRTA_{dr}$-(K_1, K_2, p) and the hybrid $LRTA\text{-}PP_{dr}$-(K_1, K_2, p). These techniques make a joint spatial/spectral processing with the aim of decorrelated and compressing the spatial and spectral dimension simultaneously. The cross-dependency of the spatial and spectral processing is guaranteed using an alternating least squares (ALS) algorithm. The spatial/spectral processing concerns: *(i)* a lower spatial rank-(K_1, K_2) approximation is performed to project the spatial dimension onto a lower subspace dimension $(K_{1,2} < I_{1,2})$; *(ii)* a spectral dimension reduction is performed to extract p components $(p < I_3)$. For this latter issue, when $LRTA_{dr}$-(K_1, K_2, p) is considered, PCA_{dr} is used and, when hybrid $LRTA\text{-}PP_{dr}$-(K_1, K_2, p) is considered, PP_{dr} is used.

Classification results are exemplified using two real-world HYDICE data and highlight the overall classification improvement when multilinear algebra-based methods are used rather than the traditional matrix algebra-based methods.

Acknowledgments

The authors thank Dr. N. Renard and A. Cailly for their helpful for the computer simulations.

References

1. Muti, D., Bourennane, S.: Survey on tensor signal algebraic filtering. Signal Proceesing Journal 87, 237–249 (2007)
2. Landgrebe, D.: Hyperspectral image data analysis as a high dimensional signal processing problem. IEEE Signal Process. Magazine 19(1), 17–28 (2002)
3. Wang, J., Chang, C.: Independent component analysis - based dimensionality reduction with applications in hyperspectral image analysis. IEEE Trans. on Geosc. and Remote Sens. 44(6), 1586–1588 (2006)
4. Ifarraguerri, A., Chang, C.-I.: Unsupervised hyperspectral image analysis with projection pursuit. IEEE Trans. on Geoscience and Remote Sensing 38(6), 2529–2538 (2000)
5. Jones, M.C., Sibson, R.: What is projection pursuit. J.R. Statist. Soc. A 150(1), 1–36 (1987)
6. Friedman, J., Tuckey, J.: A projection pursuit algorithm for exploratory data analysis. IEEE Transactionson Computers 23(9), 881–889 (1974)
7. Jimenez, L., Landgrebe, D.: Hyperspectral data analysis and supervised feature reduction via projection pursuit. IEEE Trans. on Geoscience and Remote Sensing 37(6), 2653–2667 (1999)
8. Jimenez-Rodriguez, L., Arzuaga-Cruz, E., Velez-Reyes, M.: Unsupervised linear feature-extraction methods and their effects in the classification of high-dimensional data. IEEE Transactions on Geoscience and Remote Sensing 45(2), 469–483 (2007)
9. Renard, N., Bourennane, S.: Improvement of target detection methods by multiway filtering. IEEE Trans. on Geoscience and Remote Sensing 46(8), 2407–2417 (2008)
10. Renard, N., Bourennane, S., Blanc-Talon, J.: Denoising and dimensionality reduction using multilinear tools for hyperspectral images. IEEE Geoscience and Remote Sensing Letters 5(2), 138–142 (2008)
11. Letexier, D., Bourennane, S., Blanc-Talon, J.: Nonorthogonal tensor matricization for hyperspectral image filtering. IEEE Geoscience and Remote Sensing Letters 5(1), 3–7 (2008)
12. Tucker, L.: Some mathematical notes on three-mode factor analysis. Psychometrika 31, 279–311, 66
13. Kolda, T.: Orthogonal tensor decomposition. SIAM Joural on Matrix Analysis and Applications 23(1), 243–255 (2001)
14. Harshman, R., Lundy, M.: Research methods for multimode data analysis. In: Law, H., Snyder, J., Hattie, J., McDonald, R. (eds.), vol. 70, pp. 122–215. Praeger, New York
15. Tucker, L.: The extension of factor analysis to three-dimensional matrices, pp. 109–127. Holt, Rinehart and Winston, NY (1964)
16. De Lathauwer, L., De Moor, B., Vandewalle, J.: A multilinear singular value decomposition. SIAM Journal on Matrix Analysis and Applications 21, 1253–1278 (2000)
17. De Lathauwer, L., De Moor, B., Vandewalle, J.: On the best rank-$(r_1,\ldots,r_N)$ approximation of higher-order tensors. SIAM Journal on Matrix Analysis and Applications 21, 1324–1342 (2000)
18. Comon, P.: Independent component analysis: A new concept. Signal Process. 36, 287–314 (1994)
19. Hyvarinen, A., Karhunen, J., Oja, E.: Independent Component Analysis. John Wiley Sons, Chichester (2001)

20. Chang, C., Chiang, S., Smith, J., Ginsberg, I.: Linear spectrum random mixture analysis for hyperspectral imagery. IEEE Trans. on Geoscience and Remote Sensing 40(2), 375–392 (2002)
21. Sidiropoulos, N., Giannakis, G., Bro, R.: Blind parafac receivers for DS-CDMA systems. IEEE Trans. on Signal Processing 48(3), 810–823 (2000)
22. Hazan, T., Polak, S., Shashua, A.: Sparse image coding using a 3d non-negative tensor factorization. In: Computer Vision, IEEE International Conference on Computer Vision (ICCV), Beijing, China, October 2005, vol. 4179(1), pp. 50–57 (2005)
23. Vasilescu, M., Terzopoulos, D.: Multilinear independent components analysis. In: Proc. of IEEE CVPR 2005, vol. 1, pp. 547–553 (2005)
24. Basedow, R., Aldrich, W., Colwell, J., Kinder, W.: Hydice system performance - an update. In: SPIE Proc., Denver, CO, August 1996, vol. 2821, pp. 76–84 (1996)

Comparative Analysis of Wavelet-Based Scale-Invariant Feature Extraction Using Different Wavelet Bases

Joohyun Lim[1], Youngouk Kim[2], and Joonki Paik[1]

[1] Image Processing and Intelligent Systems Laboratory,
Graduate School of Advanced Imaging Science,
Multimedia, and Film, Chung-Ang University,
221 Heuksuk-Dong, Dongjak-Ku, Seoul 156-756, Korea
joo-hyun@wm.cau.ac.kr, paikj@cau.ac.kr
[2] Intelligent Robotics Research Laboratory, Korea Electronics Technology Institute (KETI),
401-402 B/D 193, Yakdae-Dong, WonMi-Gu, Puchon-Si, KyungGi-Do 420-140, Korea
kimyo@keti.re.kr

Abstract. In this paper, we present comparative analysis of scale-invariant feature extraction using different wavelet bases. The main advantage of the wavelet transform is the multi-resolution analysis. Furthermore, wavelets enable localigation in both space and frequency domains and high-frequency salient feature detection. Wavelet transforms can use various basis functions. This research aims at comparative analysis of Daubechies, Haar and Gabor wavelets for scale-invariant feature extraction. Experimental results show that Gabor wavelets outperform better than Daubechies, Haar wavelets in the sense of both objective and subjective measures.

Keywords: Daubechies, Haar wavelets, Gabor wavelets, Feature extraction.

1 Introduction

Extraction of image feature has been an active area in the object recognition research for decades. Feature extraction is based on the local property of an image. Many existing methods utilize Harris corner detectors [1], difference of Gaussian (DoG) [2], and the behavior of local entropy [3], to name a few. The description of the detected image features is based on the properties of the associated local regions, the simplest solution being the grey-level histogram. Lowe has proposed scale-invariant feature transform (SIFT), which provides distinctive, stable, and discriminating features [1], [4].

The main advantage of using such SIFT is the simplicity due to the unsupervised nature. However, the SIFT method is practically classified into semi-supervised since it requires a certain degree of supervision. In this paper propose a more efficient approach when the image feature detector uses supervised learning. It is clear that the semi-supervised methods are the most suitable for the cases where the object instance remains the same, such as in the interest point initialization for object tracking [5], but they cannot tolerate appearance variation among different instances of the object class on the local level. In this paper, comparative analysis of wavelet transform using

D. Ślęzak et al. (Eds.): SIP 2009, CCIS 61, pp. 297–303, 2009.
© Springer-Verlag Berlin Heidelberg 2009

Daubechies, Haar and Gabor wavelets for evaluating performance of scale-invariant feature extraction.

2 Comparison of Feature Extraction Performance Using Different Wavelet Bases

The wavelet transform can selectively use various basis functions. In this section, we compare two orthogonal basis functions including Daubechies and Haar wavelets. Gabor wavelets is Gaussian enveloped basis functions that are orthogonal -like basis functions.

2.1 Feature Extraction Using the Daubechies, Haar Wavelets

The discrete wavelets transform (DWT) decomposes an input signal into low and high frequency component using a filter bank. Daubechies, Haar wavelet, which characteristics the filter bank, has important properties of orthogonality, linearity, and completeness. We can repeat the DWT multiple times to multiple-level resolution of different octaves. For each level, wavelets can be separated into different basis functions for image compression and recognition.

Fig. 1. Level-3 DWT representation of a Lena image

Figure 1, shows the result of multi-resolution expansion using Daubechies wavelets. The wavelet transform can be used to represent a two-dimensional (2D) signal by the 2D resolution decomposition procedure, where an image is repeatedly decomposed into an approximation and several detail components at each level. In Figure 1, Level-3 wavelet decomposition of a Lena image is shown. A 256x256 Lena image is first decomposed into four sub-images including one approximation (low-frequency part of the signal) and nine details (high-frequency part of the signal) images. The approximation image is again decomposed into four sub-images. In the experiment, we use Daubechies-4(db4) wavelets to obtain Level-3 wavelet decomposition [6].

In order to construct the wavelet pyramid, we decide the number of Daubechies coefficients and approximation levels. We would like to extract salient points from any part of the image where "something" happens in the image at any resolution. A high wavelet coefficient (in absolute value) at a coarse resolution corresponds to a region with high global variations. The properly chosen length of the Daubechies, Haar wavelet and the number of the approximation levels provides the optimum local key points or features [7].

2.2 Feature Extraction Using Gabor Wavelet

The Gabor wavelet transform uses a set of Gaussian enveloped basis functions that are orthogonal-like basis functions [8]. Gabor wavelets provide analysis of the input signal in both spatial and frequency domains simultaneously. In addition, the Gabor domain function provides efficient extraction edges. The Gabor wavelet $\psi_{u,v}(z)$ is defined as [9],

$$\psi_{\mu,v}(z) = \frac{\left\| k_{u,v} \right\|^2}{\sigma^2} e^{\frac{\left\| k_{u,v} \right\|^2 \left\| z \right\|^2}{2\sigma^2}} \left[e^{ik_{\mu,v}z} - e^{-\frac{\sigma^2}{2}} \right], \tag{1}$$

where $z=(x,y)$ represents the point with the horizontal coordinate x and the vertical coordinate y. The parameters μ and v define the orientation and scale of the Gabor kernel, $\| \bullet \|$ denotes the norm operator, and σ is related to the standard derivation of the Gaussian window and determines the ratio of the Gaussian window width to the wave length. The wave vector $k_{u,v}$ is defined as

$$k_{u,v} = k_{ve^{i\phi_\mu}}, \tag{2}$$

where $k_v = \frac{k_{max}}{f^v}$ and $\phi_\mu = \frac{\pi \mu}{8}$ if 8 different orientations are chosen. k_{max} repre-

sent the maximum frequency, and f^v the spatial frequency between kernels in the frequency domain.

The Gabor kernels in (1) are all self-similar since they can be generated from the mother wavelet, by scaling and rotation via the wave vector $k_{u,v}$. Each kernel is a product of a Gaussian envelope and a complex plane wave, while the first term in the square brackets in (1) determines the oscillatory part of the Direct Current (DC) value. The effect of the DC term becomes negligible when the parameter σ, which determines the ratio of the Gaussian window width to wavelength, has sufficiently large values. In most case, one would use Gabor wavelets of five different scales, v={0,....,4}, and eight orientations, μ={0,....,7}. The parameter values of wavelength,

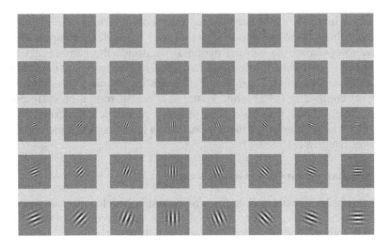

Fig. 2. Gabor function for different values representation of real part

orientation, phase offset, aspect ratio, and bandwidth are given for the Gabor function. Edge operator shows the salient feature of texture and Gabor energy operation is robust for isolated line, edge, and contours.

Fig. 3. Gabor energy filtering

3   Experiment Results

The results show that orthogonal basis function of Daubechies, Haar wavelet is in complex background and rough texture. However, for orthogonal-like basis function of Gabor wavelet, noise and rough texture of the edge are removed hence image feature can be extracted.

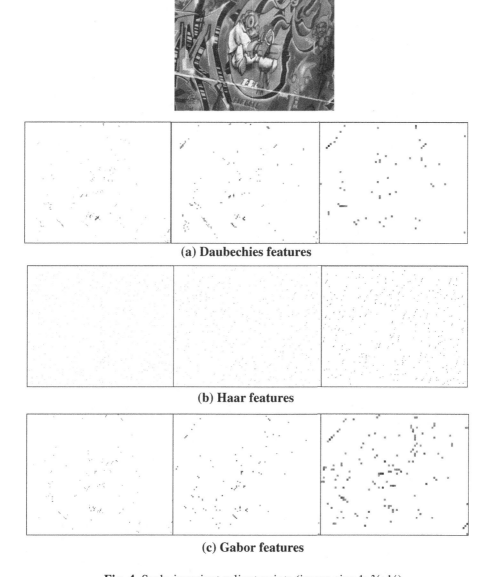

Fig. 4. Scale-invariant salient points (image size 1: ¾: ¼)

5 Discussion

In this paper, we compared of Daubechies, Haar wavelet and Gabor wavelet. Gabor wavelet respond to contours and edges. However, noise and rough texture of the edge are removed hence image feature. Gabor wavelet basis function extracts image feature more efficiently than Daubechies, Haar basis function in wavelet bases.

(a)

(b)

(c)

(d)

(e)

(f)

Fig. 5. Experimental results: (a),(d) Daubechies features, (b),(e) Haar features, (e),(f) Gabor features

Acknowledgement

This research was funded by Ministry of Commerce, Industry and Energy (MOCIE) in Korea under the Technologies Development for Future Home Appliance project, by the Ministry of Knowledge Economy, Korea, under the HNRC(Home Network Research Center)–ITRC(Information Technology Research Center) support program supervised by the Institute of Information Technology Assessment, and by Ministry of Culture, Sports and Tourism(MCST) and Korea Culture Content Agency(KOCCA) in the Culture Technology(CT) Research & Development Program 2009.

References

1. Mikolajczyk, K., Schmid, C.: Scale and affine invariant interest point detectors. Int. J. Comp. Vis. 60, 63–86 (2004)
2. Lowe, D.: Distinctive image features from scale-invariant keypoints. Int. J. Comp. Vis. 60, 91–110 (2004)
3. Kadir, T.: Scale, saliency and scene description. Ph.D. dissertation. Oxford Univ., Oxford (2002)
4. Mikolajczyk, K., Schmid, C.: A performance evaluation of local descriptors. In: Proc. Conf. Computer Vision, Pattern Recognition, pp. 257–263 (2003)
5. Kyrki, V., Kragic, D.: Integration of model-based and model-free cues for visual object tracking in 3D. In: Int. Conf. Robotics, Automation, pp. 1566–1572 (2005)
6. Bagci, U., Bai, L.: A Comparison of Daubechies and Gabor Wavelets for Classification of MR Images. In: Int. Conf. Signal Processing, Communications, Dubai, United Arab Emirates, pp. 676–679 (2007)
7. Lee, J., Kim, Y., Park, C., Paik, J.: Robust feature detection using 2D wavelet transform under low light environment. In: Int. Conf. Intelligent Computing. LNCIS, vol. 345, pp. 1042–1050. Springer, Heidelberg (2006)
8. Qian, S., Chen, D.: Discrete Gabor Transform. IEEE Trans. Signal Processing 41(7), 2429–2438 (1993)
9. Liu, C.: Gabor-based kernel PCA with fractional power polynomial models for face recognition. IEEE Trans. Pattern Analysis, Machine Intelligence 26(5), 572–581 (2004)

A Novel Video Segmentation Algorithm with Shadow Cancellation and Adaptive Threshold Techniques

C.P. Yasira Beevi[1] and S. Natarajan[2]

[1] MTech. CS&E Student, PES Institute of Technology, Bangalore, India
`yasihere@gmail.com`
[2] Professor, Dept. of IS&E, PES Institute of Technology, Bangalore, India
`natarajan@pes.edu`

Abstract. Automatic video segmentation plays an important role in real-time MPEG-4 encoding systems. This paper presents a video segmentation algorithm for MPEG-4 camera system. With change detection, background registration techniques and real time adaptive threshold techniques. This algorithm can give satisfying segmentation results with low computation load. Besides, it has shadow cancellation mode, which can deal with light changing effect and shadow effect. Furthermore, this algorithm also implemented real time adaptive threshold techniques by which the parameters can be decided automatically.

Keywords: Adaptive threshold, background registration, object extraction, shadow cancellation, video segmentation.

1 Introduction

The MPEG-4 standard has been taken as the most important standard for multimedia and visual communication and will be applied to many real-time applications, such as video phones, video conference systems, and smart camera systems. The most important function of MPEG-4 video part is content-based coding, which can support content-based manipulation and representation of video signal and random access of video objects (VO). Automatic video segmentation is the technique to generate shape information of video objects from video sequences. It is very important in a real-time MPEG-4 camera system with content-based coding scheme, since the shape information is required for shape coding, motion estimation, motion compensation, and texture coding.

Change detection based segmentation algorithms [1], threshold the frame difference to form change detection mask. Then the change detection masks are further processed to generate final object masks. The processing speed is high, but it is often not robust. The segmentation results are suffered from the uncovered background situations; still object situations, light changing, shadow, and noise. The robustness can be promoted by a lot of post-processing algorithms; however, complex post-processing will make the efficiency of less computation lost. The threshold of change detection is very critical and cannot be automatically decided. These reasons make this kind of algorithms not practical for real applications.

D. Ślęzak et al. (Eds.): SIP 2009, CCIS 61, pp. 304–311, 2009.

In this paper, a fast video segmentation algorithm for MPEG-4 camera systems is proposed. The algorithm has three modes: baseline mode, shadow cancellation mode and adaptive threshold mode. It is based on work using change detection [6]. With background registration technique, this algorithm can deal with uncovered background and still object situations. An efficient post-processing algorithm can improve segmentation results without large computation overhead. Moreover, it has a shadow cancellation mode, in which light changing effect and shadow effect can be suppressed. Furthermore, an adaptive threshold mode is also proposed to decide the threshold automatically.

This paper is organized as follows. Sections II–VI describe baseline mode, shadow cancellation mode (SC mode) and adaptive threshold mode (AT mode), respectively. The experimental results are shown in Section V. Finally, Section VI gives a conclusion of this paper.

2 Baseline Mode

Baseline mode is designed for stable situations. That is, the camera is still, and there is no light changing and no shadows. It is based on change detection and background registration technique. Unlike other change detection algorithms, the change detection mask here is not only generated from the frame difference of current frame and previous frame but also from the frame difference between current frame and background frame, which can be produced by background registration technique. Since the background is stationary, it is well-behaved and more reliable than previous frame. Besides, still objects and uncovered background problems can be easily solved under this scheme. The block diagram of baseline mode is shown in Fig. 1. There are five parts in baseline mode: *Frame Difference*, *Background Registration*, *Background Difference*, *Object Detection*, and *Post processing*.

Frame Difference
In *Frame Difference*, the frame difference between current frame and previous frame, which is stored in *Frame Buffer*, is calculated and threshold. It can be presented as

$$FD(x, y, t) = |\ I(x, y, t) - I(x, y, t-1)| \tag{1}$$

$$FDM(x, y, t) = \begin{array}{ll} 1 & if\ FD \geq Th \\ 0 & if\ FD < Th \end{array} \tag{2}$$

Where I is frame data, FD is frame difference, and FDM is Frame Difference Mask. Pixels belonging to FDM are moving pixels.

Background Registration can extract background information from video sequences. According to FDM, pixels not moving for a long time are considered as reliable background pixels. The procedure of *Background Registration* can be shown as

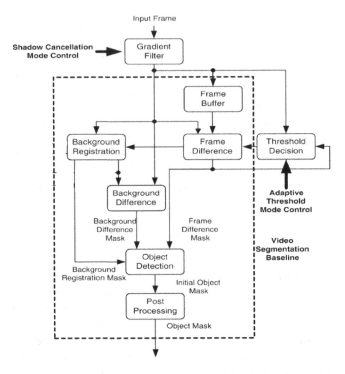

Fig. 1. Block diagram of video segmentation Algorithm Using change detection

$$SI(x, y, t)= SI(x, y, t-1)+1, \ if \ FDM=0$$
$$0, \qquad \qquad if \ FDM=1 \tag{3}$$

$$BG(x, y, t)= \quad I(x, y, t), \quad if \ SI(x, y, t)=Fth$$
$$BG(x, y, t-1), \quad else \tag{4}$$

$$BI(x, y, t)= 1, \qquad if \ SI(x, y, t)=Fth$$
$$BI(x, y, t-1), \quad else \tag{5}$$

Where SI is Stationary Index, BI is Background Indicator, and BG is the background information. The initial values of SI, BI and BG are all set to "0." Stationary Index records the possibility if a pixel is in background region.

Background difference mask: After Background Difference, another change detection mask named Background Difference Mask (FDM) is generated the operations of Background Difference, can be shown as.

$$BD(x, y, t)= | \ I(x, y, t) - BG(x, y, t-1) \ | \tag{6}$$

$$BDM(x, y, t) = \quad 1, \; if \; BD \geq Th \qquad (7)$$
$$0, \; if \; BD < Th$$

Where BD is background difference, BG is background frame, and BDM is Background Difference Mask, respectively.

Both of FDM and BDM are input into Object Detection to produce Initial Object Mask (IOM_k). The procedure of Object Detection can be presented as the following equation.

$$IOM(x, y, t) = \quad BDM(x, y, t), \; if \; BI(x, y, t) = 1 \qquad (8)$$
$$FDM(x. \; y, t), \quad else$$

Post processing

The *Initial Object Mask (IOM)* generated by *Object Detection* has some noise regions because of irregular object motion and camera noise. Also, the boundary may not be very smooth. Therefore, there are two parts in *Post-processing*: noise region elimination and boundary smoothing. The connected component algorithm can mark each connected region with a special label. Then we can filter these regions by their area. Regions with area smaller than a threshold value are removed from the object mask. In this way, the object shape information is preserved while smaller noise regions are removed. Next, the morphological close–open operations are applied to smooth the boundary of object mask.

3   Shadow Cancellation Mode

Conventional change detection algorithms usually cannot give acceptable segmentation results when light changes, or shadows exist. Shadow regions are always falsely detected as parts of foreground regions, and the whole frame will be regarded as foreground object if light changes. In these situations, shadow cancellation mode (SC mode) [8] is preferred.

Proposed Shadow Cancellation Algorithm

Based on the analysis, a shadow cancellation algorithm is proposed and is employed in shadow cancellation mode (SC mode). There are two different parts in SC mode: *Gradient Filter* and *Post-processing*.

1) Gradient Filter: The morphological gradient is chosen because of its simple operations. It can be described by given equation,

$$GRA = (I \oplus B) - (I \ominus B) \qquad (9)$$

Where I is the original image, B is the structuring element of morphological operations, $\oplus$ is morphological dilation operation, $\ominus$ is morphological erosion operation, and GRA is the gradient image. After the gradient operation, the shadow effect will be reduced, and the motion information is still kept.

2) Post-processing: After morphological gradient operation, the edges are thickened. To eliminate this effect, a morphological erosion operation should be added at the end of *Post-Processing*.

4 Adaptive Threshold Mode

The threshold is a very critical parameter for change detection based algorithms. If the optimal threshold cannot be decided automatically, these kinds of video segmentation algorithms are hardly used in real applications. Therefore, the automatic threshold decision is very important in our video segmentation system.

The proposed non-parametric algorithm[2] computes a threshold of each block of an image adaptively based on the scatter of regions of change (ROC) and averages all thresholds for image blocks to obtain the global threshold. First, the output D_n of change detection at time instant n is divided into K equal-sized blocks. Then a ROC scatter estimation algorithm is applied, where each image block W_k, k = {1, 2, ... ,K}, is marked either as containing ROC, denoted W_k^r, or not containing ROC, denoted W_k^b. The threshold T_k^b of a W_k^b is computed by a noise statistical-testing algorithm. The threshold T_k^r of a W_k^r is computed by a noise-robust thresholding method. That is, the threshold T_k of a W_k in D_n is defined as

$$T_k = \begin{cases} T_k^r & \text{of} \quad W_k^r \\ \\ T_k^b & \text{of} \quad W_k^b. \end{cases} \tag{10}$$

Finally, the global threshold T_n of a difference image D_n is

$$T_n = \frac{1}{K} \sum_{k=1}^{K} T_k \tag{11}$$

Since size and velocity of objects, noise, local changes in videos may affect the histogram of a W_k, the first moment of histogram is used to estimate the scatter of ROC making the estimation adaptive to these characteristics. This also account for noise and local changes.

The ROC in D_n are, in general, scattered over the K image blocks. Let i be a pixel in D_n that varies between 0 and 255. i is high in ROC which is caused by strong changes such as motion or significant illumination changes and is low in non ROC which is caused by slight changes such as noise or slight illumination changes. We use the first moment, m_k, of the histogram of each image block W_k as a measure for determining if an image block contains ROC. If m_k of W_k is greater than a threshold T_m, the image block is regarded as a block containing ROC, and marked as W_k^r, otherwise, it is marked as W_k^b, i.e.,

$$W_k = \begin{cases} W_k^b & : \quad m_k \leq T_m \\ W_k^r & : \quad m_k > T_m \end{cases} \tag{12}$$

To find T_m, we first compute the m_k of each block and then descending sorts the m_k values. A straight line between the first bin and the last filled bin is then drawn. T_m is selected to maximize the perpendicular distance between the line and the sorted first moment curve.

Threshold in Non ROC: As per the paper [2], the equation for finding out the threshold for non ROC is

$$T_k^b = \mu_k + c \cdot \sigma_k \tag{13}$$

Where c is a varying coefficient to improve the robustness to noise, illumination changes, or background movement. The coefficient c is determined adaptively based on the min-max ratio of m_k, $r_m = m_{min}/m_{max}$, where m_{min} and m_{max} are minimum and maximum of all m_k in D_n.r_m is a good measure to adaptively estimate c , which is defined as ,

$$c_r = \frac{1}{r_m} e^{r_m} - \alpha$$

$$c = \begin{cases} c_{max} & : \quad c_r > c_{max} \\ c_r & : \quad \text{otherwise} \end{cases} \tag{14}$$

Where c_{max} is the maximum value of c determined by Eq.14, and α is a constant experimentally set to 0.9.

Thresholding in ROC: Threshold in blocks containing ROC is determined by the following equation [2]

Where g_{fl} is the most frequent gray level in each block W_k^r , μ_k is average gray level of W_k^r and L is total number of W_k^r .

$$T_k^r = \frac{\sum_{l=1}^{L} g_{fl} + \mu_k}{L + 1} \tag{15}$$

5 Experimental Results

The performance of the proposed video segmentation algorithm is tested with many video sequences. A fair evaluation method named "fix frame number test" is used here. That is, the segmentation results of fixed frames of each sequence are picked, rather than chosen by the algorithm developers themselves. The quality of segmentation results is then evaluated subjectively. In these experiments, frame 50 and frame 100 are chosen in every sequence.

Baseline Mode: The segmentation results of sequence *Akiyo* and *Mother and Daughter* are shown in Fig. 3. These sequences are not influenced by shadow and light changing effects. The segmentation results of sequence *Akiyo* and *Mother and*

Daughter are shown in Fig. 2(a)–(d) respectively, where the still object problem can be correctly solved.

2) Shadow Cancellation Mode: The experimental results of shadow cancellation mode are presented in Fig. 4. The sequence *Claire*, which is influenced by light changing effect, can be correctly manipulated in SC mode, as shown in Fig. 3(a) and (b). In Fig. 3(c) and (d), the segmentation results of sequence *Silent Voice* are shown. It shows that the shadow effect in *Silent Voice* can be reduced, and good object mask can be generated.

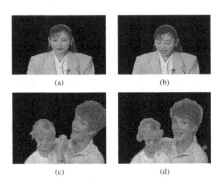

Fig. 2. Segmentation results of baseline mode. (a) *Akiyo* #50; (b) *Akiyo* #100; (c) *Mother and Daughter* #50; (d) *Mother and Daughter* #100.

Adaptive Threshold Mode: The experimental results of AT mode are shown in all the figures mentioned above because the threshold in all experiments is decided by the automatic threshold decision algorithm. The segmentation results show the proposed threshold decision algorithm is suitable for change detection and background registration based video segmentation algorithms.

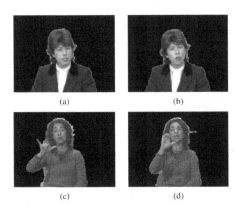

Fig. 3. Segmentation results of SC mode. (a) *Claire* #50; (b) *Claire* #100; (c) *Silent Voice* #50; (d) *Silent Voice* #100.

6 Conclusion

This paper proposes, background registration and change detection based video segmentation algorithm with shadow cancellation to analyzes light changing and shadow effects in indoor environments and a real time adaptive threshold techniques to decide the parameters automatically, This algorithm can generate segmentation results with low computation complexity and high efficiency compare to other change detection based video segmentation algorithm.

There are still limitations in the proposed segmentation system. The shadow cancellation cannot deal with parallel and strong light sources and may cause errors when the texture of background is significant. Furthermore, the decision to turn on and off each mode of the proposed algorithm is not automatic. Finally, this algorithm is designed for moving objects segmentation; therefore, for stable and accurate results, the foreground object should not be still for a long time.

References

[1] Radke, R.J., Andra, S., Al-Kofahi, O., Roysam, B.: Image Change Detection Algorithms: A Systematic Survey. IEEE Trans. Image Processing 14(3), 294–303 (2005)
[2] Su, C., Amer, A.: A Real Time Adaptive Thresholding for video change detection. In: IEEE International Conference on Image Processing (ICIP), October 2006, pp. 157–160 (2006)
[3] Chien, S.Y., Ma, S.Y., Chen, L.G.: Efficient Moving Object Segmentation Algorithm Using Background Registration Technique. IEEE Trans. on circuits and system for video technology 12(7), 577–586 (2002)
[4] Chien, S.-Y., Huang, Y.-W., Hsieh, B.-Y., Ma, S.-Y., Chen, L.-G.: Fast Video Segmentation Algorithm with Shadow Cancellation, Global Motion Compensation, and Adaptive Threshold Techniques. IEEE Trans. on Circuits and System for Video Technol. 6(5), 732–748 (2004)
[5] Gonzalez, R.C., Woods, R.E., Eddins, S.L.: Digital Image Processing Using Matlab, 2nd edn. Pearson Education, London (2006)
[6] Chien, S.-Y., Ma, S.-Y., Chen, L.-G.: An efficient video segmentation algorithm for real-time MPEG-4 camera system. In: Proc. Visual Communication and Image Processing, SPIE, Vol. 4067, 1067, pp. 1087–1098 (2000)
[7] Rosin, P.L.: Thresholding for change detection. Computer Vision and Image Understanding 86, 79–95 (2002)
[8] Haralick, R.M., Shapiro, L.G.: Computer and Robot Vision. Addison-Wesley, Reading (1992)

Considerations of Image Compression Scheme Hiding a Part of Coded Data into Own Image Coded Data

Hideo Kuroda[1], Makoto Fujimura[2],
and Kazumasa Hamano[3]

[1] FPT University, Computing and Fundamental Dep., 15B My Dinh, Cau Giay,
Hanoi, Vietnam
kuroda@fpt.edu.vn
[2] Faculty of Engineering, Nagasaki University, Nagasaki,
852-8521, Japan
makoto@cis.nagasaki-u.ac.jp
[3] Graduate School of Science and Technology, Nagasaki University, Nagasaki
852-8521, Japan
hamano@bcss.jp

Abstract. In this paper, it is considered the image compression scheme, in which a part of coded data extracted in the coding process is hidden into the other parts of coded data of own image, especially into the block address data of the best matching block within restricted blocks. The proposed scheme is able to be used in a fractal image coding in which the best matching domain block is searched, in a vector quantization in image coding in which the best matching vector is searched, and in motion compensation of moving picture in which the best matching motion vector is searched. We study each image coding method and consider the features of each coding method using the proposed scheme.

Keywords: Data hiding, Fractal image coding, Motion compensation, Vector quantization.

1 Introduction

There are some data hiding techniques. The water marking hides the personal IDs for protecting copyright[1]. The steganography hides the confidential information for confidential communication like communication data of spies[2]. The objective estimation of image quality hides the marker signals for the automatic objective estimation of image quality. By the way, from the point of technical view, any data are available to be hidden data[3].

In this paper we consider an image coding scheme using the data hiding technique. This paper is organized as follows: in Section 2, Basic principles of the proposed data hiding scheme. In Section 3, image coding methods using the proposed scheme. In Section 4 and 5, consideration is discussed and concluded, respectively.

D. Ślęzak et al. (Eds.): SIP 2009, CCIS 61, pp. 312–319, 2009.

2   Basic Principles of the Proposed Data Hiding Scheme

In the proposed image compression scheme, a part of coded data, which is extracted from the image in the coding process, is hidden into the other parts of coded data of own image. Here we focus the image coding methods in which a small block is defined as an unit of coding and decoding process like a fractal image coding, an image coding using vector quantization and a moving picture coding using motion compensation.

Fig 1 presents the proposed data hiding scheme restricting address of searching reference blocks as two bits are hidden. After inputting the pixel values of a current block, the pixel values of reference blocks for block matching and two bits hidden data "i" and "j", the hidden data(i,j) are checked. If "i" is "0", then address Ax(address of x-axis) are set even values, "1", odd values. If "j" is "0", then address Ay(address of y-axis) are set even values, "1", odd values. Only blocks having ad-dress(Ax,Ay), even/odd, are searched from the reference blocks. And the block most similar to the current block is selected as the best matching block.

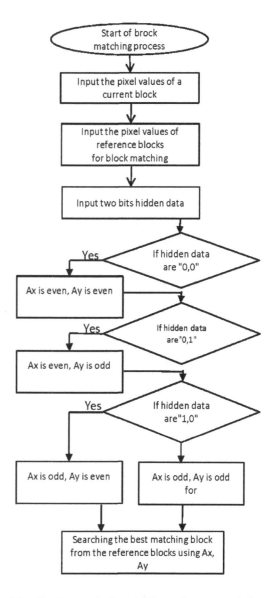

Fig. 1. Proposed data hiding scheme restricting address of searching reference blocks (as two bits hidden). Ax is the coordinate of x-axis, Ay is the coordinate of y-axis of blocks searched in the reference blocks.

3   Image Coding Methods Using the Proposed Scheme

3.1   Fractal Image Coding Method

A fractal image coding method does the block match processing at every small block by using the self-similarity in an image[4][5]. An original image for the fractal image coding is divided into range blocks (the block size is n×n) un-overlapped each other which are units of coding process, and also divided into domain blocks overlapped each other in the image, of which block size is lager than that of range block, for example 2n×2n.

First of coding process, the mean value and the standard deviation of pixel value of the range block are calculated. The range block where the value of this standard deviation is smaller than or equal to predetermined threshold value T_1 is classified into a flat block. For the flat block, only the mean value is transmitted to the decoding side. In the decoding side, the approximation reproduction is done by using only this mean value in the range block classified into a flat block.

The range block where the value of standard deviation of pixel value is more than T_1 is classified into a non-flat block. For the non-flat block, block matching processing is done by using the technique hiding data shown by Fig.1. The range block corresponds to a current block in Fig.1. Moreover, the domain blocks correspond to the reference blocks. And a part of bits of the mean value are selected as the hidden data. Odd numbers and even numbers of the address of the searched domain blocks are according to "0" and "1" of the hidden bits of the mean value, respectively.

Thus, after the restricted address is decided, the most similar domain block to the given range block is selected from the domain blocks having the restricted address as the best matching domain block.

By the coding process, for all range blocks the non-flag identification flag and the mean value are transmitted to the decoding side.

For non-flat block, the standard deviation, the address of the best matching domain block and the affine transform information, which is used for making usable block pattern much more by right and left rotation etc., are transmitted, too.

By the decoding side, first an initial image of a low resolution is generated by using only the mean value of the range block received, and this

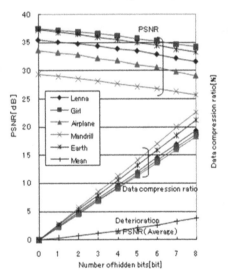

Fig. 2. Relation among number of hidden bits, PSNR, deterioration in PSNR and data compression ratio

is stored in the frame memory. The following decoding processing is repeated based on this initial image.

In the decoding processing, pixel values of all non-flat range blocks are reproduced by Equation (1) by using the transmitted various information and the initial image restored in the frame memory.

$$r_p(i,j) = \left\{ \tau d_q(i,j) - \hat{M}(\tau d_q) \right\} \frac{\sigma(r_p)}{\hat{\sigma}(\tau d_q)} + M(r_p)$$ (1)

where,

$r_p(i,j)$: The value of pixel at coordinate(i,j) in the p_{th} range block normalized

$\tau d_q(i,j)$: The value of pixel in the q_{th} domain block made small and normalized

M : mean value

$\sigma(r_p)$: Standard deviation of the p_{th} range block

$\sigma(\tau d_q)$: Standard deviation of the q_{th} domain block in the domain pool including reference blocks

$\hat{*}$ denotes the values regenerated in the decoding side.

The decoding image obtained by this reproduction is to be the following initial image, and then the decoding processing of mentioning above is done until obtaining the image of an enough resolution or the change in the image settles.

The coding and decoding experiments are carried out. Test images are five, that is, "Lenna", "Girl", Airplane", "Mandrill" and "Earth", and 256×256 all of image size. Block size is 4×4 pixels in range block, 8×8 pixels in domain block. The threshold T_1 is 4.

Fig.2 shows the relation among number of hidden bits, PSNR, deterioration in PSNR and data compression ratio. Horizontal axis is number of hidden bits[bits], and vertical axis of left side is PSNR[dB] and it of right side is data compression ratio[%].

Though the data compression ratio is different depending on a non-flat rate based on the image, the average of data compression ratio is shown by Equation (2), approximately.

$$Data\_comp \cong 2.5 \times h\_bit$$ (2)

where, Data_comp denotes data compression ratio, and h_bit , number of hidden bits.

It increases in proportion to an increase of the number of hidden bits though PSNR is different according to the image. The deterioration in PSNR on the average is shown by Equation (3), approximately.

$$Det\_inPSNR \cong 0.37 \times h\_bit$$ (3)

where, Det_inPSNR denotes deterioration in PSNR.

3.2 Image Coding Using Vector Quantization

In the image encoding process, pixel values are quantized to reduce volume of information. Some pixels values included in a block are quantized simultaneously in the vector quantization while one pixel value is quantized in the scalar quantization[6].

An original image is divided into blocks(the block size is m×m), which is an unit of coding process. This is a vector space that consists of all pixels value included in the block. The output vector to which the error margin is minimized is composed referring to all the input vectors for this vector space, and the code book that consists of the output vector group is made.

First of coding process, the input images are blocked by the off-line processing by the size of 4×4, the mean value is removed afterwards, and it normalizes it with a standard deviation. The code book of a predetermined size is made by using the LBG algorithm[7] after this processing.

Next, the input image is blocked by the size of 4×4 as well as the time of the off-line processing, and the mean value and the standard deviation are calculated. After the mean value is pulled, the input image is normalized by standard deviation in the on-line processing of encoding.

After this normalizing processing for each block, block matching processing is carried out using data hiding technique by Fig.1. The normalized input block is corresponds to a current block in Fig.1. Moreover, the vector codes in the vector code book correspond to the reference blocks. And a part of bits of the mean value are selected as the hidden data in Fig.1. An odd number or an even number of the address of the searched vector codes is according to "1" or "0" of the hidden bits of the mean value.

Thus, after the restricted address is decided, the most similar vector code to the normalized input block is selected from the vector code books having the restricted address as the best matching domain block. And the address is output as an index for the best matching vector.

The coding and decoding experiments are carried out. Test images are five, that is, "Airplane", "Earth", "Girl", "Lenna" and "Mandrill", and all of image size is 256×256. Block sizes are 4×4 pixels. The code book used in the experiment is made by using images "Airplane", "Lenna" and "Mandrill", and the size is 4096.

Fig. 3 shows the relation among the number of hidden bits, PSNR, deterioration in PSNR and data compression ratio. Horizontal axis is number of hidden bits, and vertical axis of left side is PSNR and it of right side is data compression ratio.

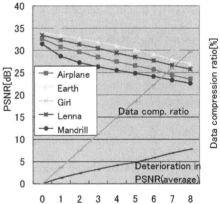

Fig. 3. Relation among Number of hidden bits, PSNR, Deterioration in PSNR and data compression ratio

The data compression ratio is shown by Equation (4).

$$Data\_comp \cong 3.6 \times h\_bit \qquad (4)$$

where, Data_comp denotes data compression ratio, and h_bit , number of hidden bits.

It increases in proportion to an increase of the number of hidden bits though PSNR is different according to the image. The deterioration in PSNR on the average is shown by Equation (5), approximately.

$$Det\_inPSNR \cong 0.95 \times h\_bit \tag{5}$$

where, Det_inPSNR denotes deterioration in PSNR.

3.3 Inter-frame Coding Using Motion Compensation

MPEG2[8] that is the motion compensated inter-frame predictive coding is studied as another example of doing the block match processing that uses the proposed data hiding technique.

Input image is blocked by small block of 16×16 which is an unit of coding process. These blocks are un-overlapped each other in the image. In the frame memory FM, reconstructed images past encoded and decoded have been accumulated. The blocks included in the FM are overlapped each other in the images from the FM. It centers on the position of the encoded current block, and the area of the length and breadth plus or minus 15 pixels in the frame memory is used as a range of the motion compensation. It searches for the block where the block match is the best to the encoded current block in one pixel from among the range of the motion compensation.

Next, it searches for the block where the block match is the best to the encoding current block in 1/2 pixels from among with oneself and the block of eight neighborhood centering on the block chosen at these intervals of one pixel. The motion vector obtained at intervals of 1/2 pixels is decided as the best motion vector.

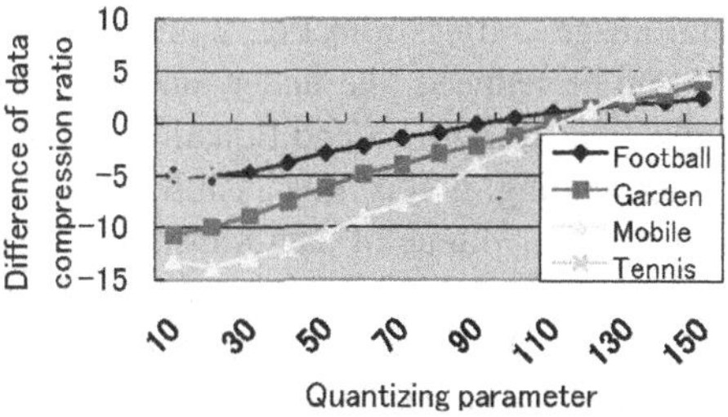

After motion compensation, the input block image is pulled the

Fig. 4. Relation between quantizing parameter and difference of data compression ratio

inter-frame predictive block image obtained by motion compensation. The obtained predictive error block image is transformed by DCT in 8×8 sizes each, and quantized.

In the MPEG2, both the best motion vector and the predictive error are transmitted to the decoding side. If the difference in block pattern between the real best motion vector and the changed motion vector by using data hiding technique are large, the volume of quantized predictive error become much more instead decrease of hidden data by using data hiding technique. In addition, the influence has a bad influence for the encoding processing of the following image frame. Then, we used the data hiding technology in the processing of the motion vector search with a small difference of the movement vector at 1/2-pixel intervals.

The coding and decoding experiments are carried out. Test images are four, that is, "Football", 125 frames and 352 ×240pixels, "Garden",115 frames and 352 ×240pixels, "Mobile", 140 frames and 352 ×240pixels and "Tennis", 112 frames and 352 ×240pixels.

Fig. 4 shows the relation between quantizing parameter and data compression ratio. Horizontal axis is the quantizing parameter, and vertical axis is the difference of data compression ratio between MPEG2 with data hiding technique and without one. When the value on vertical axis is positive, the feature of the proposed method is better than that of original MPEG2. When the value is minus, worth than that of original MPEG2. We did not obtain an good result though we limited the use of this technology to the processing with a small difference of the movement vector at 1/2 intervals.

4   Considerations

First of all, it thinks about the image quality degradation by the influence of the proposed scheme using data hiding technique. The image quality of 25dB or more is still obtained though there is a sacrifice of about 3.5dB even when the data of eight bits are hidden like understanding the introduction into the fractal image coding from Fig. 2 and Equation (3). The deterioration is large compared with the fractal as it is understood to introduce this technology into the vector quantization from Fig. 3 and Equation (5), and even when the data of four bits are buried, 25dB or more is obtained.

Next, it thinks about the data compression ratio by the influence of the proposed scheme. When the data of one bit is hidden like understanding the introduction into the fractal image coding from Fig. 2 and Equation (2), the data of 2.5% can be reduced. 3.6% can be reduced like understanding this technology is introduced into the vector quantization from Fig. 3 and Equation (4), and this is an increase of 1.1% for the fractal image coding. In this, it is a cause in the fractal image coding and the vector quantization that the volume of data transmitted is actually different. Because a lot of kinds of the block pattern are prepared to the point where the fractal image coding introduces the affine-transformation aiming at the improvement of the image quality, the volume of information originally transmitted is abundant. On the other hand, it doesn't have the same block pattern in the vector quantizing. As a result, the data reduction ratio is growing more than the fractal coding as for the vector quantizing.

MPEG2 that is encoding the motion compensated predictive inter frame coding did not achieve a good effect of the introduction of this technology because a similar block pattern was few because the number of blocks included within the range of the motion compensation was little, and the prediction error quantized due to deterioration in the motion compensation accuracy grew, and it gave the influence that deterioration in the predictive accuracy is bad for the following frame processing."

5   Conclusion

The data hiding technology was introduced, and it proposed a new image encoding technology that hides a part of image encoded data in other parts of own image. It was examined to apply this technology to the fractal image encoding, the vector quantizing, and MPEG2 that were the block processing. Experimental results showed   that it was possible to put on the fractal image coding where a lot of similar block patterns were included in the search reference block, because quality degradation was a little, and the encoding efficiency was able to be achieved . A higher data reduction was able to be achieved in the vector quantization because the volume of data transmitted of original in the vector quantizing was a little. MPEG2 did not achieve a good result because a similar block pattern was few. Future tasks are the achievement of a concrete real machine for fractal image coding.

References

1. Iwata, M., Miyake, K., Shiozaki, A.: Digital Watermarking Method to Embed Indes Data into JPEG Images. IEICE, Trans. Fundamentals E85-A(10), 2267–2271 (2002)
2. Iwata, M., Miyake, K., Shiozaki, A.: Digital Steganography Utilizing Features of JPEG Images. IEICE, Trans. Fundamentals E87-A(4), 929–936 (2004)
3. Sugimoto, Kawada, R., Koike, A., Matsumoto, S.: Automatic Objective Picture Quality Measurement Method Using Invisible Marker Signal. IEICE Trans. Information and Systems J88-D-II(6), 1012–1023 (2005)
4. Jacquin, E.: A novel fractal-coding technique for digital images. In: IEEE Int.Conf. on Acoustic, Speech and Signal Processing, M8.2, vol. 4, pp. 2225–2228 (1990)
5. Beaumont, L.M.: Image data compression of fractal techniques. BT Technology 9(4), 93–109 (1992)
6. Kim, K., Park, R.H.: Image coding based on fractal approximation and vector quantization. In: IEEE Int. Conf., on Image Processing, vol. 3(13-16), pp. 132–136 (1994)
7. Linde, Y., Buzo, A., Gray, R.M.: An algorithm for vector quantizer design. IEEE Trans. Commun. COM-28(1), 84–95 (1980)
8. MPEG Software Simulation Group, MPEG-2 Video Codec,
   http://www.mpeg.org/MPEG/video/mssg-free-mpeg-software.html

Binarising SIFT-Descriptors to Reduce the Curse of Dimensionality in Histogram-Based Object Recognition

Martin Stommel and Otthein Herzog

TZI Center for Computing and Communication Technologies,
University Bremen, Am Fallturm 1, 28359 Bremen, Germany
mstommel@tzi.de, herzog@tzi.de

Abstract. It is shown that distance computations between SIFT-descriptors using the Euclidean distance suffer from the curse of dimensionality. The search for exact matches is less affected than the generalisation of image patterns, e.g. by clustering methods. Experimental results indicate that for the case of generalisation, the Hamming distance on binarised SIFT-descriptors is a much better choice.

1 Introduction

Histogram-based methods have become increasingly popular for object recognition approaches that use a compositional model [12,14,17,4]. While in the initial Constellation Model [15] and related systems the appearance of the parts had been optimised individually, later systems (e.g. [12]) used generic code books of frequently occuring patterns. The parts of a model are often represented conveniently by high-dimensional feature vectors, e.g. in the shape of SIFT-descriptors [10]. The code book can then be created by clustering the input vectors.

Recent systems incorporate spatial [5,9] or class discriminatory [16] information, or focus on unsupervised training [7], or the meaningful combination of different kinds of local descriptors [14].

However, the effects of the high-dimensionality of the part descriptors ("curse of dimensionality") on object recognition are usually ignored, except for unrelated or general studies [2,8]. Considering that popular descriptors have tens [6] or hundreds of dimensions and that negative effects have already been reported for as few as ten dimensions [2], this topic seems highly relevant.

Therefore, this paper addresses the dimensionality of the feature descriptors in the context of object recognition. For the popular example of SIFT-descriptors, the next section demonstrates that in fact there is a curse of dimensionality. Section 3 offers a possible solution which is supported by the experimental results given in section 4.

2 Distance between SIFT-Descriptors

To perform experiments on the dimensionality of part descriptors, a data set of 200 images is sampled from ARD (a German TV network) TV news reviews

D. Ślęzak et al. (Eds.): SIP 2009, CCIS 61, pp. 320–327, 2009.
© Springer-Verlag Berlin Heidelberg 2009

of the year 2006. The images are sampled with a step width of 50 frames to ensure a minimum diversity. Then SIFT features are computed for these images with the default parameters given in the original paper [11], except for a higher contrast threshold of 10 and no initial doubling of the image size. Depending on the experiment, subsets with 100–500 000 descriptors are selected randomly.

A prerequisite for histogram-based recognition methods is a distance measure that allows for the clustering of descriptors and the comparison of an image to the trained model. Lowe [10] proposes to compare SIFT-descriptors by the Euclidean distance. More generally, the distance between two descriptors x, y can be computed via the L_p-norm

$$L_p(x, y) = \left(\frac{1}{n}\sum_{i=1}^{n}|x_i - y_i|^p\right)^{\frac{1}{p}}, \quad x = (x_1, x_2, \ldots, x_n) \quad \text{etc.} \quad (1)$$

Beyer et al. [2] explain the curse of dimensionality as a general unreliability of high-dimensional distance computations that occurs if the distance distribution becomes sparse with rising dimensionality n. Close and distant points would approach each other, thus making it more difficult to distinguish between different objects.

To test if the curse of dimensionality affects the comparison of SIFT-descriptors, sets of 1000 descriptors are created. For each set, the dimensionality of the descriptors is set to a certain value n by projecting the descriptors to a randomly chosen subspace. Then a full distance matrix is computed for each set.

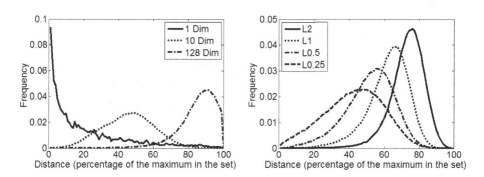

Fig. 1. Distance distributions for different dimensionalities (left chart, using the Euclidean distance) and different distance norms (right chart, for a dimensionality of 90)

Figure 1 shows histograms over the distances for different choices of the dimensionality n and the norm p (Eq. 1). Distances of zero occur trivially very often, so they are not shown in the diagrams.

It can be seen that with increasing dimensionality the maximum of the distribution moves to the right, i.e. towards the maximum distance. The distances between any two elements becomes large and the distribution becomes sparse.

This is exactly the effect reported by Beyer et al. Figure 1 (right) shows also that the effect is reduced by choosing a smaller, or even fractional norm as proposed by Aggarwal et al. [1]. However, the distance computation becomes numerically delicate for the computation of high roots. In fact, it was not possible to achieve stable results for the L0.5-norm and descriptors with more than 90 elements. The roughness of the curve for the L0.25-norm in Fig. 1 (right) is a result of the beginning instability.

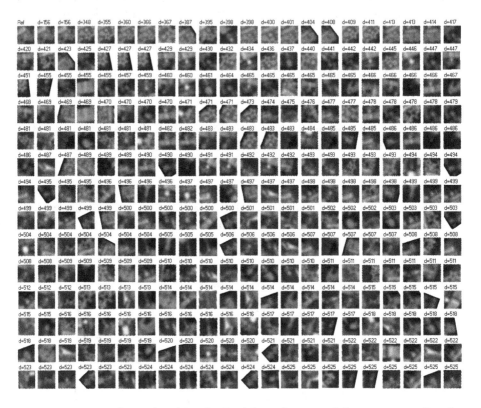

Fig. 2. Image patches ordered by the Euclidean distance of their corresponding descriptor from a reference descriptor (first patch). The distance is written on top of each patch. The maximum distance in the set is about 700.

What does this mean for the similarity of image patterns? Figure 2 shows a list of image patterns arranged according to the Euclidean distance to a reference descriptor. Generally, it appears that the image patches are only similar for minimal distances. For medium and higher distances the Euclidean distance has no visual effect on the ordering of the patterns. Also, some obviously similar patterns yield clearly different distances. However, one explanation for this may be that the SIFT operator creates multiple descriptors for a single feature if the orientation is ambiguous.

Generally, the Euclidean distance seems only appropriate for very small distances, i.e. the search for exact matches. For the goal of abstraction, a measure would be desirable that also allows for a gradual comparison of more dissimilar image regions. One possible solution is to reduce the dimensionality by sampling less than 8 gradient orientations or less than 16 grid cells in the image area. On the other hand, SIFT is generally a high dimensional method, so descriptors with less than 10 dimensions can not be expected.

3   Proposed Method

Alternatively to using a lower distance norm, Korn et al. [8] propose to reduce the dimensionality by the analysis of self-similarity of the distribution. However, a box-count plot of a set of 200 000 descriptors (Fig. 3) does not indicate the required constant Hausdorff-dimension.

Other approaches include a feature selection or a feature transform to reduce the descriptor length [3,6]. While a certain reduction can be achieved by subsuming correlated descriptor components, all dimensions represent basically the same kind of information. Since minor differences are only introduced during the descriptor normalisation, the solution seems rather application dependent.

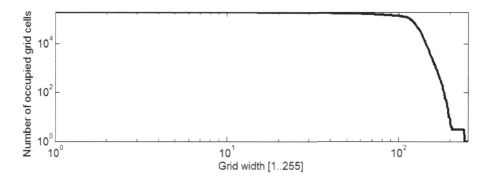

Fig. 3. Box-count plot for a set of 200 000 SIFT-descriptors

On the other hand, the box-count plot shows that even for a coarse quantisation with only two bins, i.e. a grid width of 128, almost the full diversity of the set is maintained. Since about 50% of the descriptor values are lower than twelve, this diversity must result from the few descriptor components with higher values. This observation suggests that a binarisation of the descriptor values, i.e. using only the values 0 and 1 for single descriptor components, would only minimally affect the object recognition. Though a binarisation does not reduce the dimensionality, the complexity of the problem would be reduced dramatically from 256^{128} different SIFT-descriptors to 2^{128} possibilities.

The binarisation is carried out as follows: Each of the 128 values of a SIFT-descriptor is compared to a threshold. If the value is greater than the threshold, it is set to one. Otherwise it is set to zero.

At first, a threshold of 128 is chosen dividing the domain of descriptor values in the middle. The results are however bad: About 40 percent of all SIFT-descriptors are projected to the same binary descriptor. Obviously, the distribution is not centred to the middle of the domain of descriptor values. This agrees with the already mentioned predominance of small values.

To compensate for this, the mean values of every descriptor element are chosen as thresholds. There is one threshold for every descriptor component, i.e. 128 together. Still up to 4000 descriptors result in the same binary descriptor while the majority of binary descriptors corresponds to only one SIFT-descriptor in the original set. The mapping is thus very uneven, indicating that the distribution of SIFT-descriptors is not symmetrically centred around the mean.

To compensate for this asymmetry, the medians of each descriptor element are chosen as thresholds. Now, only 20 binary descriptors occur more than ten times. The binary descriptors are thus almost uniformly distributed. The remaining inaccuracies stem from the median computation. For an original set of 100 000 SIFT-descriptors, this method results in 94 376 binary descriptors. This number also includes SIFT-descriptors that have already been identical before binarisation. Since this concerns a proportion of 4.0 percent, only 1.6 percent of original SIFT-descriptors became indistinguishable because of the binarisation.

A natural way to compare binarised descriptors is to use the Hamming-distance, i.e. the number of differing bits between two vectors.

4 Experimental Validation

The validation addresses first the information stored in the lost 1.6 percent of the descriptors. Is it possible that strongly differing patterns are spuriously combined by such a coarse quantisation? Figure 4 lists the patterns that are mapped to the

Fig. 4. Patterns from the original images that are mapped to binary descriptors A–D. No distinction is made whether the patterns are combined because the original descriptors are identical or because the binarisation leads to identical results.

same binary descriptor for four examples A–D. The groups are very homogeneous and it seems that no visually important effect is disregarded. It is in the contrary probable that different binary descriptors correspond to highly similar patterns, as for the examples B and D. For a recognition system using a code-book of descriptor prototypes, these descriptors are potential candidates for a further combination.

The next question is whether the Hamming-distance is affected by the curse of dimensionality. To answer this question, 100 000 binary descriptors are created using the component-wise medians as thresholds. Then the Hamming-distances between all pairs of binary descriptors are calculated, resulting in 10^{10} distance

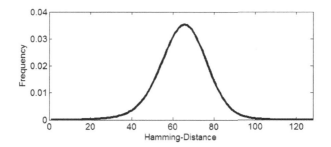

Fig. 5. Hamming-distances of binary descriptors

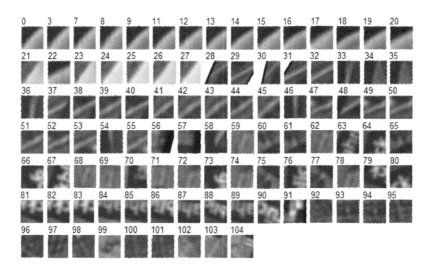

Fig. 6. A subset of the patterns sorted by the Hamming distance to the first pattern (the upper left one). The numbers give the distances. For every distance only one pattern is shown. Distances greater than 104 did not occur in the set. In comparison to the Euclidean distance, the sorting by the Hamming distance shows a softer and more continuous increase in dissimilarity towards the maximum distance. Similar patterns cluster more obviously.

values. Figure 5 shows a histogram of the distance values. Because of the descriptor length, the Hamming distances range from 0 to 128. The distance distribution is symmetric and centred around the value 64. There is no general shift towards high distance values ("sparseness"), so the curse of dimensionality does not seem to influence the descriptor comparison.

To exclude that the Hamming distance is an artificial measure without visual correspondence to the image data, it is important to visually compare image patches with low and high Hamming-distances. Figure 6 shows a list of patterns and the Hamming distances of the corresponding descriptors. The first pattern in Fig. 6 is the reference pattern. The patterns are ordered by Hamming distance. In the case of multiple patterns with the same distance, one pattern is chosen randomly.

The figure shows a clear correspondence between distance and visual appearance. Even for higher distances, neighboring patterns are visually related but also different to patterns with another distance. This makes the Hamming distance measure highly interesting for the application of clustering algorithms that aim at the generalisation of visually similar object parts. Compared to the Euclidean distance (Fig. 2), the results for the Hamming distance (Fig. 6) are obviously better.

To demonstrate the value of this method for object recognition, experiments on the Graz-2002 image data base [13] are conducted. The data base consists of 460 images of persons, 373 images of bikes and 270 images without bikes or persons. The image sets are randomly split into equally sized training and test sets. The recognition system is a KNN-classifier that uses the mean Hamming distances of single training images to a set of 1000 binary vectors with random values as a model. The classifier is enhanced by the class probabilities for each configuration within the k nearest neighbours. The system achieves a competitive accuracy of 76.7%. Averaged over all classes (person, bike, rejection), a recall of 79% and a precision of 80% is achieved. In comparison, the authors of the image data base [13] report a lower precision between 70% and 64% for a similar recall of 78–79%.

5   Conclusion

The experiments show that the comparison of SIFT-descriptors by Euclidean distance suffers from the curse of dimensionality. Though this does not seem to substantially influence the search for exact matches, it makes a gradual comparison of more dissimilar image regions unreliable. The gradual comparison however is a prerequisite for many modern compositional object recognition techniques. To solve this problem, binarised descriptors together with the Hamming distance are proposed. The experiments show that no crucial information is lost by the binarisation. Compared to the Euclidean distance, the proposed method is visually more plausible and does not show the sparseness of the distance distribution seen for the Euclidean distance. Competitive object recognition results support the use of the proposed method.

Acknowledgments. Thanks go to Arne Jacobs and Hartmut Messerschmidt for the discussions.

References

1. Aggarwal, C.C., Hinneburg, A., Keim, D.A.: On the Surprising Behavior of Distance Metrics in High Dimensional Space. In: Van den Bussche, J., Vianu, V. (eds.) ICDT 2001. LNCS, vol. 1973, pp. 420–434. Springer, Heidelberg (2000)
2. Beyer, K., Goldstein, J., Ramakrishnan, R., Shaft, U.: When Is Nearest Neighbor Meaningful? In: Int. Conf. on Database Theory, pp. 217–235 (1999)
3. Bonev, B., Escolano, F., Cazorla, M.: Feature selection, mutual information, and the classification of high-dimensional patterns. Pattern Analysis and Applications, 309–319 (February 2008)
4. Bosch, A., Zisserman, A., Munoz, X.: Scene Classification Using a Hybrid Generative/Discriminative Approach. IEEE Transactions on Pattern Analysis and Machine Intelligence 30(4), 712–727 (2008)
5. Grauman, K., Darrell, T.: The Pyramid Match Kernel: Discriminative Classification with Sets of Image Features. In: Proc. IEEE International Conference on Computer Vision (ICCV), vol. 2, pp. 1458–1465 (2005)
6. Ke, Y., Sukthankar, R.: PCA-SIFT: A More Distinctive Representation for Local Image Descriptors. In: IEEE Conference on Computer Vision and Pattern Recognition (CVPR), vol. 2, pp. 506–513 (2004)
7. Kim, G., Faloutsos, C., Herbert, M.: Unsupervised Modeling of Object Categories Using Link Analysis Techniques. In: IEEE Conference on Computer Vision and Pattern Recognition, CVPR (2008)
8. Korn, F., Pagel, B.-U., Faloutsos, C.: On the Dimensionality Curse and the Self-Similarity Blessing. IEEE Transactions on Knowledge and Data Engineering 13(1), 96–111 (2001)
9. Lazebnik, S., Schmid, C., Ponce, J.: Beyond bags of features: spatial pyramid matching for recognizing natural scene categories. In: IEEE Conference on Computer Vision and Pattern Recognition (CVPR), vol. 2, pp. 2169–2178 (2006)
10. Lowe, D.G.: Object Recognition from Local Scale-Invariant Features. In: Proc. of the International Conference on Computer Vision (ICCV), Kerkyra, Greece, September 1999, vol. 2, pp. 1150–1157 (1999)
11. Lowe, D.G.: Distinctive image features from scale-invariant keypoints. International Journal of Computer Vision 60(2), 91–110 (2004)
12. Mikolajczyk, K., Leibe, B., Schiele, B.: Multiple Object Class Detection with a Generative Model. In: IEEE Conference on Computer Vision and Pattern Recognition (CVPR), June 2006, vol. 1, pp. 26–36 (2006)
13. Opelt, A., Fussenegger, M., Pinz, A., Auer, P.: Weak Hypotheses and Boosting for Generic Object Detection and Recognition. In: Pajdla, T., Matas, J(G.) (eds.) ECCV 2004. LNCS, vol. 3022, pp. 71–84. Springer, Heidelberg (2004)
14. Varma, M., Ray, D.: Learning The Discriminative Power-Invariance Trade-Off. In: IEEE International Conference on Computer Vision, ICCV (2007)
15. Weber, M., Welling, M., Perona, P.: Unsupervised Learning of Models for Recognition. In: Vernon, D. (ed.) ECCV 2000. LNCS, vol. 1842, pp. 18–32. Springer, Heidelberg (2000)
16. Yang, L., Jin, R., Sukthankar, R., Jurie, F.: Unifying Discriminative Visual Codebook Generation with Classifier Training for Object Category Recognition. In: IEEE Conference on Computer Vision and Pattern Recognition, CVPR (2008)
17. Zhang, J., Marshalek, M., Lazebnik, S., Schmid, C.: Local Features and Kernels for Classification of Texture and Object Categories: A Comprehensive Study. In: Proceedings of the 2006 Conference on Computer Vision and Pattern Recognition Workshop (CVPRW 2006), June 17-22 (2006)

Author Index